X-Ray Astronomy

NATO ADVANCED STUDY INSTITUTES SERIES

*Proceedings of the Advanced Study Institute Programme, which aims
at the dissemination of advanced knowledge and
the formation of contacts among scientists from different countries*

The series is published by an international board of publishers in conjunction
with NATO Scientific Affairs Division

A	Life Sciences	Plenum Publishing Corporation
B	Physics	London and New York
C	Mathematical and	D. Reidel Publishing Company
	Physical Sciences	Dordrecht, Boston and London
D	Behavioural and	Sijthoff & Noordhoff International
	Social Sciences	Publishers
E	Applied Sciences	Alphen aan den Rijn and Germantown
		U.S.A.

Series C – Mathematical and Physical Sciences

Volume 60 – X-Ray Astronomy

X-Ray Astronomy

Proceedings of the NATO Advanced Study Institute
held at Erice, Sicily, July 1-14, 1979

edited by

RICCARDO GIACCONI
Harvard-Smithsonian Center for Astrophysics,
Cambridge, Mass., U.S.A.

and

GIANCARLO SETTI
University of Bologna, Laboratorio di Radioastronomia CNR,
Bologna, Italy

D. Reidel Publishing Company

Dordrecht : Holland / Boston : U.S.A. / London : England

Published in cooperation with NATO Scientific Affairs Division

Library of Congress Cataloging in Publication Data
Nato Advanced Study Institute, Erice, Italy, 1979.
    X-ray astronomy.

    (NATO advanced study institutes series : Series C, Mathematical and
physical sciences ; v. 60)
    'Published in cooperation with NATO Scientific Affairs Division.'
    1. X-ray astronomy—Congresses. I. Giacconi, R. II. Setti,
Giancarlo, 1935–   III. North Atlantic Treaty Organization. Division of
Scientific Affairs. IV. Title. V. Series.
QB472.N37 1979   522'.686   80–23948
ISBN 90–277–1156–9

Published by D. Reidel Publishing Company
P.O. Box 17, 3300 AA Dordrecht, Holland

Sold and distributed in the U.S.A. and Canada
by Kluwer Boston Inc.,
190 Old Derby Street, Hingham, MA 02043, U.S.A.

In all other countries, sold and distributed
by Kluwer Academic Publishers Group,
P.O. Box 322, 3300 AH Dordrecht, Holland

D. Reidel Publishing Company is a member of the Kluwer Group

Printed in The Netherlands

TABLE OF CONTENTS

FOREWORD

This volume contains a series of lectures presented at the 5th
Course of the International School of Astrophysics held in Erice
(Sicily) from July 1st to July 14, 1979 at the "E. Majorana"
Centre for Scientific Culture. The course was fully supported by
a grant from the NATO Advanced Institute Programme. It was
attended by about one hundred participants from ten countries.
    Since the discovery of the first extra-solar X-ray source in
the early 1960's, X-ray astronomy has played an increasingly important rôle in the study of the Universe, bringing new insight
to almost every field of modern astrophysics from stellar evolution
to cosmology. Generally speaking, this branch of astronomy is
concerned with the discovery, classification and study of "hot
matter" in the universe, including high energy non-thermal phenomena. In particular, X-ray observations appear to provide the
main, if not the only, probe to inspect regions where collapsed
objects are formed, such as the environment of neutron stars and
of black holes in the presence of matter accretion onto the objects themselves. It is significant that the first candidate
black hole (Cyg X-1) has been primarily singled out by its X-ray
emission. In the same context, it is well known that one of the
fundamental problems in modern astrophysics is the understanding
of the strong activity taking place in galactic nuclei. There is
strong evidence that, whatever the final explanation of the physics
and evolution of these nuclei, one is dealing with a central power
engine basically constituted by a massive collapsed object. Here
again X-ray observations are playing a central rôle. Quite apart
from the physics of collapsed objects, X-ray astronomy is providing
a wealth of impressive and important data on stellar coronae,
supernova remnants, normal galaxies, quasars, radiogalaxies,
Seyfert's and other active galaxies and on the properties of the
very hot intergalactic gas which has been found in clusters of
galaxies.
    A big step forward in the development of X-ray astronomy has
come from the all sky survey of UHURU, the first dedicated X-ray
satellite. The dramatic findings of UHURU have been further confirmed and enriched by the observations of other satellites and
with balloons up to very high photon energies.

R. Giacconi and G. Setti (eds.), X-Ray Astronomy, vii-viii.
Copyright © 1980 by D. Reidel Publishing Company.

Prior to the launch of HEAO-2 (Einstein Observatory), which took place at the end of 1978, several hundred X-ray sources were known. Some of these sources were identified with known classes of extragalactic objects. However, due to the limited sensitivity of the instruments, only the brightest objects (close by) were found to be X-ray emitters. The Einstein Observatory with its improved sensitivity and angular resolution (about a factor 1000) provides the first opportunity to probe the Universe in depth and to map the structure of extended X-ray sources.

When planning the School we had in mind, of course, that early results from HEAO-2 (Einstein), which was due for launch before the start of the school, may be available. We decided to dedicate a large fraction of the programme to the extragalactic aspects of X-ray observations on which HEAO-2 data could have the largest impact. Therefore, only approximately one third of the lectures have been dedicated to the discussion of galactic X-ray sources, including the pertinent results which have become available in the first six months of operation of the Einstein Observatory.

In organizing the School we also had in mind the goal of bringing together researchers and students with different specializations in astronomy to encourage a discussion of the X-ray results in light of the knowledge available in parallel domains such as optical and radio, and also in view of a better exploitation of the potentiality of the Einstein Observatory. In this respect it was, of course, a most opportune time to hold a NATO Advanced Study Institute on this subject to acquaint students and researchers, not only in X-ray astronomy, but in related fields, with the state of the art in this important branch of astronomy and with the new results made available in the early stages of the life of the Einstein Observatory.

The various aspects of the subject were covered in a series of lectures and topical seminars totalling 52 hours. Unfortunately, for a number of reasons, it has not been possible to include in this proceedings the texts of all the lectures which were delivered at the School.

We wish to express our gratitude to the Scientific Affairs Division of the North Atlantic Treaty Organization for the generous support given to the Institute. Sincere thanks are also due to Ms. Rhona Kamer for her patient typing of part of this book, and to Mr. L. Baldeschi and Mr. R. Primavera for the drawing and photographic reproduction, respectively, of a number of figures.

Special thanks are due, of course, to all lecturers and participants who contributed so much to the success of the course, and to Prof. A.Zichichi for his hospitality at the "E. Majorana" Centre for Scientific Culture.

Riccardo Giacconi, Harvard-Smithsonian Center for Astrophysics, 60 Garden Street, Cambridge, Mass. 02138, USA
Giancarlo Setti, University of Bologna, Istituto di Radioastronomia. Via Irnerio 46, 40126 Bologna, Italy.

X-RAY ASTRONOMY

Riccardo Giacconi

Harvard/Smithsonian Center for Astrophysics
Cambridge, Massachusetts USA 02138

1. INTRODUCTION AND HISTORICAL BACKGROUND

Many of the most interesting discoveries about the Universe, which
have occurred in the last few decades, have resulted from observations
of the sky in light not visible to our eye. To emphasize this point it is
only necessary to recall the discovery of the microwave background
radiation, still the most convincing evidence for the Big-Bang theory,
apart from the discovery of the recession of the galaxies 50 years ago.
The discovery of pulsars through radio observations established the
existence of neutron stars, which until then had only existed as speculative objects predicted by theoretical astrophysicists to be the end
point of stellar evolution for intermediate mass stars. Radio observations first revealed the existence of quasi-stellar objects, now generally
believed to be objects at greater distance from us than any known, and
therefore offering clues to the early stages of the formation of galaxies
and their evolution.

Important information about the world we live in is transmitted
to us in wavelengths that we cannot perceive with our eyes. From a
scientific point of view, the reasons for this mismatch are quite clear.
Our eyes have evolved to respond to a narrow range of wavelengths
which are transmitted by the Earth's atmosphere and which are useful
for high resolution imaging even with the limited aperture detectors
(the eyes) with which we are endowed. Natural physical processes,
however, give rise to characteristic radiations in a very much wider
range of wavelengths. Molecular and atomic transitions, in which only
the most weakly bound electrons are involved, give rise to radiation in

1

R. Giacconi and G. Setti (eds.), X-Ray Astronomy, 1–13.
Copyright © 1980 by D. Reidel Publishing Company.

the infrared and visible range of wavelength. Transitions of more
tightly bound electrons may give rise to increasingly energetic (or
shorter wavelength radiation) from ultraviolet to X-rays. Bodies at
temperatures of a few thousand degrees emit substantial fractions of
their radiation in the visible part of the spectrum, while bodies (or
better, gases) at millions of degrees temperature emit a negligible
fraction of their energy in the visible and most of it in the X-ray range
of wavelengths. Nuclear transitions typically result in gamma-ray
emission of even shorter wavelength. At the opposite end of the spectrum
we find that the characteristic radiation from different physical processes
extends from the infrared through the microwave to the longest radio
wavelengths, produced by the deflection of free energetic electrons in
magnetic fields.

It has become increasingly evident in the last quarter century
that in the Universe explosive, violent processes are the norm rather
than the exception in most of the events which determine the dynamics
and evolution of stars and galaxies. To cite only an example of how this
view has changed, it is interesting to consider how the study of super-
novas has convinced us that while a star uses only a few percent of its
rest mass energy in the billions of years of its evolution by nuclear
burning processes, it can liberate most of it during an abrupt gravitational collapse at the end of its evolutionary track. A violent ejection of
the outer shell of the star and the mixing of this material in the interstellar medium provide the mechanism by which the primordial hydrogen
is enriched with heavier materials. The high temperature gas which is
produced by the slowing of the ejecta into the interstellar medium emits
primarily in the X-ray range of wavelength, as does the remnant left
behind, - the pulsar.

It is a general feature of natural processes that whenever an
explosive event occurs, gases are heated to very high temperatures and/
or protons and electrons are accelerated to very high energies. The
emission of light from gases at 10 to 100 million degrees occurs
primarily in the X-ray range of wavelength, and high energy particles
emit X-rays by interaction with magnetic fields (synchrotron) or photons
(Inverse Compton).

In summary, the high energy events are dominant in the dynamics and evolution of the Universe. When high energy processes take place,
X-rays are emitted. X-rays furnish us, therefore, a particularly good
tool to study these processes.

2. EARLY DEVELOPMENT OF X-RAY ASTRONOMY

X-ray astronomy is normally understood to encompass the photon
energies between 0.1 and 100 kilo electron volts. Radiation in this
range cannot penetrate most of the atmosphere. Development of X-ray
astronomy became possible therefore only after the advent of rocket
engines capable of lifting instruments above the atmosphere for short
periods of time, or even placing these instruments into orbit as artifi-
cial earth's satellites.

At first, attention was directed to the Sun. Starting with the
early discovery by Burnight in 1948 of solar X-ray emission, Herbert
Friedman and his collaborators at the Naval Research Laboratory
carried out an extensive program of solar X-ray studies over an entire
11-year solar cycle. Attempts made by the same group to detect X-ray
emission from stars, however, failed. This was not totally unexpected
since the Sun was found to emit only a very minute part (one in a million)
of its energy in the X-ray band. If all stars emitted X-rays at the same
rate as the Sun, then their X-ray flux at Earth would be so weak as to be
undetectable with then available instruments.

Still there was a growing awareness that X-ray observations
could provide a new and potentially rewarding window in which to study
the Universe, and several groups were endeavoring to improve the existing instrumentation. Our group at American Science & Engineering, Inc.,
(which included Herbert Gursky and Frank Paolini) had succeeded, by
1962, in improving the sensitivity of the detector by 100-fold as a result
of a research program I had initiated in 1959 following a suggestion by
Bruno Rossi. After two unsuccessful attempts, we were fortunate enough
in June 1962 to discover the first source of X-ray radiation outside the
solar system, a star in the constellation Scorpio, which we called
Sco X-1.

In retrosepct Nature had been very kind to us. While the
pessimistic predictions about fluxes from normal stars appeared to hold
true, what we had discovered was the first of a new class of stellar
systems in which the X-ray emission exceeded by 1000-fold the visible
light emission. In the succeeding years a number of rocket flights by
many groups, including that of NRL, Lockheed and our own at AS&E,
confirmed our first results, increased the sample of galactic sources to
about 30 and identified the first extragalactic sources, M-87 and 3C273.
However, these rocket flights could not study the X-ray sources in sufficient detail to give us a clue to the origin of the large energy release
taking place in these systems or to the physical processes giving rise to
the observed radiation.

A solution to this problem had to await the next major step in instrumentation which was taken with the launch in 1970 of UHURU, the first X-ray satellite observatory, which I had conceived and proposed to NASA in 1963, and on which our group had been working since then. It was the first of a series of similar satellite instruments which provided most of the results in X-ray astronomy in the 1970's: the ANS of the Dutch, SAS-3 of the MIT group, Ariel V of the British, OSO-8 and Copernicus, and finally the first of the High Energy Astronomical Observatory series, HEAO-1.

Among the most important findings of UHURU which were confirmed and extended by the following missions, was the identification of many galactic sources as close binary mass exchange systems. In a series of exciting discoveries following rapidly one after the other, regular pulsations were discovered from one, then two, of the sources: Her X-1 and Cen X-3 (Figure 1). Careful study of the period of pulsation showed that their frequency was modulated by the Doppler effect due to the motion of the X-ray emitting star in orbit about a companion (Figure 2). Long term study of the period revealed that the X-ray star was increasing its speed of rotation rather than slowing down, and therefore that the energy for the emission could not be extracted from rotation, as in the case of pulsars. Gravitational energy released by the accretion of material from the companion star onto the X-ray emitting object remained as the only plausible energy source. The infall of the material was also responsible for the acceleration of the rate of spin.

Optical identifications became possible due to the improved locations, and by using both X-ray and visible light observations the mass of the X-ray emitting object could be inferred. From the change in the speed of rotation due to the infall of the material, its moment of inertia could be measured. The conclusion was that the X-ray emitting object was a neutron star of mass about 1 to 2 solar masses. This represented the first direct measurement of the mass of a neutron star. In another system, Cyg X-1, we found a compact object which exhibited flickering on time scales as short as milliseconds, but no regular pulsations (Figure 3). Following the optical identification of the system, the mass of the X-ray emitting object could be estimated to be greater than 6 solar masses. The fact that it was compact and possessed a large mass, provided the first strong evidence for the existence of a black hole.

Since these early discoveries, we have become aware that mass exchange binary systems are the source of many, and possibly all, of the most bizarre high luminosity phenomena we observe in X-rays in our galaxy. The different classes of pulsating X-ray sources, bursters, transient sources, all seem to be but varied manifestations

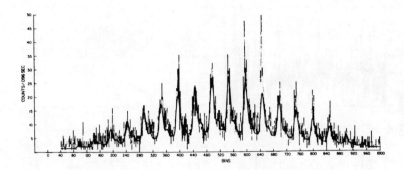

Figure 1.

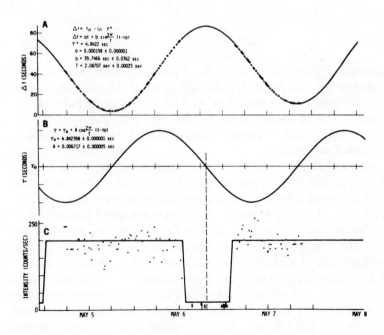

Figure 2.

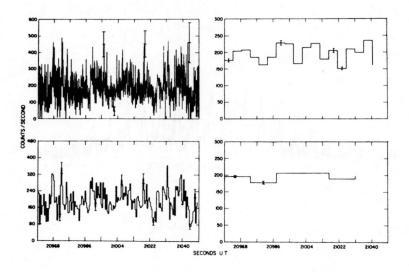

Figure 3.

of this basic underlying energy source. These systems have also been
used as a sort of astrophysical laboratory in which to study the behavior
of matter at extremely high densities as well as the properties of
neutron stars. It is in these conditions that one can hope to observe
strong general relativistic effects which in normal conditions are so
difficult to detect and study.

Another important and far-reaching discovery from UHURU
observations was that of high-temperature gas (about 10 to 100 million
degrees) filling the space between galaxies in clusters of galaxies.
These objects are the largest bound aggregate of matter known in the
Universe. We are still puzzled about their origin and evolution. We
do not know, for example, whether galaxies first formed and then coal-
esced together to form a cluster, or whether a cloud of gas already
existed as a distinct entity prior to galaxy formation. Questions related
to cluster formation are among the most probing one can ask about the
history of our Universe in the intermediate epoch between the present,
when most galaxies are formed, and the distant past when such condensations might have been just slight density fluctuations on a more or less
uniformly distributed medium. The X-ray observations here play a
unique role. Although the mass of the gas contained in a cluster is as

great as that of all the visible galaxies in it, its existence was previously unknown and it could only be revealed by X-ray measurements.

Using UHURU and other satellites up to HEAO-1 (launched in 1977 and the largest of its kind), the catalog of galactic and extra-galactic X-ray sources had expanded to more than 400. Yet the sensitivity of these observatories fell still quite short from that nec-essary to realize the full potential of X-ray astronomy. In fact we were constrained to study only the most luminous and unusual systems in our galaxy, and the most powerful and nearest of the extragalactic sources. For X-ray astronomy to join the mainstream of astronomical research it was necessary to increase the sensitivity by a large enough factor so that we could study the X-ray emission from a much broader class of objects and extend the X-ray observations of extragalactic objects, such as clusters and quasars to the most distant known objects of each class, as was required for serious cosmological research.

The reason one would not easily achieve this goal had to do with the underlying technological basis of the missions from 1962 to 1977. From the first rocket flights to the HEAO-1 launch, the basic detector had remained unchanged. It was a thin-window, gas-filled counter operating in the Geiger, or proportional, region. The detector was relatively easy to construct and its main advantage was the ability to detect each individual X-ray photon. The main disadvantages were its lack of directionality and its high background noise, produced by gamma and cosmic rays. The problem of determining the direction of the sources was solved with increasingly sophisticated arrangements of baffles which allowed us to achieve angular resolutions of $1/2^o$, in all-sky surveys. Higher angular resolutions, to 1 arc minute, could be achieved by special apertures such as the modulation collimator, in-vented by Oda, but only with a severe sacrifice in sensitivity. But the basic problem remained: in a background limited situation, the improve-ment in sensitivity only occurs proportionately to the square root of the area of the detector. HEAO-1 was already one of the largest scientific satellites ever built: approximately 20 feet high, 10 feet in diameter, and 6000 pounds of weight. Yet due to these constraints, its detectors, although 50 times larger than those of UHURU, were only 7 times more sensitive. Clearly, to make further progress we needed instruments built on entirely different principles.

3. DEVELOPMENT OF FOCUSING X-RAY OPTICS

In other wavelength intervals, such as optical or radio, it has become customary since Galileo first turned a telescope to study the sky, to use optical instruments to gather, concentrate and image the radiation from celestial objects. In X-rays this is particularly difficult because

refractive lenses absorb the radiation rather than focus it and because the efficiency of reflection of X-rays perpendicular to a mirror
surface is nearly zero. It had been discovered in the 1930's that one
could achieve high efficiency of reflection if X-rays were made to reflect from a surface at very small angles of incidence (grazing incidence)
and optical designs based on this principle were studied in the early
1950's, particularly by Wolter (Figure 4). However, no practical
application of his studies, directed to implementing an X-ray micro-
scope had been possible. This was due to the need to polish the reflecting surfaces to a very high degree, 1/1000 of the wavelength of
visible light, and the added difficulty that for microscopes this had to
be accomplished on surfaces of extremely small dimensions.

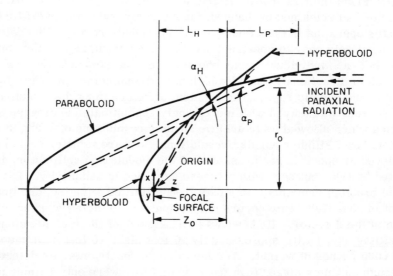

Figure 4.

     I had independently conceived of the use of grazing incidence
paraboloids in 1959, when first concerned with the need to improve
X-ray instrumentation. It was clear that focusing optics could have
practical application to X-ray astronomy and this was the subject of

a 1960 article published by B. Rossi and me. In the following decade
while the research effort on X-ray astronomy proceeded first with
rockets and then with the UHURU satellite, my group at AS&E carried
out the long-range technological development which perfected X-ray
telescopes. At each step in the development we were able to use these
techniques in a progeam of solar research which obtained the first real
X-ray pictures of the Sun in 1963 (with 1 arc minute resolution) and
culminated in 1973 with the many thousands of highly resolved (5 arc
seconds) pictures of the solar corona which the astronauts brought
back from Skylab (Figure 5). Giuseppe Vaiana had joined our group
in 1964 and assumed direction of our solar X-ray rocket program. The
analysis of the ATM data carried out under his scientific guidance un-
covered a number of new and important results on the physical processes
taking place in the solar corona.

      Observation of the Sun with this technique was considerably
simpler than the observation of stars. The flux of X-ray radiation from
the Sun is much larger at Earth than that of any star; therefore, the
telescopes could be smaller and the detection of the image could be
achieved with film. For stellar work we needed to develop larger telescopes as well as a very sensitive X-ray television camera which could
transmit information continuously to the ground over a period of years.

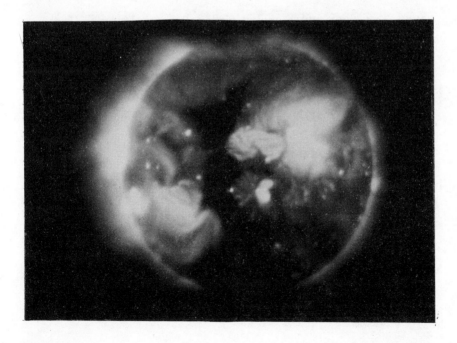

Figure 5.

The technology was sufficiently developed by 1970 to allow
NASA to seriously consider a proposal by a consortium of four institu-
tions (American Science & Engineering, Inc., Columbia Astrophysical
Laboratory, Goddard Space Flight Center, and Massachusetts Institute
of Technology for a stellar X-ray telescope mission as part of the
HEAO program. The project was finally realized with the launch on
November 13, 1978 of the Einstein (HEAO-2) Observatory.

4. EINSTEIN

The Einstein Observatory (Figure 6) is perhaps one of the most advanced
scientific spacecrafts ever flown by NASA. Schematically it consists
of a three-axis stabilized spacecraft which can be pointed on command
from the ground anywhere in the sky. The spacecraft, solar power
system, attitude control and telemetry, as well as all other housekeeping
functions were built by TRW under contract from the MSFC team led by
Freed Speer, who managed the entire HEAO Program. The scientific
payload consists of a grazing incidence telescope of 0.6 meter diameter.
Four nested paraboloids, followed by four nested hyperboloids yield a
total area of collection of 300 cm^2 at low energy (0.1 keV). At high

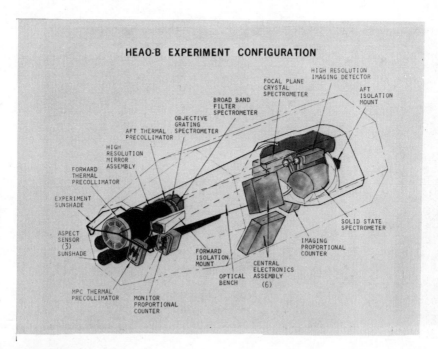

Figure 6.

energy the reflection efficiency becomes quite small effectively giving
us no useful area above 4 keV. An optical bench attaches the telescope
to a focal plane moveable platform on which several instruments are
mounted. On command one of the several instruments can be placed
at the focus of the telescope. Star sensors are used to select the target
field in orbit and on the ground for post facto determination of the
pointing direction. Perkin-Elmer constructed the mirror, Honeywell
the star sensors, and AS&E much of the structure and instrumentation
under the scientific direction of my group which in 1973 had moved to
the Harvard-Smithsonian Center for Astrophysics (CFA). While
responsibility for overall scientific management was vested in me, as
Principal Investigator, it was in fact shared with Harvey Tananbaum, of
CFA, the Scientific Program Manager. The successful design and
fabrication of the mirror, the largest and most sophisticated X-ray telescope ever built, was in large part due to the scientific guidance of
Leon Van Speybroeck, of CFA (Figure 7).

    The group at CFA was also responsible for the two imaging
instruments at the focal plane. Harvey Tananbaum, Steve Murray,
Pat Henry and Ed Kellogg developed a photon counting X-ray camera
(HRI) capable of detecting X-ray images at the full resolution of the
telescope (4 arc seconds) over a field of about 20 arc minutes. The
position of each detected photon as well as its time of arrival is trans-

Figure 7.

mitted to the ground where it is stored in a computer. The photons accumulated over a long interval of time can be used to display an X-ray
image of the sky on a television screen which is then photographed to
produce the images shown later in this article. Paul Gorenstein and
Rick Harnden, of CFA, were responsible for the design of a larger
field (60 arc minutes), lower resolution (1 arc minute) imaging
camera which uses a proportional counter to yield not only position and
time, but also rough energy information on each detected photon.
Elihu Boldt, Steve Holt and Bob Becker, of Goddard Space Flight Center,
were responsible for the development of a cryogenically-cooled Lithium
drifted silicon detector which is used as a medium resolution, high
sensitivity spectrometer (SSS). George Clark and Claude Canizares,
of MIT, developed a Bragg Crystal Spectrometer yielding the highest
available spectral resolution, but unavoidably with low sensitivity.
Robert Novick, Knox Long, Bill Ku and David Helfand, of Columbia
University, were maintly involved with software and data analysis.
The mission operation planning and implementation was carried out
under the scientific direction of Ethan Schreier, of CFA.

The data from the spacecraft are transmitted to ground sta-
tions around the world, and from there to the Einstein Control Center
at Goddard Space Flight Center, which is manned 24 hours a day by a
crew composed of GSFC, TRW and CFA engineers. The data are then
stripped and sent to the CFA, where a sophisticated data handling system
has been developed, thanks primarily to the efforts of Christine Jones,
Bill Forman, Arnie Epstein and Jeff Morris. The data stream is there
further manipulated, the instantaneous pointing direction of the spacecraft is found as a function of time and images are accumulated in the
memory of the computer. They can then be recalled and displayed on
a TV screen.

Since its launch in November 1978, Einstein has performed
with no serious mishaps and it has certainly fulfilled our expectations
and hopes with regard to increased sensitivity and angular resolution.
We have achieved detection of X-ray sources 1000-fold fainter than any
previously detected and 10 million times fainter than the first we observed, Sco X-1. This progress is comparable to the improvement in
sensitivity which has occurred in visible light astronomy ever since
Galileo first pointed a telescope to the sky more than 300 years ago.
The faintest object that can be observed by the 200 inch telescope at Mt.
Palomar is about 10 million times fainter than the faintest star which
can be detected with the naked eye.

Each day some 10 different targets are selected and observed
according to a master observing program. In the 10 months since launch,
some 3000 different fields have been observed. In each field one or
more X-ray source can be seen, most of them never previously studied.

This has resulted in an enormous amount of new information on the
X-ray sky which is just now beginning to be analyzed; the first publications on the subject will appear in the scientific literature this fall.

To the fervor and scientific excitement of the scientists at
the four institutions, primarily involved with Einstein, there is now
added that of the more than 200 guest observers, - interested scientists
from all observational branches of astronomy who have joined us, under
a NASA-sponsored program administered at CFA by Fred Seward to
utilize the data. This level of interest and participation by the astronomical community is comparable to that at any major ground-based
national observatory operating in the visible or radio range of wavelengths and is unprecedented in X-ray astronomy. This Summer
School offers the first opportunity for a discussion on the relation
between the new findings of Einstein (and those of previous X-ray
astronomy missions) to the general body of astronomical knowledge
acquired through visible light and radio observations. The results
of these discussions may have a significant impact in shaping the research program of Einstein in the remaining two years of mission life.

OBSERVATIONS OF SUPERNOVA REMNANTS WITH THE EINSTEIN OBSERVATORY

Giuseppina Fabbiano

Harvard-Smithsonian Center for Astrophysics
60 Garden Street
Cambridge, Massachusetts   02138

1. INTRODUCTION

As for almost any other aspect of X-ray Astronomy, the sensitivity, the arcsecond angular resolution and the spectral resolution achieved with the Einstein Observatory are giving new surprising
results in the observation of Supernova Remnants (SNR's). We will
concentrate here on the results obtained with the imaging experiments on the galactic SNR's.
Previous X- and γ-ray observations have given indications of
the existence of two kinds of remnants, in agreement with optical
and radio data. Namely:
1) Remnants where a pulsar is present, like the Crab Nebula and
the Vela SNR;
2) Shell-like remnants, with no obvious pulsating neutron star
present, like Cas A, Tycho, SN 1006, Puppis A, the Cygnus Loop,
the Lupus Loop, IC 443.
Except for the Crab, where the emission mechanism is clearly
non-thermal and dominated by the presence of the pulsar, the X-ray
emission from the other SNR's was understood as being generally of
thermal nature, as evidenced by fit to an exponential of the spectrum
and more by the detection of an emission feature at 6 keV that was
interpreted as thermal emission lines from iron (cf., Davison,
Culhane and Mitchell, 1976, for Tycho; Pravdo et al., 1976, for
Cas A). This thermal emission was also discovered to originate
from extended regions, roughly reminiscent of the radio and/or
optical emitting regions (i.e., Fabian, Zarnecki and Culhane, 1973;
Rappaport et al., 1978; Levine et al., 1978; Fabbiano et al., 1979).
The information from these X-ray observations, on the whole,
supported the general evolutionary picture of SNR's (Woltjer, 1972).
In particular, soft X-ray excesses present in the spectra of young

15

R. Giacconi and G. Setti (eds.), X-Ray Astronomy, 15-34.
Copyright © 1980 by D. Reidel Publishing Company.

SNR's like Cas A and Tycho were interpreted as evidence for reverse
shock emission (McKee, 1974), while the result on older SNR's,
like the Cygnus and Lupus Loops and Vela, supported the picture of
SNR's in an adiabatic state of expansion (see e.g., review by
Winkler, 1978).

The observations of SNR's with the Einstein Observatory show
that the "old" picture was quite simplistic. A considerable variety
of shapes and structures is seen in SNR's, which makes each one of
them worthy of a detailed study. We expect that these new observations will ultimately allow a more profound understanding of the
physics of the emission processes in SNR's and of the interaction
of the supernova ejecta with the interstellar medium.

In the following sections some of the SNR's observed with the
Einstein Observatory will be discussed, namely the Crab Nebula,
the Vela Nebula and pulsar, the Tycho SNR, SN 1006 and Cas A.

2. CRAB NEBULA

Figure 1 shows the Einstein HRI picture of the Crab Nebula.
This image has a spatial resolution of the order of a few arcseconds.
It bears a very close resemblance with the radio picture of the
nebula (Swinbank, 1978) and with the spatial extent of the optical
continuum. The Crab appears clearly to be a filled-in structure,
with a horizontal feature which corresponds with the region of the
main optical wisp. The pulsar can be clearly spotted. A preliminary study of the pulsar emission shows the presence of pulsations
similar in shape to the ones already reported (cf. Bradt et al., 1969
(Figure 2). We are currently analyzing the Crab pulsar data to
investigate the relative contributions of the pulsed emission and
any DC component which may originate from the hot neutron star surface.

A total of 120 ct/s was obtained from the Crab Nebula, which
is what we would expect for the typical Crab spectrum $N(E) = 9.5 \times E^{-2.08}$ ph/s cm^2 keV and an equivalent absorption column of hydrogen $N_H \sim 3 \times 10^{21}$cm^{-2} as the one measured by Charles et al. (1979)
with the HEAO-02 experiment.

3. VELA SNR

The Vela SNR is the other known X-ray emitting supernova remnant containing a pulsar. Pulsations (p $\approx$ 89 ms) have been reported
both in radio, optical and γ-ray wavelength but not in X-rays.
Previous X-ray observations (Harnden and Gorenstein, 1973) showed
an extended soft source of diameter comparable with the one of the
optical nebula (5°) and a hard 1° source centered on the pulsar
(Smith, 1978). The Einstein Observatory observed the pulsar region
with both the IPC and HRI. We are mapping the entire 5° Vela SNR
with a series of $\sim$30 IPC observations. Figure 3 (Harnden et al., 197

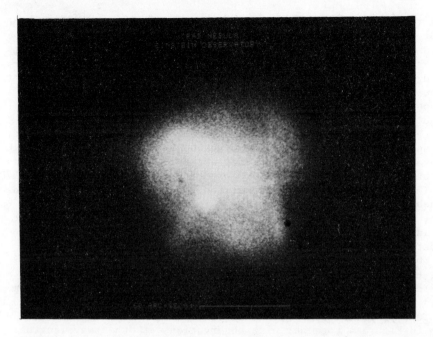

Fig. 1. HRI picture of the Crab Nebula (Tananbaum, private communication).

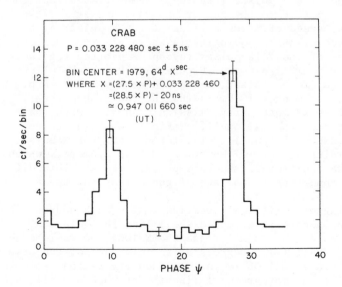

Fig. 2. Pulsations of the Crab pulsar, as seen with the Einstein
Observatory (Harnden, private communication).

shows three separate fields of ∿1 sq deg each of the Vela SNR. The
pulsar is located in the field at the bottom, where a bright extended nebula is the main visible feature. The image to the right
shows the same field in the hard band (2 - 4 keV). The bright extended nebula is absent but there is a prominent discrete source at
the position of the pulsar. The two histograms on the right show
the spectrum of the entire IPC field and a ∿2' region centered on
the pulsar respectively. The pulsar spectrum is clearly harder
than the spectrum of the nebula and is cut-off at the low energies.
This spectrum is consistent with black body spectra in the 2-3 x 10^6K
temperature range, but the fit does not rule out a syncrotron self-
absorption model like the one proposed for pulsars by Sturrock (1970).

The difference in spectra between the pulsar and the surrounding nebula might indicate that the pulsar is not directly related
to it, but rather that the pulsar happens to be located near a hot
spot in the original blast wave or along its line of sight. We do
not detect pulsations from the pulsar. Our folding analysis has
used ∿6000 s of IPC data and 2000 s of HRI data, giving 95% confidence upper limits of 1% and 2% respectively. (These percentages
refer only to the discrete source region, not to the entire 5° rem-
nant.) The IPC limit corresponds to ≤ 2 x 10^{-12}erg cm^{-2} s^{-1} at 1.5 keV
The Einstein pulsed upper limit and non-pulsed X-ray measurements are
shown in Figure 4 together with observations at other wavelengths:
radio, optical, γ-rays, and previous X-ray observations. With the
Einstein upper limit, the X-ray now falls somewhat below the optical
to γ-ray interpolation.

The HRI image of the pulsar (Figure 5) demonstrates that there
is a nebulosity connected to the pulsar of ∿2' diameter (∿1/4 pc).
(The radius of the speed of light cylinder for the pulsar is about
$1R_\odot$.) The presence of this nebulosity is shown clearly when the
surface brightness of the Vela pulsar region is plotted together
with the HRI point source response (Figure 6). Using the point
source contribution of the pulsar, we obtain a ∿1.5 x 10^6 °K black
body temperature, assuming a 10 km radius for the neutron star.

The Vela SNR shows both X-ray emission connected with the
pulsar and an extended thermal shell. It is interesting to compare
it with the Crab SNR. The angular diameter of the Crab Nebula (∿2')
corresponds at a distance of 1.7 kpc to a ∿0.7 pc diameter, which
is larger than the diameter of the Vela pulsar nebula. This difference could be connected to the age of the pulsar and the amount
of energy provided by it.

An interesting observational question that we can address is:
does the Crab Nebula have an extended shell like the Vela SNR?
Since the Crab Nebula has a similar age to SN 1006, we would expect
a ∿43 pc diameter shell (∿1.5 degrees at the distance of the Crab
Nebula) or less, since SN 1006 is in a low density region above the
galactic plane. If the integrated luminosity of such a shell is
L_x ∿10^{35}erg (1% of the total X-ray luminosity of the Crab Nebula)
probably it would have escaped detection in previous observations.

We will now discuss a different class of SNR's. These are

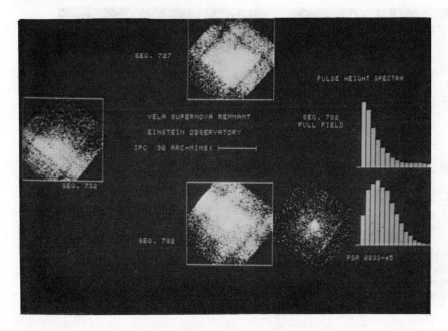

Fig. 3.   IPC observations of the Vela SNR (see text) (Harnden et al.,
1979).

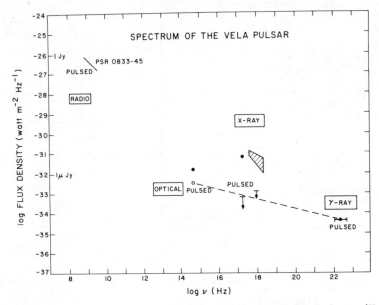

Fig. 4.   Summary of observations of the Vela Pulsar (Harnden et al.,
1979).

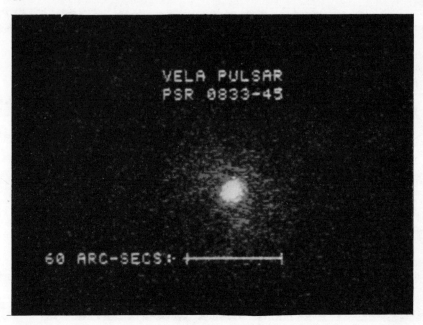

Fig. 5.   HRI image of the Vela Pulsar (Harnden et al., 1979).

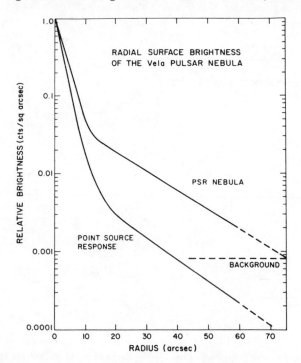

Fig. 6.   (Harnden et al., 1979).

SNR's for which, so far, no evidence of a compact remnant has been found. They all have a shell-like structure in the radio and some of them present filaments and optical nebulosities. We will discuss here the Tycho SNR, SN 1006, and Cas A. Tycho and SN 1006 appear to be remarkably similar – they both show a very close relationship between their radio and X-ray features; they both have only thin optical filaments, and their optical spectra show hydrogen Balmer emission lines only. The optical filaments lie just outside the radio/X-ray outermost contours. Cas A, instead, is a case on its own, showing an enormous complexity of features, which can be related both to optical and radio features.

4. TYCHO SNR

The spatial extent of the Tycho SNR has been measured for the first time by the HEAO-A3 modulation collimator experiment (1.5 – 5 keV) (Fabbiano et al., 1979). Figure 7 shows the annulus fit to the A 3 data superimposed on a radio map of the remnant. A region of more intense emission of about $\sim$2' extent was detected on the Northwestern side. Tycho turns out to be a very simply shaped, symmetrical remnant, for which a non-imaging experiment could get a reasonably accurate description. This can be seen from the HRI

THE RADIO OPTICAL AND X-RAY STRUCTURE
OF THE REMNANT OF TYCHO'S SUPERNOVA

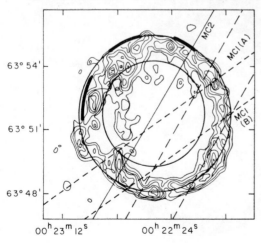

Fig. 7. The Tycho SNR as seen by the HEAO-A3 experiment (Fabbiano et al., 1979).

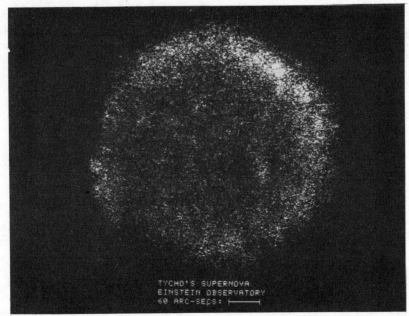

Fig. 8. The Tycho SNR as seen by the Einstein Observatory (HRI)
(Gorenstein et al., 1979).

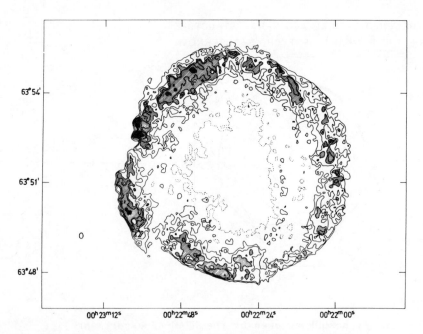

Fig. 9. Radio map of Tycho (Duin and Strom, 1975).

picture of Tycho (Figure 8) (Gorenstein et al., 1979). The SNR is
circular in appearance, showing a certain amount of limb brightening
in the North-West region and very little, almost no, internal structure. It is interesting to compare the Einstein image with a radio
map of Tycho (Figure 9). The similarity of the two images is remarkable. We see the broken shell both in the radio and in the
X-rays and we see some internal structure in both frequencies again.
An interesting feature is that the region more bright in radio is
less bright in the X-rays.
    The spectrum of Tycho in the X-rays shows the presence of
thermal radiation. It also shows the presence of a second harder
component, that led to the suggestion of reverse shock mechanism
operating in this remnant. It will be interesting to see how the
different regions of the remnant behave spectrally. An indication
will be given by the analysis of the IPC picture of Tycho which is
being carried out by the Columbia University scientists.
    From the surface brightness in the central region of the SNR

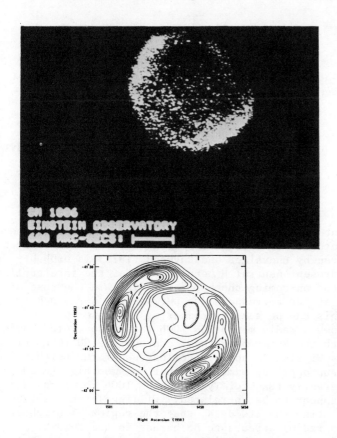

Fig. 10. Upper part: IPC image of SN 1006. Lower part: radio
map of SN 1006 (Stevenson, Clark and Crawford, 1977).

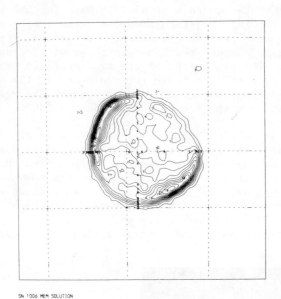

SN 1006 MEM SOLUTION

Fig. 11. Maximum entropy method X-ray brightness contours of the
SN 1006 (IPC) (Willingale, private communication).

an upper limit of $\sim 2 \times 10^6$ °K can be put on the temperature of a
compact stellar remnant of 10 km radius, radiating as a black body.

5. SN 1006

     The upper part of Figure 10 shows the IPC image of this SNR.
The optical filament is on the North-East side of the remnant, where
the X-rays are fainter. For this filament, as for the ones of Tycho,
the explanation given by Chevalier and Raymond (1978) is probably
valid that they represent neutral H nebulosities of the interstellar
medium excited by the on-coming shock front. Also, in the case of
SN 1006 there is a strict correspondence between the X-ray and
radio features. This can be seen by comparing the lower part of
Figure 10 which shows a radio map of SN 1006 (Stevenson, Clark and
Crawford, 1977) with the X-ray brightness contours of SN 1006 (Figure 11) obtained by Willingale and his colleagues of the Leicester
X-ray astronomy group applying a maximum entropy smoothing to the
IPC picture. Differently than for Tycho, in SN 1006 the X-ray and
radio intensities appear to be correlated. Willingale and collaborators also found, using the IPC data, that the region of brighter
emission (X-ray and radio) appears to be harder in the X-rays.
     Winkler (1978), based on the different age of Tycho and SN 1006,
hypothesized that the former was explainable in terms of the reverse

shock model, while the latter because of its greater age had to be
representative of a SNR in the adiabatic stage of expansion. He
interpreted the soft excess seen by SAS-3 in the X-ray spectrum of
SN 1006 as a probable indication of a pulsar remnant. The Einstein
data does not show any point-like component in the remnant. An upper limit of $\sim 1 \times 10^6$ °K can be put on the temperature of black
body radiation from a compact stellar remnant of 10 km radius.

The SSS data (Becker et al., 1979) show that the X-ray emission
of the Tycho SNR is principally thermal. Spectral high resolution
data on SN 1006 are not available as yet. It will be very important
to compare the spatial and spectral properties of these two SNR's.
The similarities between them support the hypothesis that they are
in a similar evolutionary stage, in spite of the difference in their
age. In particular, reverse shock mechanism has been suggested to
explain the two temperatures X-ray spectrum of Tycho (Coleman, et
al., 1973). If this were the case, the X-ray emission region would
lie well inside the outermost radio contour (Gull, 1973) contrary
to what we see in the X-ray picture. More debatable was the suggestion of a reverse shock mechanism operating in SN 1006, to explain the soft X-ray excess (Winkler et al., 1978), in view of the
fact that the age of this SNR would put it well in the adiabatic
stage. Here again the X-ray and radio emitting regions are fairly
well coincident. Clearly more detailed models will be needed to
explain these two SNR's.

6. CAS A

Cassiopeia-A is the youngest known SNR in our galaxy and as
such is of great interest in the understanding of the interaction
of blast wave with the surrounding interstellar medium and the mixing of the ejecta with it. The spatial resolution of previous
X-ray observations of Cas A (e.g., Charles, Culhane and Fabian,
1977) has been too coarse (~ 3 arc minutes) to permit comparison
with optical and radio features, which are typically of arc-second
angular scale. Cas A has been observed with both the HRI and the
IPC (Murray et al., 1979). The exposure time in the HRI picture
(Figure 12) is 32,819 seconds, in the IPC picture (Figure 13) is
1,660 seconds. In both pictures it is clear that the image contains complex structure, but that the general appearance of the
remnant is roughly circular. From the HRI picture we see that the
brightest regions lie in a ring-like structure and that there is an
emission plateau, extending beyond this ring, surrounding the source.
This is similar to the emission plateau observed at 6 GHz by Bell
(1977). The measured radius is 170 arcsec with a thickness of
~ 30 arcsec. Taking a distance to Cas A of 2.8 kpc (Van den Berg
and Dodd, 1970) the corresponding linear sizes are: r = 2.3 pc,
$\Delta r = 0.4$ pc.

The ring of brightest X-ray emission corresponds in general
with the radio ring with a radius of about 2 arcmin (1.7 pc).

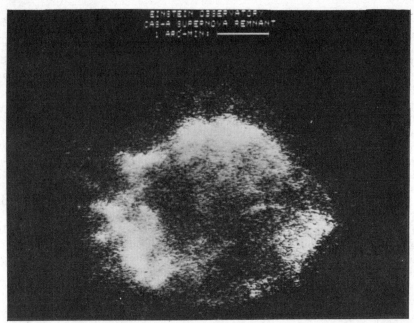

Fig. 12. HRI picture of Cas A (Murray et al., 1979).

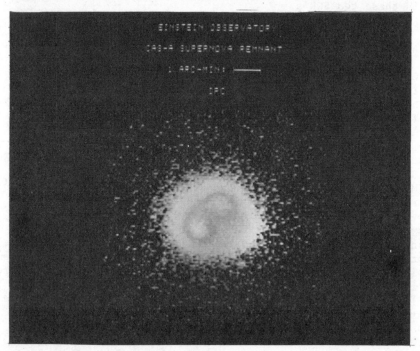

Fig. 13. IPC picture of Cas A (Murray et al., 1979).

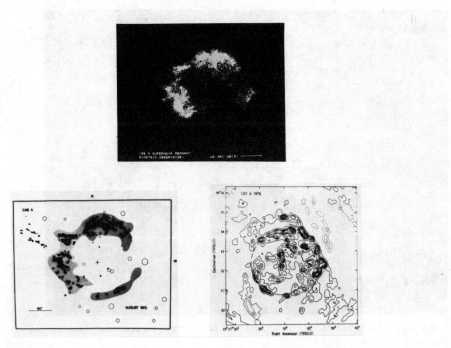

Fig. 14. X-ray, optical (Kamper and Van den Bergh, 1976) and radio
(Dickel and Greisen, 1979) maps of Cas A.

However this correlation is not always maintained on the arcsecond
scale where we find several isolated individual X-ray or radio
peaks. Similarly, with regard to optical features we find a general
correlation between the optical ring of knots and flocculi and the
X-ray emission ring (Figure 14). There appears to be a stronger
correlation between the X-ray emission and optical knots with stronger [SII] emission relative to [OIII] than with other features. The
IPC data indicate that most of the X-ray emission in the optical
nebulosity region is in the 1 to 2.5 keV band. This is the energy
range where S and Si lines are expected. The solid state spectrometer experiment has reported a very strong contribution from these
elements as well as Ca and Ar in the Cas A spectrum (Becker et al.,
1979). However there is also at least one example of an X-ray feature coincident with a quasi-stationary flocculus {2Q2, Kirshner and
Chevalier (1978)} which has primarily Hα and [NII] optical emission.
     The IPC image for Cas A also shows the features discussed above,
but at significantly lower resolution. We use the IPC data to study
spectral variations over the remnant. We have constructed images
which show the ratio of either hard to total or soft to total flux
as the function of position (Figure 15). The soft band is from 1
to 2 keV (on the right) and the hard band is from 2.4 to 3.6 keV
(on the left). The structure shown in this figure shows an east-west

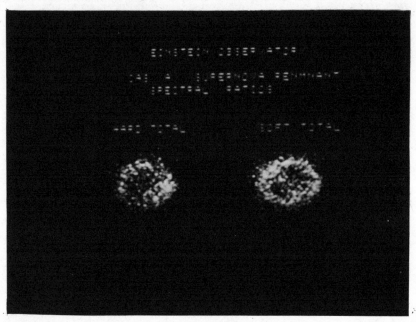

Fig. 15. Spatial distribution of spectral ratios in Cas A (IPC)
(Murray et al., 1979).

asymmetry which corresponds to differences in the radio and optical
images. The radio emission feature in the western section of Cas A
does not have a corresponding optical feature. Figure 16 shows the
radial dependence of these ratios and indicates quantitatively the
variation of the relative intensity in the 1 to 2 keV band. We
have plotted the spectral data taken from the northeast and southwest sections of the SNR (Figure 17). We observe that the spectral
distributions are significantly different. The southwest region
has more flux at high energies ($\gtrsim 1.5$ keV) and is more cut off at
lower energies.

In what follows we will discuss separately the physical implications of our data for the SN shell, the interior region and a
neutron star remnant.

I. _Shell_. From the radio measurements of Bell (1977), the
expansion velocity of the shock in Cas A is about 1700 km s^{-1} while
the optical knots have an expansion rate of $\sim$6000 km s^{-1} (Kamper
and Van den Bergh, 1976). Bell interprets this as evidence for
deceleration of the shell which requires that the mass swept up by
the shock be greater than the initial mass during the shock. In
this view, the fast moving knots are explained as a slower component
of the original ejecta which have not undergone appreciable deceleration due to their high density. The quasi stationary flocculi could
instead be due to pre-supernova ejecta and contribute to the medium
in which the main shock has been propagating.

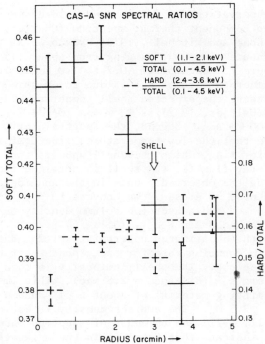

Fig. 16. A plot of the spectral ratios versus radial annulus for Cas A (Murray et al., 1979).

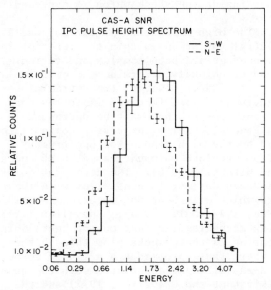

Fig. 17. The spectral data for features in the northeast and southwest of the Cas A image (Murray et al., 1979).

Various non-imaging observations of Cas A have yielded spectral data best fit by a two component thermal spectrum (Charles
et al., 1975; Davison, Culhane and Mitchell, 1976; Pravdo et al.,
1976) with the high temperature component ranging from about 3.5 to
5 keV and the low temperature component from ~ 0.7 to 1.0 keV. It
is likely that the high temperature component is due to emission
from the shell. First, because the expected shell temperature, as
given by $T_s \sim 0.8~M_p \mu V_s^2/K$ (Woltjer, 1972), is $T_s \sim 4 \times 10^7~°K$
(kT = 3.5 keV) for V_s = 1700 km s^{-1}. Second, the spectral ratio
plots of Figure 16 indicate that the low temperature, line dominated
emission is from the interior ring, while the shell appears to have
a flatter spectrum. The X-ray flux for the shell is obtained from
the HRI image by integrating the observed counts in the annulus of
radius 170 arcsec and thickness 30 arcsec. We convert this to flux
assuming a temperature of 4 keV and a hydrogen column density of
$\sim 1 \times 10^{22}$ cm^{-2} (see below). We obtain F_{shell} (0.1 - 4) $\cong$
2.3 x 10^{-10} erg cm^{-2} s^{-1} with an estimated uncertainty of 25% due
mainly to the corrections in counting rate for background subtraction
and spillover from bright regions. The shell luminosity is then
L_x (0.1 - 4) $\cong$ 2.2 x 10^{35} erg s^{-1} (distance = 2.8 kpc). This implies a density of X-ray emitting material of about 6 cm^{-3} and a
total mass in the shell of $\sim 3.5~M_\odot$. The ambient density is estimated to be $\geq$ 1.5 cm^{-3} (Bell, 1977) and the mass of swept-up material
is $\geq$ 2.2 M$_\odot$. These values indicate that Cas A is in the adiabatic
expansion phase (cf., Woltjer, 1972). The shock velocity predicted
for the Sedov similarity solution is about 2500 km s^{-1} which is
comparable with Bell's value of 1700 km s^{-1}. Also, the ambient
density is about 1/4 of that inside the shock which is expected for
a cold pre-shock gas.

II. <u>Interior Region</u>. The high resolution picture of Cas A
and the observation of spectral variations within the interior indicate that the emission processes are both complex and varied.
With our current data we cannot uniquely determine the emission
mechanism for each region of the SNR. However, the degree of correlation among X-ray, radio, and optical features indicates that
there are underlying relations present that must be addressed in
models for the object. In general, thermal processes dominate the
X-ray emission as indicated by the high resolution spectrum by
Becker et al. (1979) and from previous observations at high energies
(e.g., Davison, Culhane and Mitchell, 1976). The deceleration of
the expanding shell as discussed earlier is consistent with the
presence of a reverse shock which will heat the interior material
to X-ray temperatures (McKee, 1974; Gull, 1975). Chevalier (1975)
has also suggested a heating mechanism for fast moving knots which
is basically conductive heating as these knots plow through the
ambient medium. Examples where these processes are likely to be
occuring are given by the X-ray filament associated with the quasi-
stationary flocculus 2Q2 (Kirshner and Chevalier, 1977) and the
X-ray "jet" in the northeast region which corresponds to the jet
of high speed knots (Kamper and Van den Bergh, 1977). We have

also noted that the variation in spectrum within the SNR is characterized both by an increase in the "hardness" of the spectrum and
in the low energy cutoff. The increase in absorption is similar
to that observed by Greisen (1973) of about a factor of 2 in hydrogen column density from the east to the west $\{N_H \sim (0.5 - 1)$ x
10^{22} cm$^{-2}\}$. We note that the western structure which corresponds
to the bright radio peaks does not have an optical counterpart. If
this is not due to obscuration, it may indicate a contribution from
non-thermal processes which would require recent acceleration of
electrons to high energies. A proposed mechanism for such cosmic
acceleration has been given by Scott and Chevalier (1975). The required electron energy and estimated magnetic field strength limit
the synchrotron lifetime for these electrons to about 10 years, thus
requiring an on-going acceleration process.

In order to estimate the amount of material associated with
the X-ray emission from Cas A we have developed a simple model by
assuming basically only the thermal processes (with line emission)
and pressure equilibrium. We obtain a range of solutions consistent
with present observations which result in a range of total mass
from about 10 to 30 $M_\odot$.

III. <u>Neutron Star</u>. Supernova explosions are considered to
play an important role in producing pulsars with the Crab Nebula
being a classic example. In the case of Cas A no such object is
detected. An optical search by Kamper and Van den Bergh (1976)
gives an upper limit of $M_r \gtrsim 8^m$ for any stellar remnant near the
explosion center. We have searched the X-ray image for a possible
point source and determine an upper limit of 7.5 x 10^{-3} cts s^{-1} (3σ)
for the HRI counting rate. This corresponds to a black body temperature at 1.5 x 10^6 °K for a 10 km radius stellar remnant.

7. NEUTRON STAR TEMPERATURES

Table 1 summarizes our results on the temperature of stellar
remnants for the SNR's which we have discussed above. We have one
measurement, the temperature of the Vela pulsar (in the hypothesis
that the Vela pulsar emission is truly black body) and three upper
limits for stellar remnants in Tycho, SN 1006 and Cas A. As pointed
out by Bahcall and Wolf (1965) theories describing the cooling of
neutron star can be subjected to experimental test. The Einstein
experiment brings us to the point at which the experimental limit
on neutron star temperatures can put strong constraints on the
theory. Theories of cooling of neutron stars (see reviews by
Tsuruta 1978, 1979; Helfand and Novick, 1979 and Maxwell and Soyeur,
1979) give a minimum theoretically expected temperature for neutron
stars, which cool by the "standard" neutrino process, compatible
with the temperatures of the Vela pulsar, but above the upper limits
for Tycho, SN 1006 and Cas A. Our measurements exclude the possibility of such a star having been formed. However they are still
compatible with a pion core stellar remnant.

Table 1

Neutron Star Temperatures

SNR	T (10^6 °K)	References
Vela Pulsar	∿1.5(black body)	Harnden et al., 1979
Tycho	≤2.	Gorenstein et al., 1979
Cas A.	≤1.5	Murray et al., 1979
SN 1006	≤1.	Seward, 1979

8.   CONCLUSION

Summarizing, the Einstein observations of galactic SNR's open up a whole new chapter in the understanding of the detailed phenomena in these remnants, and pose strict upper limits on the temperature of a compact pulsar remnant. This could be both a test and a challenge to the existing theories and, on the other hand, poses a serious question on the origin of neutron stars in our galaxy. It has been hypothesized by Weiler and Panagia (1978) that pulsars would be found only in type II SN events, whose outcome would be a Crab-like or plerion type of SNR, showing flat radio syncroton spectra, filled emission regions and linear polarization, while the type I SN event would produce the shell-like, steeper radio spectra, remnants. The Einstein Observatory, and the imaging experiments of the future, while opening up more new questions, may give us the observational evidence needed to solve the puzzle.

ACKNOWLEDGMENTS

A number of people are involved in the study of SNRs with the Einstein Observatory and have substantially contributed to this paper. In particular, we thank Paul Gorenstein, Fred Seward, Rick Harnden, Steve Murray, and Harvey Tananbaum of the Center for Astrophysics and Richard Willingale and Daniel Rolf of Leicester University for communicating to us their results prior to publication and for assistance in preparing this talk. We thank Jeff Morris for helping, with his expertise in the computer display system, in the preparation of many of the figures that appear in this paper. We thank Cathy Roach for assistance in the preparation of the manuscript.

REFERENCES

Bahcall, J.N. and Wolf, R.A., 1965, Phys. Rev., 140, B 1452.
Becker, R.H., Holt, S.S., Smith, B.W., White, N.E., Boldt, E.A.,
Mushotzky, R.F. and Serelemitsos, P.J., 1979, Astrophys.J.(Letters),
234, L73.
Bell, A.R., 1977, Mon.Not.Roy.Astr.Soc., 179, 573.
Bradt, H., Rappaport, S., Mayer, W., Nather, R.E., Warner, B.,
MacFarlane, M. and Kristan, J., 1969, Nature, 222, 728.
Charles, P.A., Culhane, J.L., Zarnecki, J.C. and Fabian, A.C., 1975,
Astrophys.J.(Letters), 197, L61.
Charles, P.A., Kahn, S.M., Bowyer, S., Blisselt, R.J., Culhane, J.L.
and Cruise, A.M., 1979, Astrophys.J.(Letters), 230, L83.
Chevalier, R.A., 1975, Astrophys.J., 200, 698.
Chevalier, R.A. and Raymond, J.C., 1978, Astrophys.J.(Letters),
225, L27.
Coleman, P.L., Bunner, A.N., Kraushaar, W.L., McCammon, D.,
Williamson, F.O., Kellogg, E. and Koch, D., 1973, Astrophys. J.
(Letters), 185, L121.
Davison, P.J.N., Culhane, J.L. and Mitchell, R.J., 1976, Astrophys.
J.(Letters), 206, L37.
Dickel, J.R. and Greisen, E.W., 1979, Astron.Astrophys., 75, 44.
Duin, R.M. and Strom, R.G., 1975, Astron.Astrophys., 37, 33.
Fabbiano, G., Doxsey, R.E., Griffiths, R.E. and Johnston, M.D.,
1979, Astrophys.J.(Letters), submitted.
Fabian, A.C., Zarnecki, J.C., and Culhane, J.L., 1973, Nature, 242, 18.
Gorenstein, P., Murray, S., Epstein, A., Griffiths, R.E., Fabbiano,
G. and Seward, F.D., 1979, Bull.Am.Astron.Soc., 11, 462.
Greisen, E.W., 1973, Astrophys.J., 184, 363.
Gull, S.F., 1973, Mon.Not.Roy.Astr.Soc., 161, 47.
Harnden, F.R., Jr. and Gorenstein, P., 1973, Nature, 241, 108.
Harnden, F.R., Jr., Hertz, P., Gorenstein, P., Grindlay, J.,
Schreier, E., and Seward, F.D., 1979, Bull.Am.Astron.Soc., 11, 424.
Helfand, D.F. and Novick, R., 1979, private communication.
Kamper, K. and Van den Bergh, S., 1976, Astrophys.J.(Suppl.), 32, 351.
Kirshner, R.P. and Chevalier, R.A., 1977, Astrophys.J., 218, 142.
Levine, A., Petre, R., Rappaport, S., Smith, G.C., Evans, K.D.,
and Rolf, D., 1979, Astrophys.J.(Letters), 228, L99.
Maxwell, D.V. and Soyeur, M., 1979, 8th International Conference on
High Energy Physics and Nuclear Structure, Vancouver, August 13-18.
McKee, C.F., 1974, Astrophys.J., 188, 335.
Murray, S.S., Fabbiano, G., Fabian, A., Epstein, A. and Giacconi, R.,
1979, Astrophys.J.(Letters), 234, L69.
Pravdo, S.H., Becker, R.H., Boldt, E.A., Holt, S.S., Rothschild,
R.E., Serlemitsos, P.J. and Swank, J.H., 1976, Astrophys.J.(Letters),
206, L41.
Rappaport, S., Petre, R., Kayat, M., Evans, K., Smith, G. and
Levine, A., 1979, Astrophys.J., 227, 285.
Scott, J.S., and Chevalier, R.A., 1975, Astrophys.J.(Letters), 197, L5.
Seward, F.D., 1979, private communication.

Smith, A., 1978, Mon.Not.Roy.Astron.Soc., 182, 39P.
Stevenson, F.R., Clark, D.H. and Crawford, D.F., 1977, Mon.Not.Roy.
Astron.Soc., 180, 567.
Sturrock, P.A., 1970, Nature, 227, 465.
Swinbank, E., 1978, Mem.Soc.Astr.Ital., 49, 569.
Tsuruta, S. 1978, Preprint RIFP-344, Research Institute for Funda-
mental Physics, Kyoto University, Japan.
Tsuruta, S., 1979, this volume.
Van den Bergh, S. and Dodd, W.W., 1970, Astrophys.J., 162, 485.
Weiler, K.W., and Panagia, N., 1978, Astron.Astrophys., 70, 419.
Winkler, P.F., Jr., 1978, Mem.Soc.Astr.Ital., 49, 599.
Woltjer, L., 1972, Ann.Rev.Astron.Astrophys., 10, 129.

X-RAY SPECTRA OF SUPERNOVA REMNANTS

S.S. Holt

Laboratory for High Energy Astrophysics
NASA/Goddard Space Flight Center
Greenbelt, Maryland 20771 U.S.A.

1. X-RAY SPECTROSCOPY

Until quite recently, there has been little hard evidence for the
existence of discrete features (such as emission lines) in the
spectra of X-ray sources. The basic problem, of course, has been
that the exposure obtainable with contemporary instrumentation has
precluded the effective utilization of dispersive techniques, so
that we were limited to the use of proportional counters (with
typical resolving power $E/\Delta E$ no better than $\sim$ 5) for spectroscopic
study. Nevertheless, the high relative abundance of Fe has allowed
the detection of both thermal and fluorescent emission lines from
a variety of astrophysical contexts, even with the relatively poor
resolution available with proportional counters, so that we could
be certain that better resolution would eventually allow the
detection of emission features from atomic transitions which were
less pronounced relative to the surrounding continuum.

The model-fitting procedure we employ is a straightforward
application of Poisson statistics. The raw data from an exposure
to a celestial X-ray source are binned in pulse-height channels
(labelled with the argument E', which should be nominally relatable
to energy E); the counts accumulated in the j'th channel over the
exposure are $N_j(E')$, with a Poisson error of $\delta N_j(E')$ taken to be
$\sqrt{N_j(E')}$. An independent estimate of the background contributing
to $N_j(E')$ over the accumulation time is $B_j(E')$, with error $\delta B_j(E')$,
so that the actual "source" data accumulated are given by

$$T_j(E') = N_j(E') - B_j(E') \tag{1.1}$$

with an estimated 1σ error given by

R. Giacconi and G. Setti (eds.), X-Ray Astronomy, 35-45.
Copyright © 1980 by D. Reidel Publishing Company.

$$\delta T_j(E') = ((\delta N_j(E'))^2 + (\delta B_j(E'))^2)^{\frac{1}{2}} \qquad (1.2)$$

The only fair test of a hypothesis that a given spectral form can be adequately fitted by the data must be performed in E' (rather than E) space, because the transformation of the $T_j(E')$ to an input spectrum is non-unique. In general, the best non-dispersive spectrometers have response functions (or the probability that a photon of energy E will be recorded at a pulse-height E') with the following three characteristics:
1. A stable maximum in the response function at a pulse-height E' corresponding to E
2. A gaussian shape centered at E'-equivalent-to-E with a FWHM (= 2.36σ) such that > 50% of the response function area lies below the gaussian
3. The remainder of the response function area in a relatively featureless plateau at E' < E
More detailed discussions of the response functions of X-ray detectors may be found in the literature (e.g. Holt 1970)
    The general fitting procedure begins with the assumption of a model input spectrum S(E) which is characterized by a number m of free parameters. The expectation value for the data in the j'th bin from this model is then:

$$S'_j(E') = \int_0^\infty S(E)R_j(E,E')dE \qquad (1.3)$$

where $R_j(E,E')$ is the response function for the j'th channel. If there are a total of P channels for which equation (1.3) can be independently constructed, the quantity

$$\chi^2 = \sum_{j=1}^{P} \left[\frac{S'_j(E') - T_j(E')}{\delta T_j(E')} \right]^2 \qquad (1.4)$$

is a suitable statistic for (p-m) degrees of freedom. Several authors have recently discussed the appropriate "acceptability" criteria to apply to overall models and to the ranges over which individual parameters can be constrained (e.g. Cash 1976). In general, the important points to remember are that there is no statistical requirement to increase the number of free parameters once an acceptable fit is obtained, and that the range of acceptable parametric values will always be model-dependent. In other words, a physically interesting set of parameters will be obtained only if a physically interesting model is chosen at the outset. Before the advent of satellite experiments, the statistical quality of the data did not permit severe constraints on potentially interesting parameters (e.g. in most cases, we found that power laws and featureless thermal bremsstrahlung continua could fit the data

equally well). More recently, we are finding that just the reverse
is true (i.e. that the statistical errors are so small that most
simple model forms can be excluded, and that our ability to achieve
statistically adequate fits is limited by the availability of
detailed model calculations).

All of the data presented in this lecture will be analyzed in
this manner, but it is worth discussing alternative data pre-
sentations which appear in the literature. Because the above
approach is model-dependent, a simple model-independent manner in
which the ordinate of a figure labelled "counts" can be converted
into "incident photons" is to construct

$$
S^{(n+1)}(E) = \frac{T_j(E') \, S^{(n)}(E)}{\int_0^\infty R(E,E')S^{(n)}(E) \, dE}
\tag{1.5}
$$

where the superscripts on the input spectra refer to the trial
number. If, for the initial guess at a trial spectrum, a flat
spectrum $S^{(1)}(E)$ = constant is used, it is clear that subsequent
iterations are likely to yield a convergent value of $S^{(\infty)}(E)$ which
is independent of any a priori assumptions about the input
spectrum. A χ^2-test of $S^{(\infty)}(E)$ against trial models in E-space
may then be performed, but it is not quite as sensitive as that of
equation (1.4) because the iteration process tends to smooth out
sharp features which might be present in the true input spectrum.

A representation for the input spectrum which is quite popular
at present replaces $S^{(n)}(E)$ in equation (1.5) with the input spectrum
deduced from the fitting procedure in E'-space. The difficulty
with this representation is that the model-independency has been
removed, so that sharp features which may be incorrectly introduced into the fitting procedure will be enhanced in the data pre-
sentation. Much of the recent proportional counter data in the
literature has been displayed in this manner. A new technique
developed by Blissett and Cruise (1979) appears to offer the best
features of both (i.e. model independency and enhancement of real
features). In this approach, the data are essentially filtered
with a window function commensurate with the resolution of the
detector, so that only those sharp features which are at least as
wide as the natural detector response in E'-space will survive the
spectral inversion.

2.   DIFFUSE X-RAY PRODUCTION

The simplest model spectra are expected from those astrophysical
situations for which X-ray transfer effects in the source are
minimized. Such "diffuse" X-ray sources can be generally characterized by single interactions of electrons with electromagnetic
fields which give rise to X-ray emission. If we assume that the

electrons of interest can be approximately represented by a power
law spectrum:

$$\frac{d^2N}{dVd\gamma} \quad \alpha \quad \gamma^{-\Gamma} \text{ cm}^{-3} \tag{2.1}$$

where γ is the electron Lorentz factor, the rate at which the
electron loses energy in interactions with the field is

$$- \frac{d\gamma}{dt} \quad = \quad \frac{4}{3} \frac{\sigma_o}{mc} \rho \gamma^2 \qquad s^{-1} \tag{2.2}$$

where σ_o is the Thomson cross-section and ρ is the energy density
in the field. The characteristic "cooling time" of the process
is then simply

$$\tau \quad = \quad (\frac{1}{\gamma}\frac{d\gamma}{dt})^{-1} \quad = \quad \frac{3mc}{4\sigma_o}\rho\gamma \quad s \tag{2.3}$$

In the case of Compton interactions, the field energy density is
that in soft target photons (e.g. infa-red or optical). For
synchrotron radiation, the appropriate energy density is that in
the magnetic field, $B^2/8\pi$. Coulomb collisions can also be formally
accommodated by equations (2.2) and (2.3), but the fact that the X-ray
yield is so low ($\sim 10^{-4}$) means that the appropriate distribution
of electrons to consider is not a power law, but is instead an
equilibrium Maxwellian to reflect the fact that a large number
of Coulomb collisions are required in order to produce an X-ray
photon. The two non-thermal processes yield power-law X-ray
spectra with energy index $\alpha = (\Gamma-1)/2$, where Γ is the spectral
index from equation (2.1) in the energy range appropriate to the
electrons responsible for the X-ray production. The X-radiation
produced by Coulomb interactions has an equilibrium spectral form
given by

$$E\frac{dq}{dE} \approx 10^{-19} \text{ g } Z^2n^2 \text{ (kT)}^{-\frac{1}{2}} e^{-E/kT} \qquad \text{erg cm}^{-3} \text{ erg}^{-1} \tag{2.4}$$

The expression in equation (2.4) is the volume emissivity for bremsstrahlung only, and any line emission must be added to it in order
to determine the total X-ray emission; above $\sim 10^7$ K bremsstrahlung
dominates, while line emission dominates the cooling at lower temperatures. T is the temperature, k is the Boltzmann constant, n is
the density, Z is the effective atomic number of the plasma (~ 1.3
for cosmic abundance) and g is the Gaunt factor which, for temperatures in our range of interest, can be approximated by

$$g \approx (E/kT)^{-0.4} \tag{2.5}$$

The total bremsstrahlung luminosity per unit volume can be obtained
by integrating equation 2.4 over energy, so that an effective
cooling time can be defined as the ratio of the plasma kinetic
energy density to the luminosity per unit volume:

$$\tau \approx 10^{19} \frac{\sqrt{kT}}{n} \quad s. \tag{2.6}$$

It is important to note that the bremsstrahlung X-rays are produced by electrons of comparable energy, while the non-thermal
processes typically require electrons of much higher energy. For
Compton interactions, an X-ray of energy E will result from a
single scattering of a photon of initial energy E_o according to

$$<E> = \frac{4}{3} \gamma^2 <E_o> \tag{2.7}$$

while the analogous synchrotron case (in a field of magnitude B)
will be

$$<E> \approx 10^{-20} \gamma^2 B \quad ergs \tag{2.8}$$

The production of 10 keV photons from 1 eV photons or a 1 gauss
field would require, therefore, γ of 10^2 or 10^6, respectively.
More detailed treatments of diffuse X-ray production may be found
in many references (e.g. Blumenthal and Tucker 1974; Holt 1974).

3.   X-RAY PRODUCTION IN SUPERNOVA REMNANTS

The prime examples of diffuse X-ray production in galactic X-ray
sources are the X-ray emitting supernova remnants. The energy
source in these remnants is the kinetic energy which still remains
from the initial supernova outburst.

　　　The Crab nebula is a special case, in the sense that we have
yet to find another remnant with the same X-ray characteristics.
It contains a pulsar (i.e. a rotating magnetized neutron star)
which rotates with a period of 33ms. The fact that the pulsar is
slowing down with time was one of the pieces of evidence which
originally led to its identification with a neutron star; its
importance in the present context is that the rate at which it is
losing rotational kinetic energy matches the total luminosity of
the remnant in all energy bands. Pacini pointed out that since
the rate at which energy is converted to electromagnetic modes
scales like P^{-4} (P is the rotation period), and pulsed X-ray
luminosity created at the speed-of-light-circle scales like P^{-10},
it would be unlikely that other remnants would be found which
could either power X-ray sources at a level $> 10^{36}$ ergs/s, or

which would yield X-ray pulsations; in particular, the next-fastest rotator, that in Vela X with a period of 87 ms, does neither. It is important to note that "pulsars" are defined here as those neutron stars whose rotational energy is responsible for radio or X-ray luminosity; "X-ray pulsars" in binary systems derive their energy from other means, as will be discussed in later lectures.

We are now certain that the X-radiation from the Crab nebula remnant arises from synchrotron emission. The spectrum is well-fit by a power law out to at least 100 keV (e.g. Dolan et al. 1977) and Weisskopf et al. (1976) have measured X-ray polarization which matches that in the optical. There remain some difficulties in the theoretical considerations involved in applying the appropriate electrodynamics to efficiently transfer the rotational energy to X-ray production in detail, but both the energy source (rotational kinetic energy) and the final X-ray production mechanism (synchrotron radiation) are firmly established.

The other SNR which are X-ray emitters are quite different in all observational respects. There is no evidence whatsoever for pulsars at their centers, but we still believe that they are powered by kinetic energy; in these cases, the kinetic energy is that contained in the ejecta originally expelled by the explosion. The traditional theory of Shklovsky (1968) was first investigated in detail by Heiles (1964) for its X-ray implications, and many newer treatments may be found in the literature (e.g. Itoh 1977).

In bare outline, the post-explosive history of supernovae may be characterized by three distinct phases. In the first phase, the ejecta move supersonically out through the interstellar medium, with the attendant shock accumulating ISM material (but not so much as to effectively brake the outflow, i.e. the SNR radius increases linearly with time). When the accumulated mass becomes comparable to ejected mass, the shock slows down because of its increased inertia. It is during this phase that X-ray emission becomes important, because of both the increased mass and the lower temperatures associated with the decreasing shock velocity. When X-ray cooling becomes so large that the expanding remnant loses a significant fraction of its energy, it rapidly slows down (the third phase) and eventually breaks up into the ISM.

The second phase is often called the "adiabatic" phase because the X-ray cooling is not yet high enough to have radiated away a substantial fraction of the initial energy, and the Sedov similarity solution yields the following relations. The remnant radius R, in terms of the initial energy of the ejecta ε_0, the ISM mass density in front of the shock ρ_0 and the age of the remnant t is

$$R \approx \left(\frac{\varepsilon_0}{\rho_0}\right)^{1/5} t^{2/5} \approx \frac{5}{2}Vt \qquad (3.1)$$

where V is the present shock velocity. The temperature T of the material which has equilibrated with the shock is

$$T \; \underset{\sim}{\,} \; \frac{3}{16} \; \frac{\overline{m}}{k} \; V^2 \; \underset{\sim}{\,} \; 2 \times 10^{-9} \, V^2 \tag{3.2}$$

where $\overline{m}$ is the mean ion mass. The density just behind the shock is four times that in the unshocked ISM, and Heiles estimates the total volume emission measure of the shell to be

$$n^2 V \; \underset{\sim}{\,} \; 10 \; n_o^2 \; R^3 \tag{3.3}$$

where n_o is the number density in the ISM. The Crab will probably never be such a thermal "shell" source because the ISM density in its neighborhood is so low that such a snowplow will never be effective there. Many other remnants do exhibit this type of behavior, however, as we shall see below. It is important to note that as the X-ray luminosity of a shell source increases, we expect its temperature and shell velocity to decrease, so that it cools more even more rapidly as it progresses toward the third phase.

4. X-RAY SPECTRA OF SUPERNOVA REMNANTS

Even before the launch of Einstein, several supernova remnants were known to be thermal sources not just because they could be fit with bremsstrahlung continua, but also because thermal Fe emission at 6.7 keV could be unambiguously detected with proportional counters. Surprisingly, the temperatures to which the continua could be fit were considerably lower (several keV rather than several tens of keV) than that prescribed by equation 3.2 on the basis of the ob-served shock velocities, at least in the cases of Cas A and Tycho, and the fits further required the addition of lower temperature components to match the data below $\sim$ 4 keV. Even in the case of Fe K-emission, where the line is well separated from others and the continuum is falling rapidly so that the equivalent continuum width exceeds 1 keV, the emission feature appears only as a dis-continuity in the smooth continuum rather than a sharply defined emission line. For the emission lines expected below 4 keV from lower-Z components, there is no hint of structure in the raw data from proportional counters.

The Einstein Observatory contains a variety of dispersive and non-dispersive spectroscopic tools for the investigation of celestial X-ray sources. I will describe, here, some results from the solid-state-spectrometer (SSS), a non-dispersive device which combines high efficiency with resolution sufficient to separate individual line components, although they cannot be completely resolved. Figure 1 illustrates the capability of separating expected line features with the $\sim$ 160 eV FWHM resolution of the instrument. Not only can individual elements be easily separated, but hydrogen-like, helium-like and neutral fluorescence lines from the same elements can be separated, as well. The total field-of-view of the SSS is approximately 6 arc-minutes in diameter, and

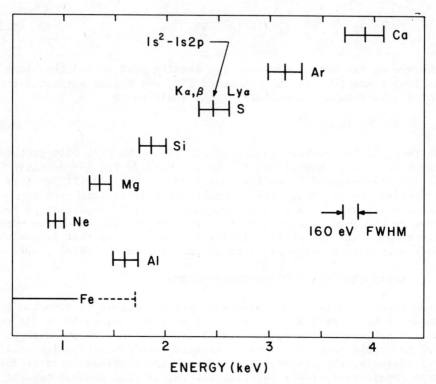

Fig. 1. Comparison of the Einstein SSS energy resolution with the
separation between line features which might be expected to be
prominent in X-ray sources.

its effective energy range is 0.8 - 4.5 keV. Additional instrumental details may be found in Joyce et al. (1978) and Holt et
al. (1979).
       A sample of the spectra obtained with the SSS from supernova
remnants is shown in Figure 2. The data displayed are all raw
pulse-height counts, compared with a solid histogram representing
a model fit, as described in Section 1. The Crab nebula is fit
with a power law only, and the others are fit with two-temperature
thermal spectra. In the latter cases, the dashed lines represents
the total contributions from elements with atomic number ≤ 10
(i.e. the large fraction of the continuum). For Cas A, where the
exposure is greatest, there is statistical significance associated
with line emission from Fe L-transitions, as well as from
K-transitions of Mg, Al, Si, S, Ar and Ca; furthermore, both helium-
like and hydrogen-like components are observed from Si and S
(although the former are at least an order of magnitude more
prominent). For Tycho, where the peak of the Si line is more than
an order of magnitude above the continuum, fully 50% of the
photons detected are in the Si and S lines.

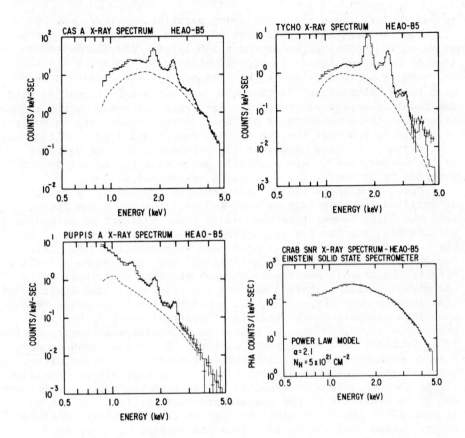

Fig. 2. Einstein SSS experimental spectra of four supernova remnants.

All of the spectral fits for the displayed thermal spectra (and those of similar remnants not explicitly displayed in Figure 2) are performed in the same manner. A high temperature component is fixed from previously obtained HEAO A-2 data, and we allow a lower temperature component, the abundances of all the line-prominent elements, and the relative normalizations of the two temperature components to be free parameters (the same abundances for both components). The models assume that the two components are each separately in collisional equilibrium (Raymond and Smith 1977,1979).

5.   DISCUSSION OF SPECTRAL RESULTS

The younger remnants (Cas A, Tycho and Kepler are all ∿ 400 years old) display spectra with marked similarities. The abundances,

in particular, exhibit the following peculiarities: Fe and Mg
are slightly less than solar, but Si, S and Ar are each in excess
of solar and appear to go in the ratio 1:2:4. The Si abundance
in both Cas A and Kepler is approximately twice solar, while that
in Tycho is approximately seven times solar. The obvious problem
with these "abundances" is that they are intimately connected with
the model assumptions. In particular, they are based on the assumptions that the abundances of all the Z ≤ 10 elements are in solar
proportions, and that the two thermal components are in collisional
equilibrium. We have no way of directly measuring the validity
of the former, and we are painfully aware that the latter is not
expected to be true for a recently shock-heated plasma.

     Pravdo and Smith (1979) have recently concluded, on the basis
of detailed interrogation of HEAO A-2 data for Cas A and Tycho,
that the "high temperature component" shows evidence of extending
all the way from the few keV which constitutes the "best fit" to
the much higher temperatures predicted from equation 3.2. This
component is clearly not in equilibrium, but its precise modelling
affects the overall abundances through the true contributions to
the high ionization state lines and the continuum. The lower temperature component (typically, 0.5 - 0.6 keV for these young remnants) is probably characteristic of material which is cooling
behind the shock, and may be closer to an equilibrium configuration.
It is probably true, therefore, that the Si and S abundances we
fit are probably the closest to being correct (consistent with a
solar lower-Z elemental distribution), since they are so prominent.
The Ar abundance is more affected by the high temperature continuum,
the Fe is sensitive to the temperature distribution, and the Mg
is coupled to the Fe because the L-lines from the latter confuse
the continuum near the K-lines from the former.

     Nevertheless, there are some consistent trends which are
evident in the young SNR spectra. Even though the emitting plasma
is an unknown mixture of ejecta, swept-up ISM and pre-explosive
stellar mass loss, there is a clear overabundance of Si-group
material relative to Fe, as in the fast-moving optical filaments
of Cas A (Kirshner and Chevalier 1977). This relative overabundance
is approximately the same in Kepler (Type I) and Cas A (Type II),
while Tycho (Type I) is about a factor of four higher. This suggests that the pregenitor for Type I SNR is a relatively low mass
star (Tinsley 1979) rather than the He star for which Arnett (1979)
has calculated pronounced overabundances in the Fe-group ejecta.
The approximately solar Fe abundances deduced from the Fe-L blends
are in agreement with the Fe-K intensities previously measured in
earlier experiments.

     Older remnants (e.g. Puppis A), which are > 10^3 years old
and have radii > 10 pc, seem to exhibit very consistent abundances
in all line-prominent elements. The ionization temperatures
deduced from the best fits are only slightly less than those for
the younger remnants, and we are presumably measuring a near-
equilibrium situation with abundances which should be approaching

solar. In fact, the best-fits require abundances which are approximately twice solar for all elements except S, for which the
abundance is twice as high. There is a strong suggestion from
both the older and younger SNR (as well as from the spectra
measured from other thermal sources not discussed in this lecture)
that the relative abundance of S called "solar" by Raymond and
Smith is about a factor of two low, i.e. that the actual solar
abundance of S relative to Si is a factor of two higher.

Remnants which are even older, such as Cyg Loop, are too cool
to be observed with the SSS. Cyg Loop is undoubtedly cooling very
rapidly via line emission (unfortunately, the oxygen emission is
below the SSS threshold), and the remnant is entering the third
(and last) phase of significant X-ray production and SNR
morphology.

Clearly, the obvious next step in the understanding of the
spectral data from SNR requires detailed modelling of the hydrodynamics and ionization balance in order to convert the apparent
abundances to true abundances. The imaging data obtained from
Einstein will be crucial to the construction of detailed models,
and the constraints imposed by both the imagery and the spectroscopy should remove a large fraction of the non-uniqueness which
has characterized such modelling in the past.

REFERENCES

Arnett, W.D., 1979, Ap. J. (Letters) 230, L37.
Blissett, R. and Cruise, A.M., 1979, MNRAS 186, 45.
Blumenthal, G.R. and Tucker, W.H., 1974, in X-Ray Astronomy
(D. Reidel Publishing Company, Dordrecht-Holland).
Cash, W., 1976, Astr. Ap. 52, 307.
Dolan, J.F., Crannell, C.J., Dennis, B.R., Frost, K.J., Maurer,
G.S., and Orwig, L.E., 1977, Ap. J. 217, 809.
Heiles, C., 1964, Ap. J. 140, 470.
Holt, S.S., 1979, in Introduction to Experimental Techniques in
High Energy Astrophysics, NASA SP-243, 63.
Holt, S.S., 1974, in High Energy Particles and Quanta in
Astrophysics (MIT Press, Cambridge, Massachusetts).
Holt, S.S., White, N.E., Becker, R.H., Boldt, E.A., Mushotzky,
R.F., Serlemitsos, P.J., and Smith, B.W., 1979, Ap. J. (Letters),
in press.
Itoh, H., 1977, Publ. Astron. Soc. Japan 29, 813.
Joyce, R.M., Becker, R.H., Birsa, F.B., Holt, S.S. and Noordzy,
M.P., 1978, IEEE Trans. Nucl. Sci. NS-25, 453.
Kirshner, R.P. and Chevalier, R.A., 1977, Ap. J. 218, 142.
Pravdo, S.H. and Smith, B.W., 1979, Ap. J. (Letters), in press.
Raymond, J.C. and Smith, B.W., 1977, Ap. J. Suppl. 35, 419.
Raymond, J.C. and Smith, B.W., 1979, in preparation.
Shklovsky, I.S., 1968, Supernova (Wiley, New York).
Tinsley, B., 1979, Ap. J. 229, 990.
Weisskopf, M.C., Cohen, G.G., Kestenbaum, H.L., Long, K.S., Novick,
R., and Wolff, R.S., 1976, Ap. J. (Letters) 208, L125.

OBSERVATIONS OF SUPERNOVA REMNANTS
IN THE LARGE MAGELLANIC CLOUD
WITH THE EINSTEIN OBSERVATORY

David J. Helfand and Knox S. Long

Columbia Astrophysics Laboratory
Columbia University
New York, New York 10027, U. S. A.

1. INTRODUCTION

As the nearest extragalactic system, the Large Magellanic Cloud
(LMC) is destined to play a major role in understanding the X-
ray emission from all normal galaxies including our own.
Classed as a prototype Sb(s)m galaxy by de Vaucouleurs and Freeman (1972), the LMC comprises several well-defined regions: (a)
a featureless, hydrogen deficient, optical bar, (b) hydrogen
rich, young spiral arms, and (c) a diffuse, hydrogen poor halo
containing numerous star clusters (Bok 1966). The mass of the
LMC is estimated to be ~10^{10} $M_\odot$ of which from 5 to 9% is thought
to be neutral hydrogen (McGee and Milton 1966). Medium energy
X-ray (2—10 keV) emission was first reported by Mark *et al.*
(1969). Above 2 keV, the X-ray luminosity of the LMC is dominated by emission from 3 to 5 point sources similar to the bright
sources found near our own galactic center (Markert and Clark
1975; Leong *et al.* 1971). Observation of fainter sources has
been hampered by source confusion and detector sensitivity limitations.
     Prior to the launch of the Einstein Observatory (HEAO 2),
very little was known about soft X-ray emission from the LMC
aside from the total 0.4—1.5 keV luminosity of ~10^{38} ergs s^{-1}
(Long, Agrawal, and Garmire 1976). Today, however, that situation has changed. After the completion of approximately half of
a series of observations of the LMC with the imaging proportional counter (IPC) on board the Einstein Observatory (Giacconi *et
al.* 1979a), approximately 40 X-ray sources have been located.
When the initial survey is complete, the entire cloud will have
been mapped to a limiting sensitivity of 3 × 10^{35} ergs s^{-1}.
More detailed observations with a sensitivity of 5 × 10^{34} ergs

R. Giacconi and G. Setti (eds.), X-Ray Astronomy, 47-59.
Copyright © 1980 by D. Reidel Publishing Company.

s^{-1} will be performed in selected regions including the optical
bar and 30 Doradus. This survey will provide us with detailed
knowledge of the luminosity function of sources in the LMC, allowing us to delineate the contribution and characteristics of
various source populations. In our initial analysis effort, we
have concentrated on data concerning previously known supernova
remnants (SNR) and SNR candidates.

X-ray observations are critical to the study of the dynamics and evolution of SNR, their impact on the energy balance of
the interstellar medium (ISM), and their role in driving the
chemical evolution of a galaxy. The interpretation of the X-ray
emission from galactic remnants has been hampered by uncertainties in distance estimates, typically a factor of 2 to 3, and by
the often large and poorly known column densities of interstellar material along the line of sight through the galactic plane.
Studies of galactic sources have also been compromised by difficulties in comparing observations carried out with different instruments in different spectral bands.

The emerging sample of SNR from the Einstein survey of the
LMC eliminates many of these difficulties. The distance is
known to 10% and the absorption, both from our Galaxy and within
the cloud, is relatively low. Thus we can produce a uniform,
luminosity-limited sample of SNR for comparison with data from
other spectral regimes, with remnants in our own Galaxy (and
other Local Group members) observed with the same instrument,
and with theoretical models of SNR evolution. The first results
of this work, presented below, have been both surprising and encouraging in each of these regards.

2. OBSERVATIONS

Several lists including optical and radio data on possible SNR
in the LMC have been compiled. Mathewson and Clarke (1973;
hereafter MC) catalogued 12 objects, based on evidence of non-
thermal radio emission, extended filamentary structure in Hα,
and strong (S II) emission. They included two additional candidates near the crowded 30 Doradus region based solely on their
radio properties. Subsequent optical studies of several of
these objects have revealed evidence for high-velocity material,
and as a result, their identification as SNR is reasonably se-
cure. Subsequently, Davies, Elliott, and Meaburn (1976; hereafter DEM) obtained a more extensive but less compelling list of
SNR candidates from Hα data alone.

All ten of the objects tabulated by MC that fall within our
partially completed survey have been detected, including the two,
N157B and N158A, whose identification as SNR was based solely on
the nonthermal nature of their radio spectra. Of those objects
we have surveyed which are unique to the DEM list, only two were
detected as weak X-ray sources. Most are evidently not SNR or,
if they are SNR, are sufficiently old that their X-ray luminos-

ities have fallen below our sensitivity threshold. Additionally, we have detected a weak source coincident with the SNR 0522-67.9. The total of 13 objects is listed in Table 1 along with X-ray positions (column 2) determined by a computer source detection algorithm (uncertainty 1') (Giacconi *et al.* 1979b), X-ray counting rates in the Einstein IPC (column 3), and radio fluxes (column 4).

For the six brightest sources, we have carried out a standard analysis of the X-ray spectra to obtain temperatures and intrinsic luminosities corrected for interstellar absorption (see Long and Helfand 1979 for a discussion of this analysis). A distance to the LMC of 55 kpc (Bok 1966) and cosmic abundances (Brown and Gould 1970) were assumed. For the remaining objects, insufficient counts were obtained for a spectral analysis, and thus, we have assumed mean spectral parameters for these sources to determine their X-ray luminosities. The results are listed in columns 2 and 3 of Table 2. Parameters for the young galac-

Table 1

Supernova Remnants Observed in the Large Magellanic Cloud

Name (1)	X-Ray Position (1950.0) R. A. Dec. (2)		X-Ray Counting Rate (counts s^{-1}) (3)	Radio Flux* [10^{-14} erg (cm^2 s)$^{-1}$] (4)
N86	$04^h56^m01^s$	$-68°43\!.7$	0.020 ± 0.010	2.2
N186D	05 00 19	-70 12.2	0.015 ± 0.007	2.2
0522-67.9	05 22 31	-67 58.5	0.014 ± 0.005	...
N(49)	05 25 21	-66 01.5	0.517 ± 0.019	2.9
N132D	05 25 32	-69 41.1	4.549 ± 0.055	25
N49	05 25 58	-66 07.1	1.03 ± 0.025	9.6
N206	05 32 43	-71 01.1	0.031 ± 0.005	2.6
No. 238	05 34 43	-70 35.8	0.027 ± 0.006	<6
N63A	05 35 45	-66 02.4	2.84 ± 0.04	11
No. 240	05 36 49	-70 42.1	0.026 ± 0.005	<2
N157B	05 38 12	-69 11.5	0.189 ± 0.015	15
N158A	05 40 36	-69 20.8	0.540 ± 0.021	9.7
N135	05 47 42	-69 42.6	0.023 ± 0.006	3.4
Tycho	...	...	19.430 ± 0.090	...

*Radio fluxes have been calculated from the 408 MHz flux densities and spectral indices quoted in MC, integrated from 10^2 to 10^4 MHz. Where no spectral index was available, a value of -0.5 was assumed. Upper limits to the 2700 MHz fluxes for source Nos. 238 and 240 were taken from catalogs and maps of McGee, Brooks, and Batchelor (1972) and Broten (1972), respectively.

tic remnant of Tycho's SN, observed and analyzed in an identical
manner, are included for comparison.

The angular resolution of the IPC is not sufficient to determine the X-ray diameters of SNR in the LMC. Therefore, until
observations of these remnants have been carried out with the
high resolution imager (HRI), we must rely on measurements of
their size obtained at other wavelengths. Unfortunately, the
optical diameters for several of the sources are not uniquely
determined. For example, according to Lasker (1978), any of
three diameters, 6, 32, and 80 pc, may apply to N132D. Only the
largest can be ruled out with the IPC. The actual sizes of N157B
and N158A are particularly uncertain since, in the absence of
well-defined optical filaments, their diameters were obtained by
MC using the notoriously unreliable $\Sigma - D$ relation. For
consistency, we have used diameters (column 4 of Table 2) derived by MC when a SNR was in their survey, and from DEM if it

Table 2

Derived Parameters for Supernova Remnants
in the Large Magellanic Cloud

Name (1)	L_{x} (0.5–3.0 keV)* (10^{36} ergs s^{-1}) (2)	kT* (keV) (3)	D (pc) (4)	t§ (10^3 yr) (5)	n_0§ (cm^{-3}) (6)	E_0§ (10^{50} ergs) (7)
N86	<0.6	0.3$^+$	30	12.0	<0.12	<1.8
N186D	<0.5	0.3$^+$	36	14.0	<0.086	<2.2
0522−67.9	<0.4	0.3$^+$	...	...	...	...
(N49)	4	0.4	10	3.4	1.4	1.1
N132D	39	0.3	6	2.4	11	1.3
N49	11	0.4$_+$	16	5.5	1.2	3.5
N206	0.40	0.3$_+$	50	20.0	0.047	3.2
No. 238	0.34	0.3$^+$	38	15.0	0.066	2.0
N63A	27	0.2$_+$	7	3.4	9.5	1.2
No. 240	0.33	0.3$^+$	40	16.0	0.060	2.1
N157B	4.2	0.8	6	1.4	2.7	0.8
N158A	14	0.6$_+$	12	3.3	1.8	3.4
N135	0.29	0.3$^+$	40	16.0	0.056	2.0
Tycho	4.4	0.4	6	2.0	3.3	0.5

*See Long and Helfand (1979) for discussion of the uncer-
tainties involved in the determination of L_{x} and kT.

†For these weak sources, the mean values of the spectral
parameters for (N49), N132D, N49, and N63A (kT = 0.3 keV; n_{H} =
3 × 10^{21} cm^{-2}) have been assumed in deriving luminosities and
the Sedov solution parameters.

§Obtained from Sedov equations.

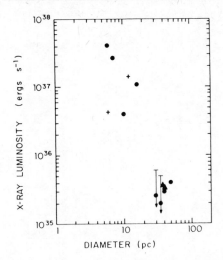

Fig. 1. Luminosities (0.5—3.0 keV) of SNR in the LMC as a function of diameter. Diameters for N157B and N158A (+) were determined from a $\Sigma - D$ relation by MC, while the rest of the diameters were optically determined (MC = filled circles; DEM =
filled triangles).

was not. [We note, however, that a recently obtained HRI image
of N49 and (N49) indicates X-ray diameters of ~25 and 40 pc, respectively, considerably larger than the optical diameters of 16
and 10 pc reported by MC.] A plot of the X-ray luminosity versus these diameters for the 13 detected sources is shown in Figure 1. There appears to be considerable range in luminosity for
SNR of the same general size. However, the trend for luminosities to decrease as the optical diameter exceeds 30 pc is evident, as it is in the case of similar studies for remnants in
our own Galaxy.

        Another quantity of interest in studying the evolution of
SNR is the ratio of X-ray luminosity L_x to radio luminosity L_R.
Seward *et al*. (1976) have discussed this parameter for galactic
remnants, but these results were compromised, as noted above, by
the indeterminate distances of many sources and the variety of
instruments used to obtain the X-ray data. Our plot of L_x/L_R
versus D for the LMC remnants is shown in Figure 2. Again,
N157B and N132D appear anomalous, with ratios of L_x/L_R that are
more appropriate for larger remnants. A determination of the X-
ray diameters for as many of these sources as possible is clearly desirable before a more detailed interpretation of these luminosity—diameter plots is undertaken. In general, though, we
again see a strong decline for this ratio toward larger diameters, reflecting the fact that the X-ray luminosity evolves much
more dramatically than does the intensity of the radio emission,
which only changes by a factor of 3 between the smallest and

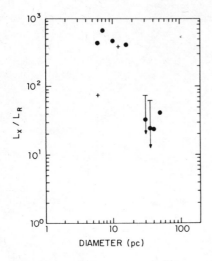

Fig. 2. The ratio of X-ray luminosity L_x (0.5—3.0 keV) to radio luminosity L_R (10^2—10^4 MHz) as a function of diameter. Symbols are as in Figure 1.

largest diameter objects.

3. COMPARISON WITH THE THEORY

As an initial step toward the interpretation of the X-ray data on SNR in the LMC and to facilitate comparison with previous work on galactic remnants (e.g., Gorenstein and Tucker 1976), we have analyzed our results within the framework of the standard blast wave picture. The Sedov solution for a symmetric blast wave propagating into a uniform-density medium allows one to solve for the expansion velocity V_b, the age t, the initial energy of the explosion E_0, and the ambient density of the ISM n_0, given the radius, X-ray luminosity, and temperature of the remnant. Assuming a mean molecular weight for fully ionized material of 0.61, one can derive the following expressions:

$$V_b = 2.69 \times 10^2 \ T_6^{1/2} \ \text{km s}^{-1} \tag{3.1}$$

$$t = 1.46 \times 10^4 \ R_{10} \ T_6^{-1/2} \ \text{yr} \tag{3.2}$$

$$n_0 = 0.4 \ R_{10}^{-3/2} \ L_{37}^{1/2} \ \Lambda_{-22}^{-1/2} \ \text{cm}^{-3} \tag{3.3}$$

$$E_0 = 0.6 \times 10^{50} \ R_{10}^{3/2} \ T_6 \ L_{37}^{1/2} \ \Lambda_{-22}^{-1/2} \ \text{ergs} \quad , \tag{3.4}$$

where the temperature T_6 is expressed in units of 10^6 K, the radius R_{10} is in units of 10 pc, the X-ray luminosity L_{37} is in units of 10^{37} ergs s^{-1}, and the band cooling function Λ_{-22} is in units of 10^{-22} erg cm^{-3} s^{-1}. Values for the band cooling func-

tion have been obtained from the work of Raymond, Cox, and Smith
(1976) for an optically thin, cosmic abundance plasma by digitiz-
ing data presented in Figure 5 of their paper and integrating
over our bandwidth 0.5–3.0 keV. The values range from Λ_{-22} =
0.04 at T_6 = 1.1 to Λ_{-22} = 0.36 at T_6 = 9.3. Heavy elements are
generally thought to be underabundant in the LMC, perhaps by as
much as a factor of 4 (D'Odorico and Sabbadin 1976, and references therein). Since Λ_{-22} enters only as the 1/2 power in equations (3.3) and (3.4), however, the error in estimating n_0 and
E_0 introduced by our assumption of cosmic abundances should be
at most a factor of 2.

The results of applying the model to our observations are
presented in the last three columns of Table 2. The relatively
narrow range of X-ray temperatures leads to blast wave velocities of typically 500 km s^{-1}; ages range from just over 1000 to
20,000 yr and are reasonably consistent with ages derived from
the observed optical expansion rates. As expected, the calculated explosion energies are not dependent upon remnant size or
age; they range from ~0.8 to 3.4 × 10^{50} ergs with a mean of 2 ×
10^{50} ergs, similar to the mean for galactic remnants (Gorenstein
and Tucker 1976). However, the ISM densities derived from the
blast wave model are highly correlated with the observed remnant
diameters as illustrated in Figure 3. The values of n_0 range

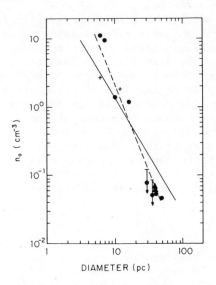

Fig. 3. Derived values of n_0 from the Sedov solutions for SNR
in the LMC as a function of SNR diameter. The solid line indicates an approximate fit to a law, $n_0 \propto D^{-5/3}$, as predicted by
McKee and Ostriker (1977). The actual slope is better fitted by
a somewhat steeper law, $n_0 \propto D^{-5/2}$ (dashed line). Symbols are
as in Figure 1.

from 0.05 to 11 cm^{-3} — a factor of ~200. This correlation between diameter and density approximated by the expression $n_0 \propto D^{-5/2}$ is clearly inconsistent with the standard blast wave picture in which n_0 should be independent of D.

McKee and Ostriker (1977) have developed a detailed theory of a supernova shock wave expanding into a tenuous ISM ($n_0 = 10^{-2.5}$ cm^{-3}), studded with small clouds ($n_0 > 10$ cm^{-3}) which have a filling factor of 0.15. The clouds have only a small effect on the dynamics of the shock. As a result, SNR ages derived from the theory differ by only 10% from the standard solution, while the values of E_0/n_0 are lower by roughly a factor of 2. X-ray emission in this model is dominated by matter which is evaporated from the cool, relatively dense clouds. The importance of the evaporating cloud material decreases as a function of time, and the derived densities are predicted to vary with radius as:

$$n_0 = n_c (R/R_c)^{-5/3} \quad , \tag{3.5}$$

where n_c and R_c are the density and radius at the time when a dense, radiative shell forms, that is, when the remnant has radiated roughly half the original SNR energy. When the standard parameters from the theory and typical SN energies ($E_0 \sim 10^{51}$ ergs) are used, R_c is approximately 180 pc.

In partial support of their model, McKee and Ostriker (1977) noted that if the data from several galactic SNR are interpreted within the context of a Sedov picture, a trend is seen toward a decreasing value of n_0 with increasing diameter. A similar trend is seen here, but over a much larger range of n_0. In Figure 3, we have plotted an eyeball fit to the data of a line with slope -5/3. The intercept of the line can be used to calculate the value of n_c, the density of the tenuous component of the ISM, of approximately $10^{-2.1}$ cm^{-3}. As noted by McKee and Ostriker, this is roughly the density of this component required to account for the diffuse X-ray background in our own Galaxy. More generally, the current observations support the conclusion that inhomogeneities cannot be ignored in explaining the evolution of SNR.

4.   COMPARISON WITH OTHER GALAXIES

In addition to providing a uniform sample of objects for the study of SNR per se, this set of remnants in a relatively unevolved irregular galaxy like the LMC, when compared with SNR in galaxies of other morphological types, may offer valuable insight into the interplay between SNR and interstellar media. Already the results are intriguing. The first surprise, of course, is the high luminosities exhibited by the LMC remnants. Although our survey of galactic SNR is far from complete, six of the LMC remnants have luminosities equal to or higher than any known galactic object excluding only the nonthermal emission of the Crab Nebula. If one naively scaled the number of SNR in a galaxy with total galactic mass, the expected number of sources at a comparable evolutionary stage in a galaxy such as ours would be ~100 — at least five should be within 3 kpc where interstellar absorption is fairly modest. There are, in fact, about 70 sources which have been detected in M31 above a luminosity of ~6 × 10^{36} ergs s^{-1} (Van Speybroeck et al. 1979), but ~9 have been identified with globular clusters. Many others are variable on short time scales, and it is estimated (Van Speybroeck 1979) that at most a dozen will ultimately be identified as SNR. This is a number comparable to that in the LMC despite a 20:1 mass ratio for the two galaxies.

Recent work by Dennefeld and Tammann (1979) indicates that the mass function and stellar birthrate per unit mass in the LMC for stars with $M \gtrsim 9\ M_{\odot}$ (likely SNR progenitors) are not substantially different from the Galaxy, and estimates of the SN rate in the LMC based on the MC catalogue are consistent with a simple scaling by galactic mass (the rates are one per 500 yr in the LMC and one per 20 to 50 yr in the Galaxy). Thus, it appears that there are not more SNR per unit mass in the LMC, and therefore more brighter ones; rather, each remnant is typically more luminous than its galactic counterpart. Although it is conceivable that the slightly lower metal abundance in the LMC may play a role in this discrepancy, it is perhaps more likely that a difference in the density and/or clumpiness of the ISM in the cloud is important. Scheduled observations of other nearby irregular and spiral galaxies (e.g., Leo II and M33) will help to establish the significance of the preponderance of high-luminosity sources in this young galaxy.

5. N49, ITS COMPANION, AND A γ-RAY BURST

While the properties of the LMC remnants taken as a class are likely to provide the most important insight into astrophysical problems associated with SNR, several of the remnants are bright enough that detailed studies of individual objects are possible. Our first HRI picture of an LMC field is shown in Figure 4. The two bright extended sources are N49 and (N49). These objects have been of particular interest since the time they were first identified as SNR. They are about 100 pc apart on the plane of the sky, yet are connected by a bridge of Hα emission extending from a break in the northwest quadrant of the optical shell of N49 through the centroid of (N49) and beyond. This remarkable morphology led MC to suggest that they were in fact physically connected and that (N49) had somehow been ejected at the time of the SN explosion. Kinematic work by Danziger and Dennefeld (1976) later showed that, since the relative radial velocity of the two remnants was small (<150 km s^{-1}), the ejection would have to have occurred within $\lesssim 3°$ of the plane of the sky.

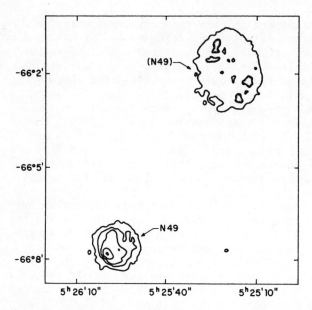

Fig. 4. The surface brightness contour map of the region containing N49 and (N49) as observed with the HRI. The map has
been smoothed by convolving the raw data with a Gaussian with a
σ of 3". The contour levels correspond to 0.025, 0.10, 0.20,
0.40, and 0.62 counts $(1' \times 1')^{-1}$ s^{-1}. The lowest contour corresponds to an equivalent flux of approximately 10^{-6} erg cm^{-2} s^{-1}
keV^{-1} in the energy range 0.5–4.5 keV.

Coupled with additional information on the velocities of the
emission regions within the two objects, they concluded that a
causal connection was most unlikely. Our detection of the pair
as distinct, bright, soft, extended X-ray sources confirms that
they are indeed simply two separate SNR. There is no evidence
in either our IPC or HRI data of any X-ray emission corresponding
to the connecting Hα filaments. It is, perhaps, simply superimposed foreground or background emission or, more speculatively,
could mark the fading shell of a very old SNR which, having induced star formation along its shock front (e.g., Herbst and
Assousa 1977), is still visible after the most massive new stars
have exploded. (At 10 km s^{-1}, the shell would have traveled
less than the diameter of either remnant in the ∿2 million year
lifetime of a 30 $M_\odot$ star and will still appear to connect the
SNR.)
     The two remnants have rather different X-ray morphologies.
(N49) is a smooth, symmetric shell with a diameter of 40 pc,
whereas N49 is only half this size and shows considerably more
structure. The general outline of the emission is similar to
that seen in an optical photograph of the object, including the
dificit of emission in the northwest corner. In detail, how-

ever, the X-ray and optical pictures (the latter being dominated by Hα emission) do not correlate well. A much better correspondence is found between the X-ray data and the Fe XIV coronal line maps of Mathewson and Dopita (1979) and Clarke (1979).
Both are tracers of the same hot (several million degrees) gas,
and so are expected to be coincident. However, the lack of correspondence with the optical emission line regions (at 10^4K) has
been used by Mathewson and Dopita as evidence in support of the
shocked cloudlet model of McKee and Ostriker (1977) cited above.
Further such detailed comparison of radio, optical, and X-ray
data on the LMC remnants map provide a welcome increase in the
rather small sample of galactic objects for which such studies
are possible.

Recently, N49 has drawn the attention of astronomers for yet
another reason. On March 5, 1979, a giant γ-ray burst was recorded by each of the nine interplanetary and Earth-orbiting satellites currently equipped with γ-ray detectors. The event had
an extremely short rise time of $\leq$250 µs (Cline et al. 1979) and
a peak flux above 30 keV of 1.5 x 10^{-3} erg cm^{-2} s^{-1} (Mazets et
al. 1979). After decaying to a level of 10^{-2} times the peak
value in a few hundred milliseconds, emission persisted for nearly a minute, exhibiting clear, double-peaked 8.1-s pulsations.
Several independent position determinations all fall within ∿1'
of the centroid of the soft X-ray emission from N49 (Evans et al.
1979). If placed at the distance of the LMC, the peak burst luminosity was ∿5 x 10^{44} ergs s^{-1}, and the total burst energy was
∿4.6 x 10^{44} ergs (Mazets et al. 1979).

With impeccable planning, we observed N49 with the IPC eight
days before the event and again five weeks later. There was no
change in the X-ray flux or spectrum to $\leq$3%, corresponding to a
limit on the change in luminosity of <3.5 x 10^{35} ergs s^{-1} at the
distance of the LMC. We note that the limit on the X-ray flux
variation is distance-independent which implies that, ∿10^6 s after the burst, the system associated with the event had a soft
X-ray luminosity <10^{-9} of that released in γ-rays. The HRI picture obtained two months after the burst shows no evidence of a
discrete point source within the remnant, although, for example,
a 10-km neutron star with a surface temperature of 3.5 x 10^6 K
would be easily detectable above the background of the remnant
(again, assuming a distance of 55 kpc). In addition, no obvious
foreground source is superimposed on the remnant emission. For
example, if the burst originated at a distance of ∿100 pc as suggested by Mazets et al. (1979), the steady X-ray flux from the
system responsible is <1.5 x 10^{30} ergs s^{-1}; i.e., less than that
expected from many normal stars (Vaiana 1979). Thus, the X-ray
observations provide some stringent constraints on any models
which may be proposed for this anomalous γ-ray event (see Helfand
and Long 1979).

6. CONCLUSIONS

We have detected 13 SNR in the course of our continuing X-ray
survey of the LMC with the Einstein Observatory. A few other
newly discovered sources with similarly soft spectra may also
turn out to be remnants when their positions are compared with
optical and radio maps, and several more objects will very likely be found from among the remaining catalogued remnants which
have yet to be observed. Several of the sources are bright
enough for detailed study of their X-ray brightness distribution
and its correlation with radio and optical properties. More importantly, though, these data offer an excellent opportunity to
compare a uniform, complete, luminosity-limited sample of SNR
with theoretical models of SNR evolution. They will also be a
useful probe with which to differentiate interstellar conditions
in, and, perhaps, to establish evolutionary trends among galaxies of differing morphological type.

ACKNOWLEDGMENTS

We would like to express our appreciation to the entire Einstein
Observatory group for bringing this superb experiment to frui-
tion. In addition, we thank Drs. T.L. Cline and L.P. Van
Speybroeck for providing information in advance of publication
and Dr. H. Van der Laan and the staff of the Leiden Observatory
for providing a most hospital environment in which to complete
the draft of this contribution. This work was supported by the
National Aeronautics and Space Administration under contract
NAS8-30753. This paper is Columbia Astrophysics Laboratory
Contribution No. 181.

REFERENCES

Bok, B.J., 1966, Ann.Rev. Astron. Astrophys., 4, 95.
Broten, N.W., 1972, Australian J. Phys., 25, 599.
Brown, R.L., and Gould, R.J., 1970, Phys. Rev. D, 1, 2252.
Clarke, D., 1979, private communication.
Cline, T.L. et al., 1979, Astrophys. J., in press.
Danziger, I.J., and Dennefeld, M., 1976, Astrophys. J., 207, 394.
Davies, R.D., Elliott, K.H., and Meaburn, J., 1976, Mem. Roy.
Astr. Soc., 81, 89 (DEM).
Dennefeld, M., and Tammann, G.A., 1979, ESO Scientific Preprint
No. 56.
de Vaucouleurs, G., and Freeman, K.C., 1972, Vistas Astron., 14,
163.
D'Odorico, S., and Sabbadin, F., 1976, Astron. Astrophys., 50, 315.
Evans, D., Klebesadel, R., Baros, J., Cline, T., Desai, U., Tee-
garden, B., and Pizzichini, G., 1979, I.A.U. Circ. No. 3356.

Giacconi, R. et al., 1979a, Astrophys. J., 230, 540.
Giacconi, R. et al., 1979b, Astrophys. J. (in press).
Gorenstein, P., and Tucker, W.H., 1976, Ann. Rev. Astron. Astrophys., 14, 373.
Herbst, W., and Assousa, G.E., 1977, Astrophys. J., 217, 473.
Helfand, D.J., and Long, K.S., 1979, submitted to Nature.
Lasker, B.M., 1978, Astrophys. J., 223, 109.
Leong, C., Kellogg, E., Gursky, H., Tananbaum, H., and Giacconi, R., 1971, Astrophys. J., 170, L67.
Long, K.S., Agrawal, P.C., and Garmire, G.P., 1976, Astrophys. J., 206, 411.
Long, K.S., and Helfand, D.J., 1979, Astrophys. J., in press.
Mark, H., Price, R., Rodrigues, R., Seward, F.D., and Swift, C.D., 1969, Astrophys. J., 155, 143.
Markert, T.H., and Clark, G.W., 1975, Astrophys. J., 196, L55.
Mathewson, D.S., and Clarke, J.N., 1973, Astrophys. J., 180, 725.
Mathewson, D.S., and Dopita, M.A., 1979, Astrophys. J., 231, L147.
Mazets, E.P., Golenetskii, S.V., Il'inskii, V.N., Aptekar', R.L., and Guryan, Yu. A., 1979, Nature, in press.
McGee, R.X., Brooks, J.W., and Batchelor, R.A., 1972, Australian J. Phys., 25, 613.
McGee, R.X., and Milton, J.A., 1966, Australian, J. Phys. Astrophys. Suppl. No. 2.
McKee, C.F., and Ostriker, J.P., 1977, Astrophys. J., 218, 148.
Raymond, J.C., Cox, D.P., and Smith, B.W., 1976, Astrophys. J., 204, 290.
Seward, F., Burginyon, G., Grader, R., Hill, R., Palmieri, T., Stoering, P., and Toor, A., 1976, Astrophys. J., 205, 238.
Vaiana, G.S., 1979, this volume, 129.
Van Speybroeck, L., 1979, private communication.
Van Speybroeck, L., Epstein, A., Forman, W., Giacconi, R., Jones, C., Liller, W., and Smarr, L., 1979, Astrophys. J., in press.

HIGH RESOLUTION X-RAY SPECTROSCOPY FROM THE EINSTEIN OBSERVATORY

P.F. Winkler[+], C.R. Canizares, G.W. Clark, T.H. Markert
C. Berg, J.G. Jernigan and M.L. Schattenburg
Massachusetts Institute of Technology
Department of Physics and Center for Space Research
Cambridge, Massachusetts 02139, USA

1. INTRODUCTION

This paper is devoted to a discussion of some results which we have recently obtained from the fourth of the principal instruments on board the Einstein Observatory: M.I.T.'s Focal Plane Crystal Spectrometer (FPCS). We shall begin with a few general remarks about X-ray spectroscopy, followed by a brief description of the FPCS instrument. The results we present here deal primarily with supernova remnants (SNRs): Puppis A and Cas A in the Galaxy, and N132D and N63A in the Large Magellanic Cloud. In addition we shall briefly discuss a member of the other class of thermal X-ray source under discussion at present; namely, to report our detection of oxygen emission from the vicinity of M87 in the Virgo Cluster.

The FPCS experiment was designed primarily for the study of lines. Numerous important spectral lines occur at energies 0.2 – 4 keV, the sensitivity range for the Einstein Observatory. In particular, the strong n=2 to n=1 resonance transitions for hydrogen-like and helium-like ions of abundant "heavy" elements from carbon through calcium lie in this energy range (see Table 1). In addition, numerous transitions downward to the n=2 level in lithium-like to neon-like ions of iron and nickel may also be observed.

One expects all of these lines as well as others to be produced in emission from an optically-thin thermal plasma at X-ray temperatures (Raymond and Smith, 1977 and references therein). Such plasma is thought to be principally responsible for the X-ray emission from most SNRs and from clusters of galaxies. Measurement of the line strengths can enable us to determine the elemental

---

+ Alfred P. Sloan Research Fellow; on leave from Middlebury
  College, Middlebury, Vermont.

R. Giacconi and G. Setti (eds.), X-Ray Astronomy, 61-71.
Copyright © 1980 by D. Reidel Publishing Company.

Table 1

Energies of n=2→n=1 Resonance Transitions

Element	Transition Energy (keV)	
	He-like Ion	H-like Ion
C	0.31	0.37
N	0.43	0.50
O	0.57	0.65
Ne	0.92	1.02
Mg	1.35	1.48
Si	1.86	2.01
S	2.46	2.62
Ar	3.14	3.32
Ca	3.91	4.11

abundances and ionization temperature. At energies $\gtrsim 1.5$ keV, lines can be resolved and identified with a resolving power $E/\Delta E \sim 15$. This could not be achieved with proportional counters, but as Holt (1980) has reported here, the Solid State Spectrometer (SSS) on the Einstein Observatory has returned beautiful data on spectral lines in this energy range. At energies below ~ 1.5 keV, thermal X-ray spectra become more complicated, due largely to the numerous L-lines from several ionization states of Fe interspersed among K-shell lines from Mg and lighter elements. In this range a resolving power ~ 100 is often necessary to identify the lines. This requires a dispersive spectrometer in combination with a focusing X-ray telescope in order to achieve sufficiently low background rates for lines to be observed – a combination of which has been realized for the first time with the Einstein FPCS.

The increased resolution also makes several notable diagnostics for plasma astrophysics accessible. These include the study of closely-spaced multiple lines, e.g., the trio of n=2 to n=1 lines in He-like ions, the relative strengths of which measure departures from ionization equilibrium (Acton and Brown,1978); the study of line profiles and satellite lines; the measurement of Doppler broadening in spectral lines. All of the latter become possible only at a resolving power of ~ 50 and are feasible for the first time with the FPCS.

2. FPCS EXPERIMENT

The FPCS is a curved-crystal Bragg spectrometer which operates at the focus of the Einstein Observatory X-ray telescope. Figure 1 shows a schematic of the instrument and illustrates its operation. X-ray from a celestial source are brought to a focus by the telescope, pass through one of several selectable apertures, and strike one of six curved-crystal diffractors. X-rays with wavelengths very

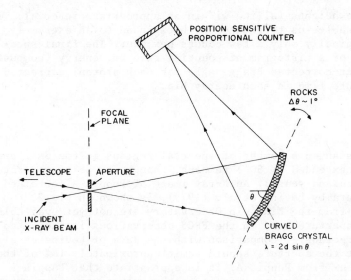

Fig. 1. A schematic diagram of the Focal Plane Crystal Spectrometer on the Einstein Observatory.

nearly meeting the Bragg condition ($\lambda = 2d \sin \theta$) for the selected Bragg angle (θ) setting of the diffractor (lattice spacing d) are reflected and astigmatically refocused onto one of two redundant, position-sensitive proportional counters. In a given observation, the crystal is rocked through a small angular range, thereby tuning the bandpass of the instrument back and forth across a limited spectral region. The instrument can be reconfigured to study an entirely different region by changing the diffractor, its angle with respect to the incident X-rays and the relative positions and orientations of the diffractor and detector. The instrument covers the range 0.2 - 3 keV, with a resolving power, E/ΔE, of 50 to 500. The FPCS is described in more detail by Canizares et al. (1979), Donaghy and Canizares (1978) and Giacconi et al. (1979).

A typical observation involves a scan width of three to five spectral resolution elements centered on some specific spectral feature and lasts 5000 to 40,000 sec. The end result of the data reduction process is a list of photons tagged with arrival time, proportional-counter pulse height, position in the proportional counter, and effective Bragg angle. The position and Bragg angle are corrected for the instantaneous source image location in the focal plane using the satellite aspect solution from on-board star trackers.

In order to reconstruct a spectrum, the data are first selected to exclude times of earth blockage, high particle background and poor aspect determination. Then further selections are made to include only those photons which: (1) are detected within the region of the detector containing the image of the source, and

(2) have pulse heights falling within an appropriate interval. The
selectivity enables us to reduce the background to an extremely
low level, typically only a few counts per hour. The final spectrum consists of a histogram of counting rate vs. energy (computed
from the aspect-corrected Bragg angle for each photon) corrected
for the exposure time in each energy bin.

3. PUPPIS A

We have scanned a number of spectral regions in the SNR Puppis
A. This is an extended ($\sim$ 50') remnant, thought to be of moderate
age (a few thousand years). An X-ray image of Puppis A, obtained
from rocked data by Levine et al. (1979), is shown in Fig.2.
(Imaging data from the Einstein Observatory are not yet available).
The 3' x 30' aperture used in the FPCS observations is indicated
by the white rectangle. Comparison with the data of Levine et al.
indicates that the aperture should admit approximately 10% of the
total X-ray flux from Puppis A. It is appropriate that Puppis A
should begin our discussion, for it is the only source from which
an X-ray line has been detected (O VIII Lyman α, by Zarnecki and
Culhane,1977) throughout the rather disappointing history of X-ray

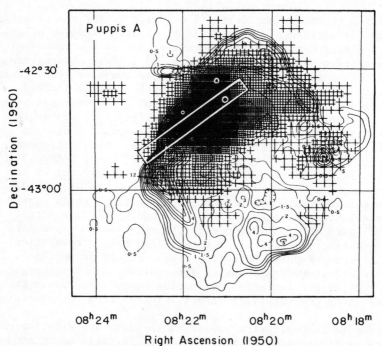

Fig. 2. X-ray Image of Puppis A (0.7 - 1.5 keV) from Levine et al.,
(1979), superimposed on the 408 MHz radio map of Green (1971).
White rectangle indicates the FPCS aperture.

crystal spectroscopy prior to the Einstein Observatory. The FPCS
results show Puppis A to have a rich line-emission spectrum below
1 keV.

Fig. 3 shows the Lyman α line of O VIII at 653 eV, as detected
in 40,000 sec of exposure with the FPCS. The line flux is 0.101 ±
0.009 photons $cm^{-2}s^{-1}$ and the width is 7 eV (FWHM), consistent
with the resolution of the instrument for an extended source at
this energy. This may be compared with the flux of 0.6 ± 0.3 $cm^{-2}s^{-1}$
from the entire Puppis A remnant measured by Zarnecki and Culhane
(1977); see also Zarnecki et al.(1978). The two results are in
reasonable agreement when we take into account that the FPCS aperture included only about 10% of the total Puppis A flux.

There is a second much weaker line visible in Fig. 3, centered
at 665 eV. We identify this line as being due to the $1s^2(^1S)$ -
$1s3p(^1P)$ transition in O VII, at energy 665.6 eV. (This is the
only likely candidate among tabulated lists of spectral lines by
Raymond and Smith (1977) or Kato (1976)). The line flux is (15 ±
5) x 10^{-3} photons $cm^{-2}s^{-1}$, about 6 times fainter than the O VII
Ly-α line. The continuum level in Fig. 3 is due almost entirely
to charged particle events. For the true continuum flux from
Puppis A within our aperture, we can place an upper limit of 0.8
photons $cm^{-2}s^{-1}$ keV^{-1} (3σ) at about 650 eV.

Since this lecture was given, we have also detected the Ly-β
line from O VIII. The results of our preliminary analysis of quick-
look data on this line are shown in Fig. 4. Once analysis of these
data are complete, we will have a measurement of the Ly-α/Ly-β
ratio, which will give a direct measurement of the electron temperature in this emitting region. For an equilibrium plasma in
which the dominant processes are electron collisional excitation
of ions, followed by radiative de-excitation, the (photon) emissivity ratio for Ly-α/Ly-β is given by

$$\frac{F_{2,1}}{F_{3,1}} = \frac{\Omega_{2,1}}{\Omega_{3,1}} e^{(E_{3,1}- E_{2,1})/kT_e} \tag{1}$$

where $\Omega_{2,1}$ and $\Omega_{3,1}$ are the effective collision strengths (including branching ratios) for Ly-α and Ly-β respectively, and the
energy difference $E_{3,1} - E_{2,1}$ = 122.4 eV (see, for example, Kato
1976). The simple prediction of Equation 1 must be modified somewhat to include contributions to the emissivity from di-electronic
recombination; Raymond and Smith (1977) include this effect in
their calculations.

We have now detected in the Puppis A spectrum emission lines
due to O VII, O VIII, Ne IX, Ne X and Fe XVII. These lines will
enable us to investigate several additional diagnostic for the
X-ray emitting plasma. For example, comparison of the $1s^2$ - $1s3p$
transition with the $1s^2$ - $1s2p$ (resonance) line in O VII will give
a second measure of T_e. The ionization balance can be measured by
comparing lines in O VII vs. O VIII and in Ne IX vs. Ne X. And

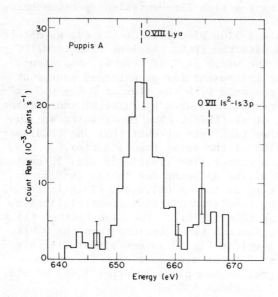

Fig. 3. An FPCS scan of portion of the Puppis A spectrum including the lines O VIII Lyman α (654 eV) and O VII $1s^2(^1S)$ - $1s3p(^1P)$
(666 eV).

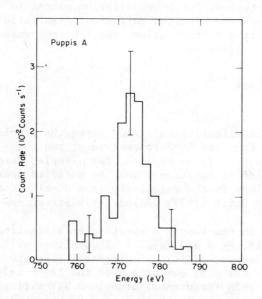

Fig. 4. FPCS scan of the O VIII Lyman β line (775 eV) in Puppis A.

departures from ionization equilibrium can be measured by studying
the relative strengths of resonance, forbidden, and intersystem
lines in the n=2 → n=1 transitions for Helium-like ions (O VII and
Ne IX), as discussed by Acton and Brown (1978).

4. CAS A

Cassiopeia A is the youngest known SNR in the galaxy, with
an age of about 300 years, and above 1 keV it is the brightest SNR
X-ray source except for the Crab. Chevalier and Kirshner (1979)
have reported highly anomalous abundances in the fast-moving optical knots, suggesting that these are composed largely of processed stellar material ejected in the supernova event. The investigation of X-ray line emission from Cas A is of great interest.
Holt (1980) has presented data on this source from the Einstein
SSS which show it to have a rich line spectrum above 1 keV. The
SSS results are discussed in further detail by Becker et al.,(1979).

As Holt (1980) and Becker et al.,(1979) have pointed out, the
extent of departures from equilibrium are extremely important for
understanding the elemental abundances in Cas A. In a recently
shock-heated plasma, the ionization temperature increases more
slowly than the electron temperature (T_e), resulting in an over-
population of lower ionization states relative to the equilibrium
population at temperature $T = T_e$. We would expect such a situation
in Cas A unless the plasma density is quite high (Itoh,1977). At
an equilibrium temperature $T = 7.3 \times 10^6$ K (kT = 0.63 keV) obtained

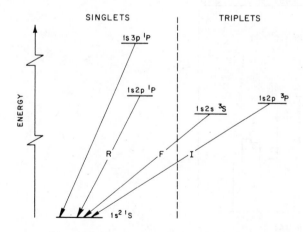

Fig. 5. Simplified term diagram for helium-like ions, showing
transitions to the ground state from n=2 and n=3 states. For sulfur,
the energies of the n=2→n=1 trio are: Resonance (R), 2460 eV;
Forbidden (F), 2427 eV; Intersystem (I), 2446 eV.

by Becker et al. as the best fit to SSS data, the helium-like
ions of Mg, Si, S, Ar and Ca are the dominant ionization state.
(In fact this temperature is determined largely from their observation of strong He-like lines relative to H-like ones). But if
the ionization lags the electron temperature significantly, there
could be a large number of Li-like ions present, and the heavy-
element abundances would then be even larger than those reported
by Becker et al.

Departures from ionization equilibrium can be measured by
comparing the relative strengths of resonance (R) forbidden (F)
and intersystem (I) lines in the n=2 → n=1 complex for He-like
ions (Acton and Brown,1978). These lines are shown in the simplified term diagram of Fig. 5. At equilibrium, the R line is dominant, and the ratio G ≡ (F+I)/R < 1. But for a plasma which is
rapidly ionizing, the triplet excited states of He-like ions may
be preferentially populated due to inner-shell ionization of Li-
like ions. This results in enhancement of the F and I lines relative to the R. A plasma which is strongly recombining from H-like
states will also enhance the triplet population and thus the F
and I lines.

We have observed this complex for He-like ions of both Si
and S in Cas A with the FPCS. The 6' circular aperture was used,
which admits virtually the entire X-ray flux from the remnant.
Preliminary results for S XV are shown in Fig.6. While the trio
of lines has not yet been well resolved, it is clear that the R
line is stronger than the F; their relative strength is consistent

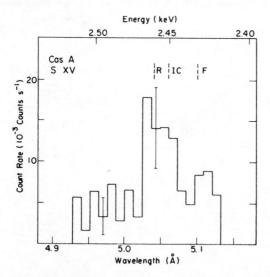

Fig. 6. FPCS scan of the S XV line complex near 2.4 keV in Cas A.
(Data from only a small segment of the extended source image are
included). Note that the resonance (R) line is stronger than the
forbidden (F).

with that at thermal equilibrium. A similar result is obtained
for Si XIII. We are led to the qualitative conclusion that there
is no excessive departure from ionization equilibrium, in that
the number of Li-like ions is small.

5. LARGE MAGELLANIC CLOUD SNRs

Helfand (1980) and Long and Helfand (1979) have reported the
detection of at least ten SNRs in the Large Magellanic Cloud as
strong X-ray emitters based on data from the Einstein IPC. In
light of their discovery, we have carried out a search for X-ray
lines in the two brightest remnants, N132D and N63A, with the
FPCS. For each of the two sources three regions of the X-ray spectrum corresponding to known lines were scanned: O VIII Ly-α (653
eV), Fe XVII (826 eV) and Ne IX (921 eV).

Preliminary results indicate that all three lines are detected
in both sources. In order to maximize the probability of a detection, these observations were carried out by rocking the Bragg
crystal over a narrower range than normal. This amounts to scanning
only a narrow region of the spectrum near the peak of the line,
resulting in high exposure to the line itself and little exposure
to the adjacent continuum. For each of the lines we have detected
a statistically significant source flux in a narrow bandwidth centered on the line. A smooth continuum spectrum would in each case
fail by a large factor to give the narrow-band flux we have observed.

For example, in N132D we scanned a range of only 9 eV centered
about 653 eV, and detected 22 source photons in 6000 sec. (The
source is clearly imaged in the detector; it appears at the 4σ
level above the non-X-ray background of only 6 counts in the same
time interval). If the flux measured by the Einstein IPC were
smoothly distributed in energy according to the exponential spectrum of Long and Helfand, we would have detected only about one
source photon in our FPCS observation. The much larger rate we have
observed must be attributed to O VIII Lyman α photons and indicates
a line flux of 0.023 $\pm$ 0.006 photons cm^{-2}s^{-1}. This corresponds to
a minimum luminosity of 8×10^{36} erg s^{-1} in this line, where we
have taken the distance as 55 kpc with zero interstellar absorption.
At least 20% of the X-ray luminosity from N132D must be a single
line.

6. THE VIRGO CLUSTER

Finally, we take a giant step from relatively nearby SNRs to
M87 and the Virgo Cluster. The X-ray emission from clusters of
galaxies is believed to stem predominantly from a hot intracluster
medium enriched with heavy elements (cf. Gursky and Schwartz,1977).
The Virgo Cluster has two X-ray components: a hot (T $\sim 10^{8}$ K) com-

ponent extending throughout much of the cluster, and a cooler
$(T \sim 3 \times 10^7$ K) "halo" component centered on M87. The morphology
and spectrum of the Virgo/M87 source are discussed in this volume
by Forman (1980) and Mushotzky (1980), respectively. The bright
"halo" component around M87 is probably due to an increased concentration of intracluster gas in the local gravitational potential
well of the massive galaxy (Bahcall and Sarazin,1978; Mathews,1978;
Mathews and Bregman,1978). Emission lines from ionized iron have
been detected with proportional counters at 6.7 keV (Serlemitsos
et al.,1977; Mushotzky et al.,1978) and at $\sim$ 1.1 keV (Fabricant
et al.,1978; Lea et al.,1979). Fabricant et al. further find that
the 1.1 keV iron-line emission is extended over a region at least
as large as that of the 3 keV continuum component.

In FPCS observations of the central M87 source we have detected Lyman α emission from O VIII, with a flux of 0.013 $\pm$ 0.003 photons cm^{-2}s^{-1} from a region countaining $\sim$ 25% of the M87 "halo"
source. For an equilibrium plasma, O VIII is an abundant species
only at lower temperatures than the 3 keV which is characteristic
of the halo source. On the other hand, at temperatures $\sim 10^7$ K
where O VIII, is abundant the iron will be in lower ionization
states than the Fe XXV and Fe XXVI required for K-line emission
at 6.7 keV. This leads us to conclude that the plasma surrounding
M87 is not isothermal; the O VIII emission is coming from cooler
material than that responsible for the iron-line emission. Upper
limits we have obtained for lines from Fe XVII, Fe XXIII, and Fe
XXIV support this picture. (An alternative explanation can be con-
cocted if we have an isothermal plasma with the O/Fe abundance ratio enhanced by a factor $\sim$ 7 times its solar value, but such extreme enhancement is untenable in current nucleosynthesis models
(Arnett,1978; DeYoung,1978)). Our results are described in detail
by Canizares et al.(1979,1980).

The most likely explanation consistent with the data is that
the central portion of the M87 "halo" source contains still cooler
material with a dominant temperature either near 10^7 K or below
$\sim 2 \times 10^6$ K. The hydrostatic, adiabatic models reviewed by Bahcall
and Sarazin (1978) are hotter rather than cooler in the interior,
and thus seem not to be applicable to this source. On the other
hand, several authors have explored the likely situation that the
gas surrounding M87 is accreting into the massive galaxy, and that
the accretion rate is in fact controlled by the rate of radiative
cooling in the central region (see, for example, Mathews and Bregman,
1978). If we take the detailed model of Mathews and Bregman with
an accretion rate of 30 M$_\odot$ yr^{-1} we obtain an estimate for the O
VIII line luminosity that is in good agreement with our observed
value. Our observations thus lend strong support to the picture of
radiatively regulated accretion in the vicinity of M87.

Acknowledgements: We are very grateful to our many colleagues at
the MIT Center for Space Research who have contributed to the success of the FPCS experiment and to other members of the Einstein

Observatory consortium for their continuing efforts. The project
has been supported by NASA contract NAS 8-30752. P.F.W. acknowl-
edges the support of the Alfred P. Sloan Foundation.

REFERENCES

Acton, L.W., and Brown, W.A., 1978, Astrophys.J., 225, 1065.
Arnett, W.D.,1978, Astrophys.J., 219, 1008.
Bahcall, J. and Sarazin, C.,1978, Astrophys.J., 219, 781.
Becker, R.H., Holt, S.S., Smith, B.W., White, N.E., Boldt, E.A.,
Mushotzky, R.F., and Serlemitsos, P.J., 1979, Astrophys.J.
(Letters), 234, L65.
Canizares, C.R., Berg, C., Clark G., Jernigan, J.G., Kriss, G.,
Markert, T.H., Schattenburg, M., and Winkler, P.F., 1980, High-
lights in Astronomy, Vol. 5, 657; presented at IAU, Montreal, 1979.
Canizares, C.R., Clark, G.W., Markert, T.H., Berg, C., Smedira,
M., Bardas, D., Schnopper, H., and Kalata, K., 1979, Astrophys.J.
(Letters), 234, L33.
Chevalier, R.A., and Kirshner, R.P., 1979, Astrophys.J., 233, 154.
DeYoung, D.S., 1978, Astrophys.J., 223, 47.
Donaghy, J., and Canizares, C., 1978, IEEE Trans., NS-25, 459.
Fabricant, D., Topka, K., Harnden, Jr., F.R., and Gorenstein, P.,
1978, Astrophys.J. (Letters), 226, L107.
Forman, W.,1980, this volume, 181.
Giacconi, R. et al., Astrophys. J., 230, 540.
Green, A.J., 1971, Aust. J. Phys., 24, 773.
Gursky, H., and Schwartz, D.A., 1977, Ann. Rev. Astr. Astrophys.,
15, 541.
Helfand, D.J., 1980, this volume, 47.
Holt, S.S., 1980, this volume, 35.
Itoh, H., 1977, Publ. Astron. Soc. Japan, 29, 813.
Kato, T., 1976, Astrophys.J. Suppl., 30, 397.
Lea, S.M., Mason, K.O., Reichert, G., Charles, P.A., and Riegler,
G., 1979, Astrophys.J. (Letters), 227, L67.
Levine, A., Petre, R., Rappaport, S., Smith, G.C., Evans, K.D.,
and Rolf, D., 1979, Astrophys.J. (Letters), 228, L99.
Long, K.S., and Helfand, D.J., 1979, Astrophys.J. (Letters), 234,
L77.
Mathews, W.G., 1978, Astrophys.J., 219, 413.
Mathews, W.G., and Bregman, J.N., 1978, Astrophys.J., 224, 308.
Mushotzky, R.F., 1980, this volume,171.
Mushotzky, R.F., Serlemitsos,P.J., Smith, B.W., Boldt, E.A., and
Holt, S.S., 1978, Astrophys.J., 225, 21.
Raymond, J.C., and Smith, B.W., 1977, Astrophys.J.Suppl., 35, 419.
Serlemitsos, P.J., Smith, B.W., Boldt, E.A., Holt, S.S., and
Swank, J.H., 1977, Astrophys.J. (Letters), 211, L63.
Zarnecki, J.C., and Culhane, J.L., 1977, Mon.Not.Roy.Astr.Soc.,
178, 57P.
Zarnecki, J.C., Culhand, J.L., Toor, A., Seward, F.D., and Charles,
P.A., 1978, Astrophys. J.(Letters), 219, L17.

X-RAYS FROM THE SURFACE OF NEUTRON STARS

Sachiko Tsuruta

Department of Physics, Tokyo University, Hongo, Bunkyo-ku
Tokyo, Japan

1. INTRODUCTION

Observations of supernova remnants with the Einstein Observatory were discussed by Drs. Fabbiano and S. Holt. Holt (1979) concentrated on line emission, while Fabbiano did so on continuum radiation both from the extended nebula regions and from point sources. She reported on observations of possible point sources in four supernova remnants. In this paper, we report on the results of the attempts to observe such point sources in five more supernova remnants, as well as nine radio pulsars, which have already been made with the Einstein Observatory. We discuss what could come out of careful point source investigations of the Crab pulsar and the Vela pulsar with the Einstein Observatory and other X-ray space programs scheduled for the near future (AXAF, etc.). Emphasis is placed on the possible theoretical implication of the outcome of such investigations.

In the next section, we summarise the observational status as of October 1979, and in the last section we explain possible theoretical implications of these observations.

2. OBSERVATIONS

As of October 1979, nine supernova remnants have been observed with the Einstein Observatory (Fabbiano, 1979; Fabbiano et al., 1979; Gorenstein et al., 1979; Harnden et al., 1979; Helfand et al., 1979; Long, 1979; Murray et al., 1979; Novick et al., 1979). They are listed in Table 1. Among these, the first five (Cas A, Tycho, Crab, SN1006, and Kepler) are historical supernova remnants (SNR), while the last four (W28, G350-18, Vela X, and G22.7) are older SNR.

73

R. Giacconi and G. Setti (eds.), X-Ray Astronomy, 73-87.
Copyright © 1980 by D. Reidel Publishing Company.

S. TSURUTA

Table 1

Some characteristics of supernova remnants observed by the HEAO-B, as of Oct. 1979, for possible detection of the thermal X-ray emission from neutron stars, as explained in the text.

Supernova Remnant	Distance (kpc)	Diameter (pc)	Age (years)	Temperature (10^6 °K)
Cas A	3.0	2	306	≤1. - 1.5
Tycho	3.0	6	400	≤1.4- 2.2
Crab	1.7-2	1	925	≤2-3 (1.-1.5?)
SN1006	1.2	4.4	972	≤0.8- 1.
W 28	1.3	10	3400	≤1.3- 2.
G350-18	4.0	35	8000	≤1.5- 2.5
Vela X	0.5-1	20-40	10000	∿1. - 1.5
G22.7-0.2	4.8	35	10000	≤1.7- 2.7

Estimated distances, diameters and ages are given wherever known (see, e.g., Woltjer, 1972 and Tsuruta, 1979a). The last column gives the 3σ level upper limit to the surface temperature of neutron stars, assuming the presence of neutron stars in these supernova remnants.[†] For the Vela pulsar, it may give the actual surface temperature, if the discrete source found at the site of the Vela pulsar includes the thermal radiation from the star, with no pulsation present (Fabbiano, 1979, for details). The observed upper limits for the other sources range from levels of approximately ∿0.8 to 3 million degrees. In order to compare these values with the latest theoretical estimates, the observed values are shown in Figure 1, together with cooling curves.[††] The supernova remnants are numbered in the order of their estimated ages. The crosses indicate the remnants in which pulsars have not been discovered. The circles show the remnants with pulsars, namely the Crab pulsar (3) and the Vela pulsar (7).

Investigation of a possible point source in the Crab by the Einstein Observatory has not been completed yet, as of October 1979

[†]These values of the temperatures (Table 1, the last column) correspond to the interstellar density n_0 = 0.3 - 1 cm^{-3} and radius of the star R = 10 - 17 km. Blackbody radiation is assumed. The uncertainties due to n_0 and R are included approximately within the size of the crosses and circles in Figure 1. No point source is detected at Kepler's SNR by the HRI (Helfand, 1979), but the upper limit to the surface temperature has not been deduced yet.
[††]The temperatures in Figure 1 refer to the equivalent values as measured by an observer on the earth. Similarly, the time refer to those measured by the observer's clock.

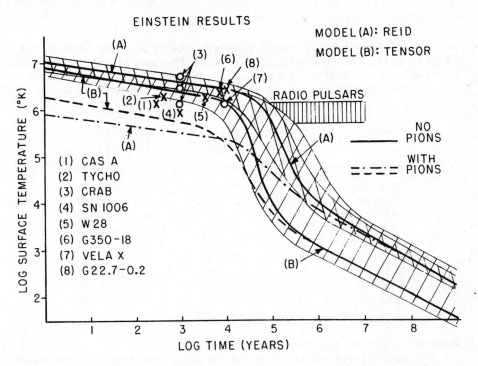

Fig. 1.   Comparison between surface temperatures predicted from the
latest theoretical work and preliminary results from the Einstein
Observatory and earlier observations. The shaded stripes (A) and
(B) are regions where theoretically estimated surface temperatures
are expected to lie in the absence of pions, for models (A) and (B).
Effects of mass, magnetic fields, superfluidity, and equations of
state (nuclear and strong interactions) are included. The shaded
stripes are wider than the figure shown in Tsuruta (1979a), because
the additional effects of variation of mass and magnetic fields and
crude estimates of general relativistic thermodynamics are also included. The solid curves are the best theoretical estimates without
pions. The dashed curves show the maximum effect of the presence
of pion condensates on cooling. For the solid and dashed curves,
we used $M = 1.3 M_\odot$ and $H = 5 \times 10^{12}$ gauss. The temperatures and
times refer to values at infinity. For further details, refer to the
main text.

(Tananbaum, 1979; Seward, 1979). In principle, under favourable
circumstances, it is estimated that the upper limit to the surface
temperature may be lowered to $\sim 1 - 2 \times 10^6$ °K (Seward, 1979). That
point is indicated in Figure 1 as the lowest of the three circles
marked as (3). The upper two circles (3) indicate the previous observation of the upper limits by the lunar occultation experiments
(Wolf et al., 1975; Toor and Seward, 1977). Within the next few
years, it is expected that other near-by supernova remnants will be

Table 2

Some characteristics of radio pulsars observed by the HEAO-B, as
of Oct. 1979, for possible detection of the surface radiation from
neutron stars. Distances, periods, and the 3 σ upper limits to
the surface temperatures are shown. (Taken from Helfand et al.,1979.)

Pulsar	Distance (pc)	Period (sec)	Temperature $(10^6$ °K$)$
0656+14	310	0.385	< 0.37 - 0.52
1529+28	640	1.125	< 0.54 - 0.88
1952+29	240	0.427	< 0.34 - 0.43
2327-20	320	1.644	< 0.38 - 0.54
0031-07	440	0.943	< 0.43 - 0.73
0149-16	560	0.833	< 0.51 - 0.86
1237-25	370	1.382	-- --
1706-16	160	0.653	< 0.30 - 0.36
1642-03	160	0.388	< 0.29 - 0.35

observed with the Einstein Observatory.

The Columbia group is planning to check the possible existence
of point thermal X-ray sources at the locations of ∿35 radio pulsars
by the IPC of the HEAO-B (Helfand et al., 1979; Long, 1979; Novick
et al., 1979). They were selected because they seem to have a
higher probability of lending themselves to observation (closer,
younger, etc.). So far, nine sources have been observed (Helfand
et al., 1979). The results are summarised in Table 2 (Helfand et
al., 1979), with distances, periods, and 3σ level upper temperature
limits (the last column), for a neutron star of 1.3M$_\odot$ and 16 km,
and with n_0 = 0.3 and 1 cm^{-3}. The general relativistic correction
on surface temperatures is included. (That is, the surface temperature values refer to the actual temperatures on the stellar surface.)
These upper limits are in the range of ∿0.3 - 1 x 10^6 °K. The age
of most of these sources is somewhat uncertain, though we expect
that most of them lie in the range of ∿10^5 - ∿10^7 years (e.g.,
Taylor and Manchester, 1977). By the end of the next one to two
years, it is expected that all of these sources will be observed.

In the next section, we discuss the possible theoretical implications of the above studies, after briefly summarising the
theoretical background.

3. COMPARISON WITH THEORIES

Soon after the first discovery of the extraterrestrial X-ray
source in the early 1960's by Giacconi et al. (1962, 1963), it was

found through detailed cooling calculations (Tsuruta 1964, Tsuruta
and Cameron 1966) that neutron stars of $\sim 10^6$ years or younger should
emit thermal X-rays of detectable strength ($L_x \geq 10^{33}$ergs sec^{-1})
with corresponding surface temperature $\gtrsim 10^6$ °K, if one neglects the
effects of superfluid nucleons and strong magnetic fields (of
$\gtrsim 10^{12}$ gauss). Soon after the discovery of pulsars, it was found
that these pulsars should have strong magnetic fields, of at least
$\sim 10^{11} - 10^{12}$ gauss at the surface level. A more direct measurement
of the surface magnetic fields was given by the group in the Max-
Planck-Institut (Trümper et al., 1978) for the X-ray binary source
Her X-1, to be $\sim 5 \times 10^{12}$ gauss, assuming that the observed line at
~ 50 keV is a cyclotron line. Also it seems that both theory and
observation support the presence of superfluid nucleons in neutron
stars (see e.g. Pines et al., 1974; Tsuruta, 1974). By including
the effect of superfluidity (and superconductivity) and magnetic
fields of comparable strength ($H \geq 10^{12}$ gauss), it was found that
neutron stars should cool faster, with $\leq 10^5$ °K for $\geq 10^5$ years
(Tsuruta et al., 1972; Tsuruta, 1974). That means, older pulsars
of $\geq 10^5$ years should be too cold to be detected (at reasonable astronomical distances), unless additional heating mechanisms are at
work (Tsuruta, 1974). However, the theory predicted that we should
observe a detectable amount of X-rays from younger pulsars such as
the Crab pulsar and the Vela pulsar, because the estimated surface
temperatures of these younger pulsars were $\geq 10^6$ °K and their luminosities were $\geq 10^{33}$ ergs sec^{-1}. The above conclusion is based on what
we call the "standard scenario". That is, all important neutrino
emissivities were included except those involving pions. However,
as early as 1965, Bahcall and Wolf (1965a,b) pointed out that if
there are substantial amounts of pions present in a neutron star,
the star will cool so fast that it will be virtually undetectable.

     Due to recent progress in various theoretical aspects of neu-
tron star physics, we can now make better estimates of the cooling
curves. The latest theoretical results are summarised in Figure 1.
The surface temperatures are shown as a function of time (after
supernova explosion). The curves (A) and (B) refer to two different
models which are based on two different "realistic" nuclear potentials
which give the softest and the hardest equations of state, respectively. Before going further into the explanation of Figure 1, we shall
briefly describe how these curves are obtained.

     (a) Cooling Curves. After a supernova explosion, an isolated
neutron star (without an extra heat source) just keeps cooling, losing its thermal energy. Therefore, choosing t=0 as the moment of
the explosion, the stellar temperature is obtained as a function
of time t by the following equation:

$$t = \int_{U_o}^{U} \frac{dU(T_c)}{\overline{L}_\nu (T_c) + \overline{L}_\gamma (T_e)} \tag{1}$$

Here, t is the local time after the explosion, U is the thermal
energy, T_c is the central temperature of the star, T_e is the surface temperature of the star, and $\bar{L}_\nu$ and $\bar{L}_\gamma$ are the average rate of
loss of the residual energy by the amount dU through neutrinos es-
caping from the interior and through photons escaping from the surface, respectively. We need a relation between the central core
temperature T_c and the surface temperature T_e, in order to solve
the above cooling equation (1). This is obtained by carrying out
integration of basic stellar structure equations (general relativistic hydrostatic equations and the energy transport equation) from
the surface into the central isothermal core (where temperature
gradients vanish), together with the equation of state, the opacity
equation and information on the composition. We refer to our earlier paper (e.g. Tsuruta, 1974) for further details on the basic
method of calculations.

(b) Effects of Nuclear Forces. Nuclear forces enter the problem
through the equation of state when we integrate the basic stellar
structure equations through the interior. Such integrations give
unique combinations of acceptable values of the mass, radius and
density of physically realistic stable neutron stars. The problem
of nuclear forces in neutron stars is still unsolved. This is be-
cause near and above the nuclear density ($\rho_n \simeq 3 \times 10^{14}$ gm/cm^3), the
data from terrestrial laboratory experiments and the two-body ap-
proach may no longer be valid. According to the latest review of
the neutron star problems in general, Baym and Pethick (1979) argue
that at and above the nuclear density repulsive forces could become
effective and that then the equation of state would be hardened.
That is, pressure would be higher at a given density. The main outcome is that a harder equation of state will give a lower central
density and a larger radius and outer crust, for a given mass star.

The arguments given by Baym and Pethick are quite convincing,
but it may be that the dispute can be settled through observation.
For instance, if the observed surface temperature of the Vela neutron star turns out to be about one to two million degrees, this
may favour the argument that already near the nuclear density, simple
application of terrestrial experimental data and the two-body approach
fail and repulsive forces should become effective, though more careful studies, both in theory and observation, are desirable, before
we can make a definite conclusion. It is encouraging to learn that
the Vela pulsar data from the Einstein Observatory is currently in
the process of being analysed very carefully, and the more conclusive result is expected to come out soon (Tananbaum, 1979; Gorenstein, 1979).

In our model (A), we used the conventional Reid potential which
is expected to best fit terrestrial scattering experiments (see
e.g. Baym et al., 1971). In our model (B), we used the tensor model
of the Illinois group (Pandharipande et al., 1976), where the above
effect of the repulsive force is taken into account. The solid
curves (A) and (B) represent our best current theoretical estimates
(as of October 1979), for the "standard" cooling of our nuclear

models (A) and (B), respectively, with the typical neutron star
mass of M = 1.3M$_\odot$ and the surface magnetic fields of
H = 5 x 10^{12} gauss. For the neutrino emissivities, we used the
latest work by Friman and Maxwell (1979) and Soyeur and Brown (1979).
Both the core and crust neutrino processes are included. Superfluid corrections have been applied wherever appropriate (to nucleon
specific heats and the URCA and nucleon bermsstrahlung neutrino
emissivities), in the manner described in Tsuruta (1979a,b). For
model (A), the central density ρ_c = 3.5 x 10^{15} gm/cm^3, the radius
R $\simeq$ 8 km, and the outer heavy ion crust has a thickness $\Delta R \simeq$ 0.5 km,
while for model (B), we get ρ_c = 4.2 x 10^{14} gm/cm^3, R $\simeq$ 16 km, and
$\Delta R \simeq$ 8 km.

The main difference between the structure of these two models
is because of the fact that model (B) has a harder equation of state
due to the repulsive tensor term in the nuclear force, which lowers
the central density, increases the radius and enlarges the outer
crust (for a given mass star). The effect of these differences is
obvious when we examine Figure 1. Namely, model (B) cools faster
than model (A). (This outcome is consistent with a general trend
already found in earlier theoretical studies – that is lower density,
larger stars generally cool faster (e.g. Tsuruta, 1974)). The major
reason can be traced to the fact that neutrons cease to be in a
superfluid state when densities get sufficiently high, above
$\sim 10^{15}$ gm cm^{-3}. Therefore, the major portion of the central neutron
core of model (A) consists of normal (non-superfluid) neutrons,
while all of the central core of model (B) consists of superfluid
nucleons. One major effect of superfluidity is to reduce the internal thermal energy of superfluid particles when the temperature
becomes less than the critical temperature, and the less the thermal energy content the cooler the star (see equation (1)). Other
reasons are found in our earlier paper (Tsuruta, 1979a).

(c) Other Effects. In Figure 1, the shaded stripes around
the solid curves take into account the various uncertainties in
our theoretical estimates, most importantly the uncertainties in
our estimates of the superfluid energy gaps, neutrino emissivities
and internal energies (see e.g. Maxwell, 1979; Tsuruta, 1979a).
Also included are the effects of changing stellar masses and magnetic fields. Here we assumed that the mass is not much less than
$\sim 0.5 M_\odot$. The range of the magnetic fields included in our striped
regions is $\sim 10^{12}$ - $\sim 10^{13}$ gauss. We believe that these are reasonable assumptions both theoretically and observationally. (See,
e.g., Gunn and Ostriker, 1970; Ruderman, 1972; Trümper et al., 1978
for the magnetic fields, and Bahcall, 1978; Joss and Rappaport, 1976;
Arnett, 1977; Bethe et al., 1979 for masses.) Neutron stars in
some SNR may not be strongly magnetised, and even in pulsars the
surface area outside the polar regions may not be strongly magnetised (Ray, 1979). The cooling curves for the zero field case for
all "physically realistic" models lie within the combined shaded
regions for t $\leq$ 10^5 - 10^6 years, generally above the shaded regions
at the older ages. The shaded stripes also include the effect of

possible reduction of photon emissivity for magnetic neutron stars
(Itoh, 1975; Brinkmann, 1979).   In our treatment of superfluidity,
we assumed, following nuclear experts (e.g.,Tamagaki and Takasuka,
private communication), that the p-wave superfluid gap does not go
beyond $\sim$1.2 MeV. As mentioned earlier, model (A) is representative
of the softest and model (B) is representative of the hardest equation of state, among relatively "realistic" models of nuclear forces.
Therefore, we may conclude that theoretical values for the "standard" scenario most likely lie somewhere within the combined
striped regions (A) and (B)[†].

   (d) <u>Pion Cooling</u>. The above results were obtained for what we
call the "standard" cooling scenario, where all effects are included in the absence of pions. The dashed curves indicate the
maximum effects of pion cooling, with the same mass and magnetic
fields, $M = 1.3M_\odot$ and $H = 5 \times 10^{15}$ gauss, as the solid curves. We
used the most recent estimates of neutrino emissivities in the
presence of pion condensates (Maxwell et al., 1977; Kiguchi, 1977).
These are the maximum effects, because it was assumed that the
nucleon-pion transitions are complete, while that assumption depends
on the ratio of the central density of the star to the critical pion
density ρ_π(the density above which pions appear). For instance,
there should be no pions if this ratio is less than 1. So far, the
best estimate gives $\rho_\pi \approx 6 \times 10^{14}$ gm cm^{-3} ($\approx 2 \rho_n$). This means that
the central core of model (A) will be almost completely in the
form of pion condensates, while there should be no pions in model (B)

   (e) <u>Comparison with Other Cooling Calculations</u>. In the last
few years, several other authors also studied the cooling of neutron stars. By using an analytic approach, Brown (1977) obtained,
as the surface temperature of the Crab neutron star, a value of
4.4×10^6 °K, which is within the range of our theoretical estimates for the "standard" cooling (the shaded stripes). Maxwell
(1979), using the better neutrino emissivities of Friman and
Maxwell (1979) (which we also used), obtained several cooling curves.
His results qualitatively agree more with our model (B) than our
model (A). This is to be expected because his density is closer
to that of our model (B). His lowest cooling curve, however, is
unrealistically low. The major reason is that he overestimated the

---

[†]Because of the incompleteness of our theories, we can not exclude
the possibility of a theoretical curve lying outside these striped
regions (e.g., when the stellar mass is unrealistically low,
$\leq 0.2M_\odot$); however, it seems rather doubtful that such a physically
unrealistic situation will actually take place.
   Brecher and Barrows (1979) have recently carried out calculations for the cooling of quark stars. These stars, they found,
tend to have a higher heat capacity but cool faster, resulting in
predicted temperatures similar to the "standard" cooling scenario
without pions (the shaded stripes in Figure 1).

effect of neutron superfluidity on the total energy.[†] Very recently, Maxwell and Soyeur (1979) obtained cooling curves for the
"standard" curve lying close to our model (B) and their pion curve
lying roughly within our two dashed curves (A) and (B). Ray (1979)
obtained surface temperatures for both the "standard" and pion
cooling. His treatment of the pion cooling seems more satisfactory.
However, like Maxwell, Ray used an oversimplified neutron star model
of constant low density (=nuclear density) with R = 10 km and
M = 1.3$M_\odot$. The qualitative agreement between our results is still
satisfactory for all except his estimate of the lowest temperatures
for the "standard" cooling. These values fall seriously below the
lowest boundary of the shaded stripes in Figure 1. The major reason is that he chose Maxwell's lowest curve, which is physically unrealistic[†], and moreover in using it he applied an incorrect method
to obtain his zero field values[††]. (More detailed comparisons and
discussions are given elswhere (Tsuruta, 1979b), for this report is
meant as a brief summary.)
       (f) Comparison with Observations. In figure 1, the two upper
circles (3) show the previous upper limits to the surface temperature of the Crab neutron star set by the lunar occultation experiments which were obtained by assuming R = 10 km (Wolf et al., 1975;

[†] Maxwell assumed R = 10 km, M = 1.3$M_\odot$ and density = 2.4 x 10^{14} gm
cm^{-3} = constant. However, any self-consistent neutron star model
of the same radius (10 km) should have higher densities (at least
$\sim 10^{15}$ gm cm^{-3} or higher - see, e.g., Baym and Pethick, 1979), and
because the electron specific heat depends on density as $\rho^{4/3}$ he
underestimated the electron thermal energy. This effect becomes
significant when T << T_n^c (= the neutron superfluid critical temperature), which is the case of his lowest curve when t $\gtrsim$ 100 years.
For his lowest curve, he assumed that T_n^c = 3.2 x 10^9 °K, which corresponds to a p-wave neutron superfluid gap energy $\Delta_n(p)$ = 2.3 MeV,
an unrealistically high value. In fact, according to Tamagaki and
Takatsuka (private communication), it is almost impossible to get
$\Delta_n(p)$ $\gtrsim$ 1 MeV, according to conventional nuclear physics.
[††] Ray simply assumed that the surface temperature T_e decreases (at
a given age) when the magnetic field H decreases, because
$\Delta T\{=T_c(\text{core temperature}) - T_e\}$ increases with decreasing H. However, the above statement is correct only when the neutrino processes dominate over the photon radiation as the cooling agent. When
the photon radiation dominates the cooling the reverse is true,
while they become comparable when neutrino and photon emissivities
are comparable (see, e.g., Tsuruta, 1974, 1979b). At the age of
Cas A, Crab, etc., in which we are interested, ($\gtrsim$ 100 years), the
star is definitely already cooling by photon radiation in his model
IIa (and Maxwell's lowest curve). Therefore, in this case, his
T_e at H = 0 should be higher than T_e with the strong H, contrary to
his prediction. In fact, the exact relationship between the T_e with
and without H can be found only by actually carrying out the stellar
structure integration.

Toor and Seward, 1977). The earlier one, the upper circle, is
4.7×10^6 °K, which is consistent with both models (A) and (B).
The later one, the lower circle, is 3×10^6 °K, which is safely
within the shaded stripe of model (B). For model (A), this circle
was safely beneath the lower boundary, when the redshift effect on
T_e was not taken into account. The surface temperatures in Figure
1 refer to the equivalent values as measured by an observer on the
earth (=infinity). Since photon radiation will be redshifted under
strong gravitational fields, the observed temperature on the earth
will be lower than the actual temperature at the star. This effect
is negligible for low density stars (e.g., our model (B)), but it
becomes significant for high density stars (e.g., our model (A)).
In fact, when the stellar mass becomes close to the maximum limit,
$\sim 1.4 M_\odot$ for the Reid type model, the redshift approaches $\sim 30\%$.
Without the redshift correction, the cooling curve of $1.4 M_\odot$ star of
the Reid type lies above the solid curve (A) with $M = 1.3 M_\odot$. However, with the redshift correction applied, it comes below the
solid curve (A), in fact, near the lower boundary of the shaded
stripe (A) in Figure 1. In our present calculations, we included
only a very rough estimate of the effect of general relativity on
thermodynamics. In a more complete treatment, general relativity
should be included in the heat transport equation, as it was done
for the hydrostatic equation. Such an equation, however, as far
as we know, is yet to be derived. The specific heats and neutrino
emissivities also should include general relativistic corrections.
Because of these uncertainties, the exact location of the lower
boundary of the stripe (A) is somewhat uncertain. The effect of
these uncertainties, however, is negligible for the lowest boundary
(=the boundary of the stripe (B)), because the general relativistic
effects are negligible for the fastest cooling model (B)(with the
lowest mass).

      The circle (7) for the Vela pulsar lies safely within the
shaded stripe (B). It is below the shaded stripe (A). From this,
we may be tempted to conclude that observations favour harder equations of state. That is, the deviation from the two-body approach
and laboratory scattering data already occurs and some of the repulsive effects already enter near the nuclear density. The profound implications of such conclusions to fundamental problems in
nuclear and particle physics are quite obvious. However, since
this point (the circle (7)) is rather close to the lower boundary
of model (A)(though it is beneath it), more quantitative, careful
examinations are desirable, both theoretically and observationally.
If the surface temperature (not just an upper limit) of the Crab
neutron star is measured, knowledge of this temperature may give
an invaluable handle on nuclear force problems and the outcome
may be far reaching. In any case, we strongly emphasize the importance of further careful spatial spectral studies of the point
sources in the Vela and the Crab.

      In Figure 1, the crosses indicate the upper limit to the sur-
face temperatures obtained from the Einstein Observatory for

supernova remnants in which pulsars have not yet been discovered.
The cross (1) stands for the Cas A, the cross (2) for the Tycho
SNR, etc. We see that the observed upper limits are safely below
the lowest boundary of the combined striped regions (A) and (B)
for Cas A and SN1006[†]. (It is near the boundary for Tycho.) One
obvious interpretation is that there are no neutron stars in these
SNR. Another interpretation is that there exist pion stars in
these SNR. Maxwell and Soyeur (1979) reached a similar conclusion.
Ray (1979), using his minimum temperatures, concluded that the
"standard" scenario is consistent with observation. However, as
already noted in the previous section, his minimum temperatures
are unrealistically low. (For further details see Tsuruta, 1979b.)
     Ray (1979), however, raised an interesting and important point.
He pointed out that for a certain class of neutron star models
which have large outer crusts (e.g., the Illinois tensor model),
the time scale of thermal conduction through the crusts may be
sufficiently long, so that thermal equilibrium may not be reached
before the elapse of a few thousand years or so. If so, the surface temperatures should be higher than the values obtained by
using the standard method of cooling calculation as explained earlier. However, if the pion critical density ρ_π is indeed $\approx 2\,\rho_n$
$\approx 6 \times 10^{14}$ gm cm^{-3} (e.g., Weise and Brown, 1975; Maxwell et al.,
1977), such low density, large crust stars most likely will not
contain pions at all. This definitely applies to our model (B).
In general, the higher the stellar density, the thinner the outer
crust. Therefore, if the stellar density is high enough so that
pions are present, such a neutron star may more likely possess a
thinner crust, and the above problem may not be applicable. These

---

[†] Besides the reasons given in previous footnotes, we expect the
surface temperatures of these young neutron stars to be higher
than the estimates shown in Figure 1, if the conduction time scale
is as long as the value estimated by Ray (1979).
     The lowest boundary of the theoretical cooling curves should
be compared with the lower end of the crosses and circles which
show observed values, because both correspond to the stellar radius
R = 17 km. Using the recent work of Brinkman (1979), we estimated
that the deviation of photon emissivity under strong magnetic
fields from the blackbody approximation, can be as large as a factor of 5 or so. This, however, increases the estimate of the observed temperatures only by a factor of ~ 1.5, due to the slow $L^{1/4}$
dependence of the surface temperatures. If we include this effect
in the observed upper limits for Cas A and SN1006, their values
still lie beneath the lowest theoretical boundary. Brinkman (1979)
obtained the above result assuming that the stellar surface will
most likely be in a liquid or solid state. We feel this is a
resonable assumption (Flower et al., 1977; Tsuruta, 1975, 1979a).

arguments are, however, at this stage still speculative, and there is no question that a more careful theoretical study of this interesting problem is desirable.

Even though the possibility of the presence of a pion star is not excluded for the SNR such as Cas A and SN1006, we may call our attention to another interesting possibility, which has already been raised by Fabbiano (1979). That is, some type of supernova explosions (e.g., Type I) may not produce neutron stars at all. The morphological and structural studies of various SNR may already suggest such a conclusion. In this case, however, the comparison of recent pulsar and supernova statistics may point to the need for neutron star formation mechanisms other than through supernova explosions (e.g., Helfand et al., 1979; Taylor and Manchester, 1977; Tsuruta, 1979b).

At this point, it may be appropriate to remind ourselves of the incompleteness of our theories. Until recently, it seemed that it would not be of much sense to carry out neutron star cooling calculations more quantitative than the existing ones. However, from our earlier discussions, it is obvious that further theoretical studies will be desirable, because through the Einstein Observatory we can now have more precise observed values (e.g., the upper limits to the surface temperatures) through which our theories can be better tested. For instance, a better treatment of general relativistic thermodynamics (as explained earlier) will make it possible to more accurately define the lower boundary of the model (A). Also, there may be some room for improvement in the problem of thermal conductivity and its effect on opacity. An improved treatment of this problem in the absence of magnetic fields may somewhat increase the difference between the interior and surface temperatures. On the other hand, the effect of magnetic fields is to decrease this difference (Tsuruta, 1975, 1979c). These problems are currently under investigation. Due to the various uncertainties mentioned above, we wish to present our present conclusion only as a tentative one. However, our preliminary results of these considerations (Tsuruta, 1979b) indicate that our present conclusion will not be seriously changed.

According to some of the pulsar radiation theories (e.g., Cheng and Ruderman, 1977, 1979; Arons and Scharlemann, 1979), weak and soft X-rays may be detected from some older radio pulsars. The upper limits to the surface temperature of these radio pulsars which can be checked through the Einstein Observatory are indicated by the vertically shaded region in Figure 1. So far, all results for the nine sources observed with HEAO-B by the Columbia group have been negative (see Table 2). In one or two years time, all of the about three dozens of pulsars in the HEAO-B programs will have been observed. The outcome promises to be very interesting, because whether they are negative or positive, the results may provide useful (in some cases perhaps severe) constraints on various theoretical models for many aspects of neutron star structure and behaviour - crust-core coupling, starquakes, glitches, and

other dynamical behaviour, and radiation mechanisms of pulsars.
Most recent general reviews of these heating, non-equilibrium, and
pulsar radiation aspects of the problem are found, e.g., in Helfand
et al., 1979; Lamb et al., 1978; Manchester and Taylor, 1977;
Ruderman, 1979; Tsuruta, 1979a.

ACKNOWLEDGEMENTS

I thank many of my colleagues, especially the following per-
sons, for helpful discussions: Drs. D.J. Helfand, G. Fabbiano, R.
Novick, F.D. Seward, H. Tananbaum, P. Gorenstein, G.E. Brown,
A. Chanan, O. Maxwell, D.Q. Lamb, K.S. Long, P. Joss, J. Trümper,
A. Ray, K. Van Riper, and G.R. Caughlan. I am indebted to some
of them, especially D. Helfand, for kindly supplying me with their
preliminary results before publication. I thank Professors
R. Giacconi and G. Setti for their hospitality at the Erice
Summer School of X-Ray Astronomy. This paper was prepared during
my stay at Harvard Center for Astrophysics, Columbia Astrophysics
Laboratory, Institute of Astronomy in Cambridge, Max-Planck-Institut
in München, and Montana State University. My thanks are due to
Professors R. Kippenhahn, R. Novick, R. Giacconi, A.G.W. Cameron,
G. Field, R. Swenson, and M.J. Rees for their hospitality.

REFERENCES

Arnett, W.D., 1977, Astrophys.J., 218, 815.
Arons, J., and Scharlemann, E.T., 1979, Astrophys.J., 231, 854.
Bahcall, J.N., 1978, Ann. Rev. Astr. Ap., 16, 241, and references
quoted therein.
Bahcall, J.N., and Wolf, R.A., 1965, Phys. Rev., 140B, 1445 and
1452.
Barrows, A., 1979, Ph.D. Thesis, MIT, Mass.
Baym, G., and Pethick, C., 1979, Ann. Rev. Astron.Astrophys., 17,
in press.
Baym, G., Pethick, C., and Sutherland, P., 1971, Ap.J., 170, 299.
Bethe, H.A., Brown, G.E., Applegate, J., and Lattimer, J.M., 1979,
to be published in Nucl. Phys. A.
Brecher, K., and Barrows, A., 1979, to be published.
Brinkmann, W., 1979, MPI Preprint.
Brown, G.E., 1977, Comments Ap. Spa. Phys., 7, 67.
Cheng, A., and Ruderman, M.A., 1977, Astrophys.J., 216, 865.
Cheng, A., and Ruderman, 1979, to be published.
Fabbiano, G., 1979, This volume, 15.
Fabbiano, G., Doxsey, R.E., Griffiths, R.E., and Johnson, M.D.,
1979, Astrophys.J.(Letters), submitted.
Flowers, E.G., Lee, J.F., Ruderman, M.A., Sutherland, P.G.,
Hillebrandt, W., and Müller, E., 1977, Astrophys.J., 215, 291.
Friman, B.L., and Maxwell, O.V., 1978, Preprint NORDITA-78/15 (revised).

Giacconi, R., Gursky, H., Paolini, F.R., and Rossi, 1962, Phys. Rev. Lett., 9, 439.

Giacconi, R., Gursky, H., Paolini, F.R., and Rossi, 1963, Phys. Rev. Lett., 11, 530.

Gorenstein, P., Murray, S., Epstein, A., Griffiths, R.E., Fabbiano, G., and Seward, F., 1979, Bull.Am.Astron.Soc., 11, 462.

Greenstein, G., 1975, Astrophys.J., 200, 281.

Gunn, J.E., and Ostriker, J.P., 1970, Astrophys.J., 160, 979.

Harding, D., Guyer, R.A., and Greenstein, G., 1978, Astrophys.J., 222, 991.

Harnden, F.K. Jr., Hertz, P., Gorenstein, P., Grindlay, J., Schreier, E., and Seward, F., 1979, Bull.Am.Astron.Soc., 11, 424.

Helfand, D.J., 1979, private communication.

Helfand, D.J., Chanan, G.A., and Novick, R., 1979, Nature, in press.

Holt, S., 1979, This volume, 35.

Itoh, N., 1975, Mon.Not.Roy.Astr.Soc., 173, Short Communication, - 1P.

Joss, P.C., and Rappaport, S., 1976, Nature, 264, 219.

Kiguchi, M., 1977, Prog.Theor.Phys., 58, 1766.

Lamb, F.K., Pines, D., and Shaham, J., 1978, Astrophys.J., 224, 969, and 225, 582.

Long, K.S., 1979, Proc. HEAO Science Symp., NASA, Marshall Center.

Manchester, R.N., and Taylor, J.H., 1977, Pulsars, San Francisco: Freeman.

Maxwell, O.V., 1979, Astrophys.J., 231, 201.

Maxwell, O.V., Brown, G.E., Campbell, D.J., Dashen, R.F., and Manassah, J.T., 1977, Astrophys.J., 216, 77.

Maxwell, O.V., and Soyeur, M., 1979, Preprint CTP No. 801.

Murray, S.S., Fabbiano, G., Fabian, A., Epstein, A., and Giacconi, R., 1979, Astrophys.J.(Letters), 234, L69.

Novick, R., 1979, private communication.

Novick, R., Helfand, D.J., Chanan, G.A., Ku, W.H.M., and Long, K.S., 1979, Bull.Am.Astron.Soc., 11, 445.

Pandharipande, V.R., Pines, D., and Smith, R.A., 1976, Astrophys. J. 208, 550.

Pines, D., Shaham, J., and Ruderman, M.A., 1974, Physics of Dense Matter (D. Reidel Publishing Co.), 189.

Ray, A., 1979, Ph.D. Thesis, Columbia University, N.Y.

Ruderman, M.A., 1972, Ann.Rev.Astron.Astrophys., 10, 427.

Ruderman, M.A., 1979, Ann.N.Y.Acad.Sci., in press.

Seward, F., 1979, private communication.

Soyeur, M., and Brown, G.E., 1978, Preprint CTP No.745; submitted to Nuclear Phys.A.

Tananbaum, H., 1979, private communication.

Taylor, J.H., and Manchester, R.N., 1977, Astrophys.J., 215, 885.

Toor, A., and Seward, F.D., 1977, Astrophys.J., 216, 560.

Trümper, J., Pietsch, W., and Voges, W., 1978, Astrophys.J. 219, L105.

Tsuruta, S., and Cameron, A.G.W., 1966, Can.J.Phys., 44, 1863.

Tsuruta, S., 1964, Ph.D. Thesis, Columbia University, N.Y.

Tsuruta, S., 1974, Physics of Dense Matter, (D. Reidel Publishing Co. 209.

Tsuruta, S., 1975, Ap.Spa.Sci., 34, 199.
Tsuruta, S., 1979a, Physics Reports, in press.
Tsuruta, S., 1979b, in preparation.
Tsuruta, S., 1979c, Proc.Intern.School of Phys., Enrico Fermi
Course LXV, 635.
Tsuruta, S., Canuto, V., Lodenquai, J., and Ruderman, M., 1972,
Astrophys.J., 176, 739.
Weise, W., and Brown, G.E., 1975, Phys.Lett., 58B, 300.
Wolff, R.S., Kestenbaum, H.L., Ku, W., and Novick, R., 1975,
Astrophys.J.(Letters), 202, L77.
Woltjer, L., 1972, Ann.Rev.Astr.Ap., 10, 129.

X-RAY SPECTRA OF GALACTIC X-RAY SOURCES

S.S. Holt

Laboratory for High Energy Astrophysics
NASA/Goddard Space Flight Center
Greenbelt, Maryland   20771 U.S.A.

1.   BINARY X-RAY SOURCES

With the exception of a few supernova remnants, we presently
believe that all the X-ray sources in the galaxy with $L_x \gtrsim L_\odot$
arise in binary systems, one member of which is a degenerate
dwarf, a neutron star, or a black hole. The energy source for the
X-radiation in such a system is the gravitational energy liberated
when mass is transferred to the surface of the degenerate star,
i.e.

$$L_x = \eta G \frac{M_x \dot{M}_x}{R_x} \tag{1.1}$$

where R_x and M_x are the radius and mass of the compact X-ray star,
$\dot{M}_x$ is the rate at which mass is accreting onto the object, and η
is the efficiency with which this gravitational potential energy
may be converted to X-rays (a typical value might be ~ 0.1). The
requirement that the star be compact is an obvious consequence of
the total gravitational potential energy liberated by a proton
falling from infinity to the surface of a 1 $M_\odot$ star:

$$G \frac{m_p M_\odot}{R_x} \approx 2 \frac{R_\odot}{R_x} \quad keV \tag{1.2}$$

Since the radiation yield in bremsstrahlung is so low, it is clear
that we must have $R_x << R_\odot$ for efficient X-ray production. A
typical white dwarf radius of 10^8 cm will yield ~ 1 MeV in equation
(1.2), and a neutron star will yield 100 times more.
   Of the catalogued varieties of binary X-ray sources, we are

R. Giacconi and G. Setti (eds.), X-Ray Astronomy, 89-102.

certain that white dwarfs are the compact objects in cataclysmic
variables, and that neutron stars are present in the X-ray "pulsars"
Note that the latter objects have unfortunately been given the same
name as the central source in the Crab nebula; it is true that both
types are rotating magnetic neutron stars, but those in binary
systems are actually acquiring rotational kinetic energy rather
than expending it. There are many other historic classifications
of sources (e.g. "soft" transients, bursters, bulge sources, halo
sources) for which arguments have been made for both dwarf and
neutron star compact objects, with the latter generally being
favored. In addition, at least three objects (Cyg X-1, Cir X-1
and V861 Sco) have been tentatively identified with black holes.
Of these, the strongest case can apparently be made for Cyg X-1,
where a compact object with a mass which exceeds that possible for
either a neutron star or a white dwarf seems to be required. The
argument for Cir X-1 is circumstantial, and based on similarities
in short-term temporal variations with Cyg X-1. The case for V861
Sco now appears to have been based on a misidentification.

   The remainder of this lecture will be devoted to the spectroscopic properties of broad classes of galactic sources, and will not
substantially address the very basic evolutionary considerations
which a detailed description of these objects requires. The reader
is directed to the lectures of Prof. van den Heuvel in this volume
for an excellent review of the elements of our present understanding of the fundamental nature of binary X-ray sources.

   Spectroscopically, the prime a priori characteristic which
distinguishes galactic binaries from the diffuse sources discussed in previous lectures is that the photon transport in the
source can no longer be ignored. In the cases of supernova
remnants and clusters of galaxies, the assumption of direct trans-
mittal of the X-ray photons to our detectors, with only the interstellar medium intervening, was a good one. Here, however, the
X-ray photons are liable to suffer a considerable number of in-
elastic collisions before escaping the source region, so that we
do not expect to be able to measure the production spectrum
directly.

   In most accretion scenarios, the infalling material cannot
fall directly to the surface of the compact star because of the
angular momentum that it carries. The material must, therefore,
spiral into the star, so that the probability of an accretion disk
forming around the object is quite high (e.g. Pringle and Rees
1972). X-radiation produced by viscosity in the disk will be
multi-temperature as it will arise from differing disk positions,
and the optical depth for Thomson scattering will be >> 1. Even
if the emission comes from close to the stellar surface (as we
might expect near the magnetic poles of X-ray pulsars), obscuration
by material in the disk or at the Alven surface will almost certainly result in high Thomson optical depths for most geometries.
In fact, the very condition which limits the luminosity of compact
X-ray sources, the Eddington condition

$$L_x \lesssim 10^{38} \frac{M_x}{M_\odot} \tag{1.3}$$

is that imposed by the balance of the gravitational and radiation
pressures on the infalling material, where the latter is dominated
by Thomson scattering.

In summary, the same compactness which allows binary sources
to be powerful X-ray sources often prevents the direct observation
of the initially produced X-radiation via electron scattering in
the source region itself. Emission signatures (such as thermal
emission lines) will, therefore, be largely washed out by the transport of the photons out of the source. A large number of treatments of this scattering have appeared in the literature, the most
recent (and complete) being that of Sunyaev and Titarchuk (1979).
Typically, smooth power-law spectra are expected over limited
dynamic ranges, with the indices determined primarily from the
Thomson optical depth and the scattering electron temperature,
and only secondarily from the input X-ray spectrum if the scattering depth is high enough.

2.   CATACLYSMIC VARIABLES

These binary systems, with a white dwarf as the compact object,
include the recurrent and classical novae as well as the "polar"
systems such as AM Her. They suffer much less from Thomson scattering than most other binary X-ray sources, possibly because they
radiate considerably below their Eddington-limited luminosities.
They possess accretion disks from which most of the optical variation is observed (particularly that associated with the "hot
spot" where the accreting material is channelled into the disk),
but the disk temperatures are much too cool to be responsible for
the X-ray emission. The X-rays are presumed to arise from material
which is directly falling to the stellar surface. Fabian, Pringle
and Rees (1976) have calculated that effective X-ray temperatures
of $\sim 10^8$ K should be observed from material which is being
spherically accreted onto white dwarfs, and the spectra of
cataclysmic variables observed with proportional counter experiments onboard OSO-8 and HEAO-2 have detected such thermal components from Am Her, U Gem, SS Cygni and EX Hydrae. It is
important to note that AM Her (which has the highest X-ray temperature of the four) is the only one for which direct evidence for
X-ray production at the dwarf stellar surface exists. The hard
AM Her X-rays exhibit modulation consistent with self-eclipsing of
the magnetic poles, with subsequent reflection and fluorescence
from adjacent regions. The others are completely consistent with
optically thin bremsstrahlung spectra, and the presumption of
spherical accretion is based only upon consistency with the
observed temperature.

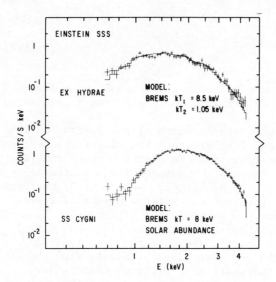

Fig. 1. Einstein SSS
experimental spectra of
two cataclysmic variables.

    Figure 1 displays some raw data from the Einstein SSS from
two of these systems. The effective temperature of the $\sim 10^8$ K
continuum is not sensitively measured by this instrument, because
its dynamic range is in the region of the spectrum where the
continuum is quite flat. The temperatures chosen were determined
from higher energy data from HEAO A-2 and, as can be observed
from the solid traces representing the model inputs folded through
the detector response as described in the lecture devoted to SNR
spectra (Holt, this volume), the fit to the SS Cygni data is quite
good. The EX Hydrae data, even though their statistical quality
are poorer, cannot be fit by the high temperature component alone.
Virtually all of the structure in the EX Hydrae spectrum can be
reconciled with an additional component (at the $\sim$ 1% level) of
lower temperature emission which is characterized by the strong
resonance emission lines from He-like Si and S. SS Cyg and U Gem
(in outburst) have exhibited intense components at even lower
temperatures ($\lesssim$ 100 eV) which are pulsed on timescales of seconds.
    Kylafis and Lamb (1979) have generalized the problem of
accretion onto a white dwarf to carefully consider cases for which
the accretion rate approaches the Eddington-limited conditions.
They find a peculiar relation between the apparent X-ray temperature
and the luminosity which is double-valued, and Copernicus data
from Cyg X-2 are apparently in agreement with this relation
(Branduardi 1977). Using these calculations, the measurement of
the apparent temperature and flux at the Earth fixes the distance
to the source and, hence, allows the determination of the absolute
luminosity and even the mass. The only difficulty with the model
is the identification of sources for which it might be applicable;
in particular, recent data (Cowley, Crampton and Hutchings 1979)
suggest that the proto-typical case for the analysis, Cyg X-2, is

almost certainly a neutron star. It is unfortunate that this,
one of the few model calculations which allows the inference of
detailed physical information about the source from its spectrum
(i.e. mass and absolute luminosity), may not have any unambiguous
candidates to which it can be applied.

3. BULGE SOURCES

The large fraction of galactic X-ray sources have no pronounced
spectral features (e.g. emission lines or sharp spectral breaks).
They can generally be fit with a bremsstrahlung continuum at a
temperature of several keV; deep exposures from satellite experiments may require deviations from a single temperature continuum,
but the required deviations are often non-unique. If we exclude
supernova remnants, X-ray pulsars and cataclysmic variables, the
great majority of the others fall into this spectroscopic category.
     Without additional information, such as we might obtain from
the determination of binary periods, for example, we have no real
handles on the nature of the compact objects or the binary system
masses or separations. Sco X-1 and Cyg X-2, for example, which
have always been characterized as similar based upon X-ray spectra
and variability, have turned out to be in binary systems with very
different periods: less than one day, and more than one week,
respectively. Many of the others may be associated with the low
mass ultra-short period systems recently suggested by Joss and
Rappaport (1979). The remainder of this section will be devoted
to the phenomenological description of the typical spectral
characteristics of all the sources which are not known to be SNR,
cataclysmics or X-ray pulsars, which include "soft" transients and
bursters as well as "steady" X-ray sources. As their similarities
far outweigh their differences, it is likely that most of them
have a common origin in a neutron star accreting mass from a
binary companion.
     After the approximately thermal nature of the overall emis-
sion, the next most striking similarity in the X-ray emission is
the considerable variability in intensity. Even before the consideration of "burst" or "transient" emission episodes, in which
the emission level may change by orders of magnitude, almost all
of the sources are variable by at least a factor of two or so on
timescales of days or more. Furthermore, this variability seems
to be correlated, on the average, with luminosity. This was
first detected by the UCL group for Sco X-1 (White et al. 1976)
but it appears to be generally (or at least typically) true. In
Figure 2 the ratio of "hard" to "soft" X-rays are plotted for
two variable sources, and there is a clear correlation between
spectral hardness and intensity level, i.e. the sources appear
"hotter" when they are more luminous. It is worth noting, however, that Cyg X-2 has, on at least one occasion, exhibited a departure from this monotonic behavior, as mentioned in Section 2.

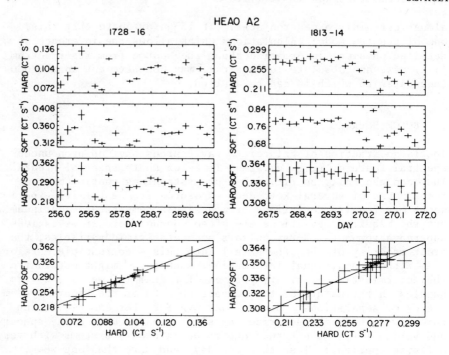

Fig. 2. HEAO A-2 "hardness ratios" for two bulge sources, indicating the correlation between spectral hardness and intensity.

The detection of unambiguous Fe-K features from diffuse X-ray sources and X-ray pulsars has led to the search for such features in the other sources, with quasi-negative results. The "quasi" is an important qualification, because while all searches for narrow line features have been unsuccessful (some with upper limits to the line equivalent widths as low as $\sim$ 2 eV), there are persistent indications of very broad Fe features with equivalent widths in excess of a few hundred eV. The nature of these indications are reductions the χ^2 in the fitting procedures (rather than actually discernable features), but such broadened line emission might well be expected from objects which are observed through a large Thomson optical depth. Typically, the overall spectra can be consistently reconciled with this assumption.

The last spectral feature worthy of note in this section is the signature of Type I bursts. The Type II bursts of the "rapid burster" are not spectrally distinct from the average properties of the sources discussed here, nor do they appear to exhibit any spectral evolution within a burst. Type I bursts, on the other hand, have a pronounced black body character which differs considerably from the average source behavior. Furthermore, this spectrum appears to cool just like a black body throughout the burst (i.e. the radius of the burst region remains constant, at a value consistent with that of a neutron star). This spectral

peculiarity, first noted by Swank et al. (1977), has been used by
van Paradijs et al. (1979) to argue that the burst sources are
neutron stars, as the Type I bursts are consistent in all respects
with He-burning episodes on neutron stars as calculated by Joss
(1979).

There are some "anomalous" sources which do not fit neatly
into the two broad categories I have defined: the "pulsars"
discussed in the next section, and the "bulge sources" described
here. The prime black hole candidate Cyg X-1, for example, exhibits a power law spectrum which is remarkably constant in both
slope and intensity with time in the range 1-100 keV, at least
during the extended "low" states which constitute the large
fraction of the source history. Even though Cyg X-1 exhibits
considerable microstructure on timescales < 1 sec, on timescales
> 1 hour it is one of the most stable sources in the galactic
catalog. This must clearly represent a true steady-state condition, and Sunyaev and Trumper (1979) have recently modelled the
emission as arising from the superposition of Thomson-scattered
components of an accretion disk around a black hole. In marked
contrast, the second black hole candidate Cir X-1 is erratic on
any timescale. Its spectrum often appears to be similar to that
of the bulge sources, but the absorption at low energies is considerably more variable than most, and there is clear evidence
for Fe-K emission which is more than merely inferential. The
similarity between the two black hole candidates in temporal
microstructure does not extend to the X-ray spectrum; this may
not be a strong yes-no discriminator for the presence of a black
hole, however, as the spectrum is likely to be dominated by the
geometry far from the central mass.

No discussion of spectra of galactic X-ray sources can be
complete without at least a mention of Cyg X-3. Like Cyg X-1, its
spectrum can be generically characterized as bimodal with a "high
state" which can be roughly fit with a black body at a few keV
and a "low state" which can be better fit by a much flatter power
law which extends to higher energies (interestingly, the total
X-ray luminosity in both cases is the same close-to-Eddington-
limited luminosity for a unit solar mass).   The prime characteristics of this source are an enormous column density (virtually
no emission below $\sim$ 1.2 keV can be detected through such a column),
and the most pronounced Fe-K emission line yet observed
(Serlemitsos et al. 1975).   The Fe line alone would easily rank
among the 50 brightest (in apparent X-ray magnitude) sources in
the sky. This line arises from fluorescence of the material responsible for the large column density around the hot source embedded in the cloud, and the SSS has recently detected S fluorescence, as well. There is a reproducible asymmetric temporal modulation of 4.8 hours ascribed to the system binary period, and with
all these temporal and spectral clues we still cannot unambiguously pin down the source geometry (i.e. whether the thick material

is a stationary shell or a wind from the companion) or even the
nature of the compact object.

4.   X-RAY PULSARS

Those sources which have periodic modulation of their X-ray emission on timescales ≤ 1000s have been termed "X-ray pulsars", even
though they are quite different physical systems than the Crab
nebula.   The Crab pulsar is the only true X-ray "pulsar", as it
is losing rotational kinetic energy with time.   All of the dozen
or so others (which are the ones I shall discuss here), are magnetized neutron stars which are increasing their rotational kinetic energies, because they are accreting material with angular
momentum in the same sense as that in which they are rotating.
Unlike the sources discussed in section 3, the X-ray pulsars have
several spectral characteristics in common which are not shared
by the bulk of the galactic X-ray catalog.
      Perhaps the most obvious of these is the very hard ($0 \leq \alpha \leq 0.5$)
energy index of the phase-averaged power-law spectrum.   A second
characteristic is the sharp cutoff of this spectrum above $\sim$ 20 keV.
Only X Persei, which is less luminous by at least two orders of
magnitude than the others, does not appear to exhibit such an
extreme spectral shape.   It should be noted here, as well, that
fully half the sample of X-ray pulsars are associated with "hard"
or "recurrent" transient sources, where the pronounced emission
episodes are believed to arise from episodes of enhanced accretion
onto the neutron stars.
      Other spectral characteristics of the X-ray pulsars are vari-
able low energy absorption, marked spectral variations throughout

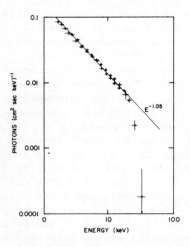

Fig. 3.   Her X-1 spectrum obtained in a rocket flight.

the pulse cycle, unambiguous Fe emission, and the possibility of cyclotron emission or absorption in at least two (and possibly all) of the sources. All of these spectral characteristics are interrelated, in the sense that they can be "explained" with a number of assumptions which is smaller than the number of spectral characteristics noted.

     If we assume that the X-ray emission from these sources originates close to the neutron star from material which is funnelled into the polar regions by the magnetic field, all of the characteristics of the X-radiation may further be assumed to arise from geometrical considerations, i.e. "beaming" of some sort in the production process itself, and/or transport of the X-radiation through the infalling material. Basko and McCray and their coworkers have devoted considerable effort to the details of a model which has had some success in achieving consistency with the data from Her X-1, which is based on the presence of a semi-opaque shell of material at the Alven surface. This shell may allow an almost unobscured view of the source region for part of the pulse cycle, and is responsible (via Thomson scattering) for the reprocessing of X-radiation from the beamed radiation which intercepts the more opaque portions of the shell. The Fe-K emission is fluorescence from the shell. Pravdo and his coworkers have been responsible for most of the detailed spectral results from OSO-8 and HEAO A-2 with which the model has been refined.

     Figure 3 is an early Her X-1 spectrum obtained from a rocket flight which has been "inverted" in accordance with equation 1.5 of the lecture on SNR spectra in this volume. The two types of

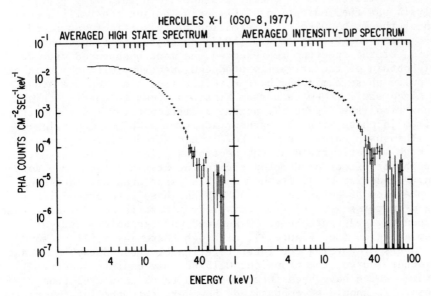

Fig. 4. OSO-8 experimental spectrum of Her X-1.

points displayed represent data from two different proportional
counter systems. Clearly evident from the figure is the hard
power-law spectrum and sharp cutoff at higher energies. Less
evident is the case for substantial Fe-K emission near 6.5 keV,
although the points from both detectors are systematically above
the best fit near that energy (and, it should be noted, the spectral inversion technique depresses such features if they are not
assumed to be present in the input spectrum). Figure 4, taken
from the OSO-8 satellite, should dispel any doubt that a substantial Fe-K feature is present in the Her X-1 spectrum. Both
of the spectra displayed are raw counts, and the Fe-K feature is
quite statistically significant in the left-hand spectrum, even
though it does not appear obvious to the eye. The right-hand
spectrum was obtained during an "absorption dip", and the fact
that the Fe emission is absorbed less than its surrounding continuum makes the feature quite obvious. Such fluorescent Fe-K
emission, with equivalent widths $\geq$ 200 eV, are typical of X-ray
pulsar spectra. For Her X-1, this equivalent width implies that
the assumed opaque fraction of the obscuring shell is at least
0.5 (Pravdo 1978).

The study of spectra at differing pulse phase can sometimes
offer unique insight into the system. The most elementary form
of "pulse-phase-spectroscopy" is the display of the pulse profile
in different energy bands. The first attempt at this sort of display was from the rocket data shown in Figure 3; this is shown in
Figure 5, where the residuals of the pulse shape at differing
energy bands compared to the 3-6 keV template are displayed. The
only clear indication of energy dependence in this very early
analysis was the indication of hardening during the falling portion of the main pulse. Confirmation of this effect was found in
OSO-8 data, and Pravdo et al. (1977) suggested that this hardest
portion of the spectrum represented the most unobscured view we
could obtain of the primarily produced spectrum, with the bulk of
the pulse phase dominated by the softer radiation reprocessed at
the Alven shell. New data from Einstein seems to confirm this
supposition (see Figure 6), as the radiation below 1 keV has its
minimum in phase with the hardest spectrum at higher energies.

The sharp cutoff at $\sim$ 20 keV is almost certainly related to
the high magnetic field at the neutron stellar surface, either
through its effect on Compton scattering or via cyclotron absorption. In addition, J. Trumper and his associates have detected
the presence of emission near 50 keV and have interpreted its
origin in terms of cyclotron emission. There is now no doubt that
the source is detectable at 50 keV and undetectable at slightly
lower energies (see Figure 4), although there may still be some
controversy regarding the interpretation of these data. At least
one other source, the transient 0115+63, may exhibit spectral
features which have been interpreted in terms of a cyclotron
origin. It should be emphasized, however, that the interpretation
of features which are phase-dependent (as is the case with the

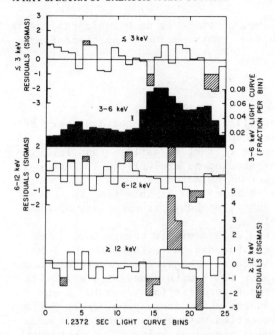

Fig. 5. Her X-1 data of Fig. 3 sorted into 25 pulse phase bins
and 4 energy bins. Using the 3-6 keV light curve as a template,
the residuals of the other three light curves are plotted (those
> 1σ are shaded for emphasis).

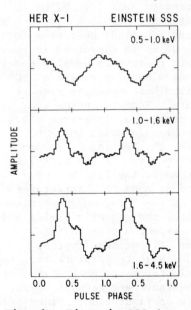

Fig. 6. Einstein SSS data from Her X-1 in three energy ranges as
a function of pulse phase.

0115+63 feature) is non-unique, as there is no unambiguous way to
determine the correct "unpulsed" contribution to subtract from
the measurement.

5. STARS

The advent of the HEAO-1 satellite resulted in the discovery of
X-ray emission from a large number of late-type RS CVn stars, and
HEAO-2 is increasing the sample to include earlier (and later)
stars, as well. The SSS has studied several of the late-type
systems, and has measured thermal emission features from many of
them. The only analysis which is mature at this time is that of
the nearby Capella system, but it probably is exemplary of what
the RS CVn spectra will yield.
     The data cannot be fitted with a single temperature (even
with the abundances variable), but can be fitted with two temperatures with variable abundances. The single temperature best
fit is at a temperature of $\sim 8 \times 10^6$ K with abundances which are
near-solar, while the two-temperature fit (at $\sim 6 \times 10^6$ K and
$\sim 40 \times 10^6$ K) requires abundances which are approximately 3 times
solar. The data can also be fitted with a distribution of temperature components which has a peak just below 10^7 K, but which
extends from $\sim 5 \times 10^6$ K to $\sim 5 \times 10^7$ K. Figure 7 displays the
data fitted with the best-fit two-temperature model described
above, the same less Fe, and the two temperature components
separately. Clearly, the emission lines of Fe(L), Mg(K), Si(K)
are constraining any fits to the spectrum to peak just below 10^7 K
to require some emission up to at least 3×10^7 K.
     Most coronal models derive the heat input to the corona from
acoustic pressure from the photosphere which is converted in the
transition region to heat the gas. This process is usually sorely
pressed to heat the corona to temperatures much in excess of 10^6 K,
so that the Capella measurement described here requires some rethinking of the heat input to such stars. Rosner, Tucker and
Vaiana (1978) have suggested that the coronae of late-type stars
may be composed of magnetic loop structures in which the hot
plasma is contained; in such a scenario, the heating is accomplished via the conversion of magnetic energy. RS CVn stars, in
particular, are more starspot-active than is the Sun, but even
the solar corona appears to be composed of such loop structures.
     In a loop model, the temperature of the X-ray emitting plasma
depends only on the pressure (and, hence, the magnetic field) and
the size of the loop, so that we would expect a peak in the
emission measure distribution characteristic of the "typical" loop,
but we should not be surprised to find a high temperature tail to
this distribution arising from atypical structures on the stellar
surface. The spectrum observed was constant over several hours,
so that these loops constitute a stable configuration. Since the
X-ray emitting gas is much too hot to be gravitationally bound to

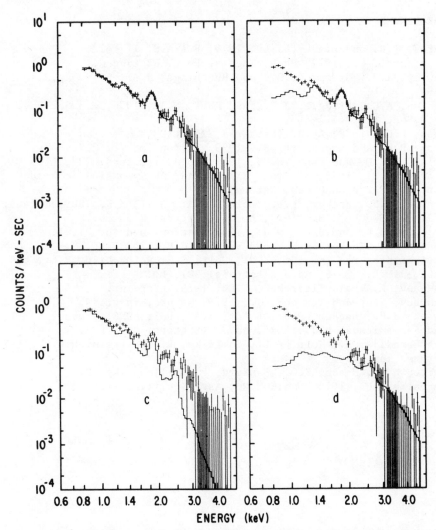

Fig. 7. Einstein SSS data from Capella fitted with a two-component collisional equilibrium model (a), the same with Fe abundance set equal to zero (b), and the lower temperature (c) and higher temperature (d) components alone.

the star, the magnetic confinement in the loops prevents catastrophic mass loss from the corona. It is interesting to note that Ayres and Linsky (1979) have recently concluded from independent evidence obtained with the IUE satellite that the GO secondary of the Capella system is very similar to the starspot-active component of RS CVn systems.

REFERENCES

Ayres, T.R. and Linsky, J.L., 1979, B.A.A.S. <u>11</u>, 472.
Branduardi, G., 1977, Ph.D. Thesis, Univ. of London.
Cowley, A.P., Crampton, D., and Hutchings, J.B., 1979, Ap.J. <u>231</u>, 539.
Fabian, A.C., Pringle, J.E., and Rees, M.J., 1976, Mon.Not.Roy. Astr.Soc. <u>175</u>, 43.
Joss, P.C., 1978, Ap.J. (Letters) <u>225</u>, L123.
Joss, P.C. and Rappaport, S., 1979, Astron.Ap. <u>71</u>, 217.
Kylafis, N.D. and Lamb, D.Q., 1979, Ap.J. (Letters) <u>228</u>, L105.
Pravdo, S.H., 1978, in Proc. XXI COSPAR/IAU Symposium in X-ray Astron. (Pergamon Press, Oxford).
Pravdo, S.H., Becker, R.H., Boldt, E.A., Holt, S.S., Serlemitsos, P.J. and Swank, J.H., 1977, Ap.J. (Letters) <u>215</u>, L61.
Pringle, J.E. and Rees, M.J., 1972, Astron. and Ap. <u>21</u>, 1.
Rosner, R., Tucker, W.H., and Vaiana, G.S., 1978, Ap. J. <u>220</u>, 643.
Serlemitsos, P.J., Boldt, E.A., Holt, S.S, Rothschild, R.E., and Saba, J.L.R., 1978, Ap.J. (Letters) <u>201</u>, L9.
Sunyaev, R.A. and Titarchuk, L.L., 1979, preprint.
Sunyaev, R.A. and Trumper, J., 1979, Nature, in press.
Swank, J.H., Becker, R.H., Boldt, E.A., Holt, S.S, Pravdo, S.H., and Serlemitsos, P.J., 1977, Ap.J. (Letters) <u>212</u>, L73.
van Paradijs, J., Joss, P.C., Cominsky, L. and Lewin, W.H.G., 1979, Nature, in press.
White, N.E., Mason, K.O., Sanford, P.W., Ilovaisky, S.A., and Chavalier, C., 1976, Mon.Not.Roy.Astr.Soc. <u>176</u>, 91.

STUDY OF HIGH LUMINOSITY X-RAY SOURCES IN EXTERNAL GALAXIES (M-31)

Riccardo Giacconi

Harvard/Smithsonian Center for Astrophysics
Cambridge, Massachusetts USA 02138

1. INTRODUCTION

The study of individual galactic X-ray sources in external galaxies is clearly interesting on many counts. It allows the study of a homogenous source population whose distance is well determined, which is rarely the case in our own galaxy. This in turn permits the determination of luminosity and source distributions functions with known and correctable bias. The relationship between source location and features of the parent galaxy, such as nuclear bulge, spiral arms, dust lanes, etc. can be studied.

In the LMC the X-ray sources were found to have higher average luminosity than sources in our own galaxy. This was interpreted by Clark (1978) to be related to the higher metal content of the galaxy. The authors argued that the sources are Eddington limited in luminosity and that this limit is higher for accreting sources in which the accreting gas is richer in high Z elements. This line of reasoning then opens the possibility of studying the metalicity content and therefore the evolutionary state of galaxies by studying their X-ray source content.

The great improvement in observational capabilities brought about by the launch of the Einstein Observatory, has made it possible to begin this type of investigation.

It was quite natural that one of our first targets would be M-31, the spiral galaxy in Andromeda. Leon Van Speybroeck, who heads the group at CFA which is responsible for this work, points out

R. Giacconi and G. Setti (eds.), X-Ray Astronomy, 103-113.

that M-31 is the first galaxy which was studied with a visible light telescope by Marius in 1612 (Van Speybroeck, 1979).

The integrated X-ray emission from M-31 was detected by
UHURU (Forman et al, 1978), Ariel (Cooke et al, 1978) and by Margon
(1974) in soft X-rays. The source was at the limit of sensitivity for
these instruments and its position correspondingly poorly known, as
shown in Figure 1.

2. OBSERVATIONAL RESULTS

The angular size of M-31 is so large (about 2°) that several exposures
of the largest field of view imaging detector (IPC) on Einstein are
necessary to image it all.

Figure 2 shows the trace of the field of view of the detectors for the two exposures obtained with the IPC as well as that obtained
with the HRI superimposed on the optical image of the galaxy.

Figures 3 and 4 show the X-ray images obtained in the IPC
exposure (approximately 30,000 seconds each).

In the central image one can observe a confused region
corresponding to the center of the galaxy surrounded by some 20 individual sources. Several sources can also be seen in the more northern
exposure.

Figure 5 shows an X-ray image of the central region obtained with the HRI in which some 45 individual sources can be resolved.
The exposure time was approximately 30,000 sec. The individual
sources are detected using computerized procedures based on a sliding
box algorithm. Maximum likelihood techniques are used to establish
position and intensity of each source. Typical systematic position
errors for HRI and IPC sources are of 2 arc seconds and 1 arc minute.

Threshold intensities for the HRI and IPC detectors corres-
pond to luminosities at M-31 of 9.1 and 4.4×10^{36} erg sec^{-1} in the
band 0.5 to 4.5 keV. The sources are too weak to measure the spectrum
directly with the IPC. The intensity is computed assuming a thermal
bremsstrahlung with temperature of 5 keV with corrections for absorption in our galaxy corresponding to $N_H = 8 \times 10^{20}$ cm^{-2}. The quoted
values of the intensity are relatively insensitive to the choice of
spectral shape or parameters, but we consider the quoted absolute
intensities to be uncertain to about 50%.

HRI and IPC observed fluxes from a region within 5 arc
minutes of the nucleus are equal within about 10% which is well within
uncertainties due to calibration or corrections due to the choice of
spectral parameters.

This consistency argues strongly against the existence of
a large number of fainter sources undetectable in the HRI, but giving a
contribution to the IPC fluxes. The central region luminosity in the

band 0.5 to 4.5 keV is of 1.24×10^{39} ergs/sec. The total luminosity of all detected sources is 2.3×10^{39} erg sec^{-1} which is consistent with previous measurements of M-31 luminosity.

3. PROPERTIES OF M-31 SOURCES.

We have compared the X-ray source positions with those of known supernova remnants, globular clusters and other interesting objects. We find a coincidence with the supernova in region 521 of Baade and Arp (1964). The X-ray luminosity of this object is of 2.2×10^{37} ergs sec^{-1}. We do not detect emission ($L_X \leq 10^{37}$ erg sec^{-1}) from Andromedae 1885.

Five of the HRI X-ray sources coincide with the position of globular clusters in the list of 60 clusters (included in the HRI field) by Sargent et al (1977) to within 5 arc seconds. Eight of the IPC sources are also found within 1 arc minute of the globular clusters, when only 4 accidental coincidences are expected. We estimate therefore that 9 out of the 237 globular clusters which are in our fields are X-ray sources. This ratio is similar to that found in our galaxy (Grindlay, 1979), where 7 out of 150 globular clusters have X-ray luminosity greater than $5 - 10 \times 10^{36}$ erg sec^{-1}.

The position of the bright optical source at the center of M-31 coincides within a few arc seconds with one of the X-ray sources. Given the density of sources in the central region, we cannot exclude that this is a purely accidental coincidence. If the X-ray source should truly represent the emission from the galactic nucleus, the M-31 nucleus at $L_X = 10^{38}$ erg sec^{-1} would be 10^3 times brighter in X-rays than the nucleus of our galaxy. On the other hand, our galactic nucleus is much brighter in radio (15 times at 1415 MH$_Z$, van der Kruit(1972)) than M-31. This might suggest that the X-ray emission process may be thermal, possibly powered by gas accretion of a massive object. Further study of this source may help decide whether it is just another binary system or truly galactic nucleus emission.

We find it useful to divide the X-ray source population into three groups: globular clusters, inner bulge and population I objects. Inspection of Figure 2 shows the association of the X-ray sources far from the central region with spiral arm features. This can be better appreciated by comparing the source positions with the HI map by Emerson (1976), Figure 6. Population I sources (not associated with the central bulge, or identified with globular clusters) are shown as filled circles. The correlation with HI features is striking. This is not surprising since X-ray sources are presumably binaries in which one of the stars has evolved to the supernova stage. This means that the progenitor of the X-ray binaries must be a massive binary system

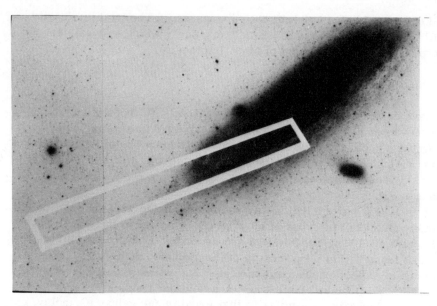

Figure 1

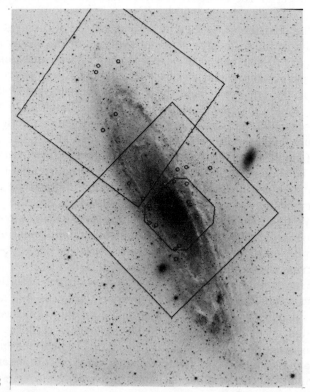

Figure 2

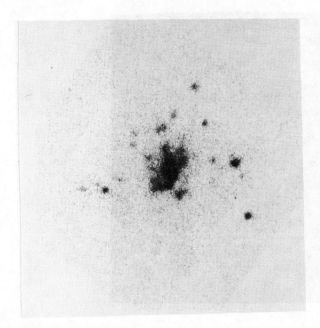

Figure 3

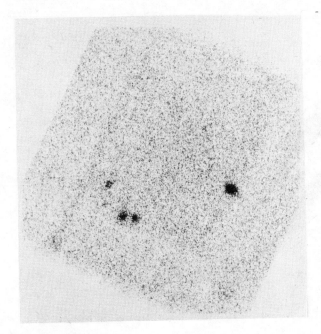

Figure 4

Figure 5

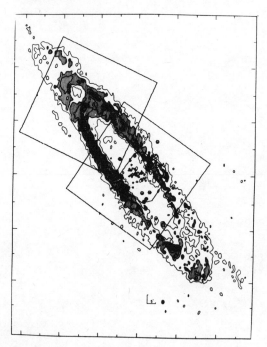

Figure 6

and therefore X-ray sources should be associated with indicators of
stellar formation and young massive star regions. In M-31 such
indicators of Population I activity, such as OB associations, HI, H II,
CO and dust regions are located in a ring at about 9 pc from the
nucleus, as shown by the work of many authors (Emerson, 1974, 1976
and 1978; Baade and Arp, 1964; Berkhuijsen, 1977; and Coombes et
al, 1977a and b).

The position of X-ray sources in the inner part of M-31
are shown on an enlarged picture in Figure 7. While some of the
sources clearly belong to the inner arms, there is a group of about 18
sources which are concentrated in the central 2 arc minutes of the
nucleus, a region corresponding to about 400 pc. These sources,
which we designate as inner bulge, are on the average more luminous
than the Population I sources. The information on luminosity is summarized in Figure 8.

It is interesting to compare the results with the study by
Clark et al (1978) on the luminosity of sources in our galaxy and in the
Magellanic Clouds. Taking into account the different energy band used
by Clark, it becomes clear that the luminosity of sources in our
galaxy and in M-31 is quite similar.

The most striking difference, however, has to do with the
distribution of what we have normally considered bulge sources in our
own galaxy with those in M-31. The sources which we consider
bulge sources in our galaxy span 40°. However, 400 pc at our galactic
center subtend only 2.3°. There are in fact only 3 sources in the 4U
Catalog (Forman et al, 1978) within 2.3° of the galactic center compared with about 20 in M-31. It is therefore clear that the density of
inner bulge sources in M-31 is much greater than in our galaxy. Nor
can absorption to our own galactic center have prevented us from seeing these sources. The temperature derived from a spectral fit to the
aggregate X-ray emission from the M-31 inner region yields a temperature of 5 keV, sufficiently high that the 3 keV typical absorption expected toward our galactic center should not prevent us from seeing similar
sources.

It is interesting that the 400 pc radius that defines the inner
bulge X-ray sources is the distance from the M-31 nucleus where Rubin,
Ford and Kumar (1973) find a local minima in the emission line velocity.
This region may constitute a distinct feature of M-31. Rubin and Ford
(1970) assign 0.015 of the M-31 mass to this region and attribute most
of it to low mass, late-type stars. Yet our finding is that this region
is responsible for about 1/3 of the X-ray emission of M-31.

It is interesting to relate these findings to some of the
evolutionary models which have been discussed for globular clusters and
inner bulge sources by Van den Heuvel and others (see Van den Heuvel

Figure 7

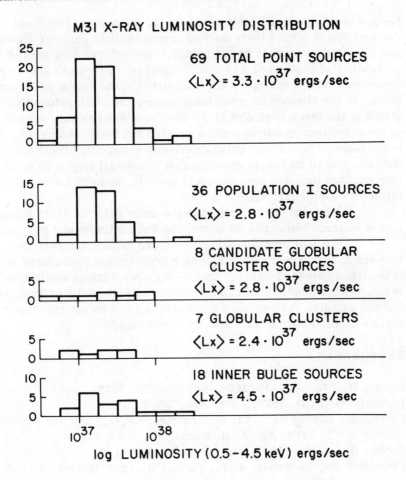

Figure 8

lecture in this book for bibliography). In particular, low mass
capture binaries have been invoked to explain both globular clusters
and inner bulge sources in our galaxy. The computation for the
probability of capture of a compact object by a low mass star is strongly
dependent on the assumed stellar densities in the region under consideration. In the absence of other information, the stellar densities determined in the inner region of M-31 have been used as representatives
of the conditions in our own galaxy. This is clearly incorrect. If the
bulge sources in our own galaxy were due to capture processes, we
would expect to find them concentrated in a small region as in M-31.
The fact that they are not, seems to point to the need for a quite
different evolutionary model.

In summary, we find in the inner bulge of M-31 a population of sources which has no corresponding analog in our own galaxy.

It will be interesting to explore further the relation
between X-ray properties and other morphological features of M-31.
It is already clear, however, that X-ray observations similar to
these, when extended to a representative sample of different types of
normal galaxies, will give us a powerful tool to investigate their
stellar populations and evolutionary conditions.

REFERENCES

Baade, W., and Arp, H. 1964, Ap. J., 139, 1027
Berkhuijsen, E. M. 1977, Astr. Ap., 57, 9
Clark, G., Doxsey, R., Li, F., Jernigan, F.G., and van Paradijs,
          J. 1978, Ap. J. (Letters), 221, L37.
Cooke, B.A., et al 1978, M.N.R.A.S., 182, 489
Coombes, F., Encrenaz, P.J., Lucas, R., and Weliachew, L.
          1977a, Astr. Ap., 55, 311
          -1977b, Astr. Ap., 61, L7
Emerson, D.T., 1974, M.N.R.A.S., 169, 607
          -1976, M.N.R.A.S., 176, 321
          -1978, Astr. Ap., 63, L29
Forman, W., Jones, C., Cominsky, L., Julien, P., Murray, S.,
          Peters, G., Tananbaum, H. and Giacconi, R., 1978 Ap. J.
          Suppl., 38, 357
Grindlay, J. E. 1979, private communication
Margon, B., Bowyer, S., Cruddace, R., Heiles, C., Lampton, M.,
          and Troland, T. 1974, Ap. J. (Letters), 191, L117
Rubin, V., and Ford, W.K., Jr. 1970, Ap. J. 159, 379.
Rubin, V., Ford, W.K., Jr. and Kumar, C.K. 1973 Ap. J., 181, 61
Sargent, W.L.W., Kowal, C.T., Hartwick, F.D.A., van den
          Bergh, S. 1977, A.J., 82, 947.

van den Heuvel, E. P. J. 1978, in Proc. International School of Physics
"Enrico Fermi", Course LXV, ed. R. Giacconi and
R. Ruffini (Amsterdam: North-Holland), p. 828
van der Kruit, P. C. 1972, Ap. Letters 11, 173
Van Speybroeck, L., 1979 Ap. J., 234, L45

X-RAY SOURCES AND STELLAR EVOLUTION

E.P.J. van den Heuvel

Astronomical Institute, University of Amsterdam,
the Netherlands, and
Astrophysical Institute, Vrije Universiteit Brussels,
Belgium

1. INTRODUCTION AND SUMMARY

The Einstein Observatory has observed in M31 the same types of
strong point X-ray sources that we know in our own galaxy and in
the Magellanic Clouds. Their numbers are about 3 times larger (76
vs. 26 sources with $L_x > 5 \times 10^{36}$ ergs/sec) presumably due to the
about 2.4 times larger mass of M31. Like in our galaxy, the
sources in M31 appear - apart from a few supernova remnants - to
fall into two broad categories (van Speybroeck et al. 1979),
viz.:
- a population I group, concentrated near the plane of the
  galaxy, and more or less evenly distributed throughout the
  system;
- a group of bright bulge sources, strongly concentrated towards
  the center of M31. To the latter group also belong 8 or 9
  globular cluster sources. In M31 the central concentration in
  the second group seems to be stronger than in our galaxy, which
  might be related to the known stronger central concentration of
  the bulge population in this system (cf. Spinrad and Taylor
  1971).

As the bulk of the X-ray emission from normal spirals like M31
and our own galaxy appears to be due to a collection of point
sources, the understanding of the X-ray fluxes from normal
spiral galaxies requires a knowledge of the nature and origin of
these point sources.

All our present knowledge about the nature of the point
sources is derived from the study of the X-ray sources in our
galaxy and in the Magellanic Clouds. For this reason, I will in
this paper discuss the current ideas about the nature and origin
of the galactic X-ray sources.

R. Giacconi and G. Setti (eds.), X-Ray Astronomy, 115-127.
Copyright © 1980 by D. Reidel Publishing Company.

2. TYPES OF STRONG GALACTIC X-RAY SOURCES

I will not consider here the few young supernova remnants, but restrict myself to the remaining strong point sources ($L_x \sim 10^{35.5}$– 10^{38} ergs/sec).

As pointed out by Jones (1977), Ostriker (1977) and Maraschi et al. (1977) the strong galactic sources can, according to their X-ray spectra be divided into two broad groups, denoted here as group I and II, which appear to differ in a number of physical aspects, as listed in table 1. (The existence of these two groups was already suggested by Salpeter (1973) on the sole basis of the space distribution of the sources.) The sources in group I have relatively hard X-ray spectra (bremsstrahlungsfit gives $T > 10^8$ K), whereas the spectra of group II sources are considerably softer ($T \leq 3 - 8 \times 10^7$ K).

All the pulsating binary sources exclusively belong to group I, and the correlation between hard spectra and pulsation is so good that it can be used as a criterion for selecting (prospective) pulsar candidates. With only one or two exceptions the optical counterparts of group I sources are luminous early-type stars and the ratio of optical to X-ray luminosity in group I is $L_{opt}/L_x > 1$. The sources in group I are strongly concentrated towards the galactic plane and clearly the majority of them belongs to extreme population I. (The only exception is 4U 1626-67 which has a very low optical luminosity and is far outside the plane; on the other hand, Hercules X-1, although also outside the plane, clearly belongs to population I, as it has a 2 $M_\odot$ A-star companion; its position outside the plane must be due to the large kick which it received in the supernova explosion that created this neutron star (cf. Sutantyo 1975a).)

Table 1. The two groups of strong (> $10^{35.5}$ ergs/sec) galactic X-ray sources (cf. Jones 1977; Maraschi et al. 1977; Ostriker 1977)

Group I	Group II
- Hard X-ray spectra ($T > 10^8$ K);	- Softer spectra ($T = 3.7 \times 10^7 - 8.6 \times 10^7$ K);
- Often pulsating;	- Non-pulsating;
- Luminous (massive) early-type optical counterparts: $L_{opt}/L_x > 1$;	- Always optically faint (blue) counterparts: $L_{opt}/L_x < 10^{-2}$;
- Concentrated near the galactic plane ($|z| < 150$ pc) like extreme Population I;	- Concentrated towards galactic center, like old disk population (8 sources in glob.clusters);
- Probably less than 50 in the galaxy (a dozen with $L_x > 5 \times 10^{36}$ ergs/sec).	- Probably more than a hundred in the galaxy (16 with $L_x > 5 \times 10^{36}$ ergs/sec).

On the other hand, the sources in group II never show
pulsation, and always have intrinsically faint optical counterparts, i.e.: $L_{opt}/L_x < 10^{-2}$. To group II belong: (i) strong
galactic center sources (also called "bulge sources"), (ii) the
globular cluster souces, (iii) the steady sources associated with
bursters and also (iv) the low-mass X-ray binary Sco X-1, which
has a photometric and spectroscopic period of $0^d.78$ (Cowley and
Crampton 1975). Notice that the subclass (iii) overlaps with the
subclasses (i) and (ii). According to their space distribution
the galactic center sources and the bursters belong to a rather
old population, concentrated in the bulge of the galaxy; the
globular cluster sources belong to extreme population II. Therefore, the group II sources appear to belong mostly to an old
population with ages of order $5 - 13 \times 10^9$ yrs.

As mentioned in the introduction, the space distribution of
the sources in M31 suggests that a similar division into two
population classes exists in that galaxy, although the bulge
sources in M31 show a stronger central concentration.

3. THE NATURE AND ORIGIN OF GROUP I SOURCES

The nature of these sources – generally neutron stars in massive
binary systems – seems presently well understood. The generation
of the X-rays is here due to the transfer of matter from the
atmosphere of a normal massive star to a magnetized obliquely
rotating neutron star. For the detailed physics of the accretion
process and the generation of the X-rays I refer to the general
literature (i.e.: Giacconi and Gursky 1974; Giacconi and Ruffini
1978; Papagiannis 1977). Also the general outlines of the
evolutionary history of these systems seems presently well understood and has been extensively reviewed elsewhere (for details
see van den Heuvel 1974, 1976, 1977, 1978). It appears that the
massive X-ray binaries represent a brief stage in the evolution
of all normal massive close binaries with a primary star mass
larger than about 15 $M_\odot$ (cf. also Tutukov and Yungelson 1973).
Figure 1 shows as an example the main stages of the evolution of
a representative system, which started its evolution with components of 20 $M_\odot$ and 8 $M_\odot$, as calculated by de Loore et al. (1975).
[In this picture the evolution during stages of mass transfer is
assumed to be "conservative", i.e. it is assumed that the total
mass and orbital angular momentum of the system is conserved. It
is well known now that these assumptions are too simplified and
that considerable loss of mass and angular momentum from the
system may occur during mass transfer stages, as was shown by
Flannery and Ulrich (1977) and Kippenhahn and Meyer-Hoffmeister
(1977). Such a loss of mass and angular momentum is certainly
required to explain the short binary periods of systems like
Cen X-3 (P = $2^d.087$) and LMC X-4 (P = $1^d.48$), but it is not expected to strongly affect the duration of the main evolutionary

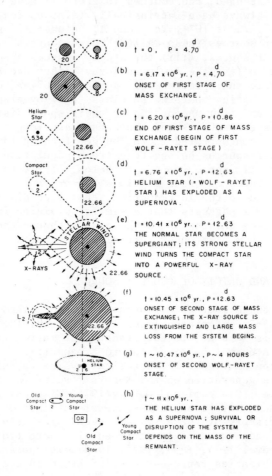

(a) $t = 0$, $P = 4.\overset{d}{7}0$

(b) $t = 6.17 \times 10^6$ yr., $P = 4.\overset{d}{7}0$
ONSET OF FIRST STAGE OF
MASS EXCHANGE.

(c) $t = 6.20 \times 10^6$ yr., $P = \overset{d}{1}0.86$
END OF FIRST STAGE OF MASS
EXCHANGE (BEGIN OF FIRST
WOLF - RAYET STAGE)

(d) $t = 6.76 \times 10^6$ yr., $P = \overset{d}{1}2.63$
HELIUM STAR (= WOLF - RAYET
STAR) HAS EXPLODED AS A
SUPERNOVA.

(e) $t = 10.41 \times 10^6$ yr., $P = \overset{d}{1}2.63$
THE NORMAL STAR BECOMES A
SUPERGIANT; ITS STRONG STELLAR
WIND TURNS THE COMPACT STAR
INTO A POWERFUL X-RAY
SOURCE.

(f) $t = 10.45 \times 10^6$ yr., $P = \overset{d}{1}2.63$
ONSET OF SECOND STAGE OF MASS
EXCHANGE; THE X-RAY SOURCE IS
EXTINGUISHED AND LARGE MASS
LOSS FROM THE SYSTEM BEGINS.

(g) $t \sim 10.47 \times 10^6$ yr., $P \sim 4$ HOURS
ONSET OF SECOND WOLF-RAYET
STAGE.

(h) $t \sim 11 \times 10^6$ yr.,
THE HELIUM STAR HAS EXPLODED
AS A SUPERNOVA; SURVIVAL OR
DISRUPTION OF THE SYSTEM
DEPENDS ON THE MASS OF THE
REMNANT.

Figure 1. Stages in the evolution of a massive close binary, in the conservative case. It is assumed that the supernova explosion of the original primary leaves a 2 $M_\odot$ compact star. The last three stages are somewhat speculative. The dashed vertical lines indicate the position of the center of gravity of the system.

phases as depicted in figure 1 (cf. van Beveren et al. 1979). As a guideline for distinguishing the main evolutionary phases and their duration, the "conservative" picture of figure 1 is, therefore, adequate.] The system of figure 1 starts with two upper main-sequence stars of 20 $M_\odot$ and 8 $M_\odot$ in an orbit of period 4.70 days. After about 6 million years the 20 $M_\odot$ star evolves off the main-sequence, expands to fill its critical Roche surface and begins to deposit mass onto its companion (stage b). In an interval of only about 30 000 years, nearly 15 $M_\odot$ of material is transferred onto the secondary, leaving a Wolf-Rayet or helium star of about 5 $M_\odot$ and a main-sequence star of about 23 $M_\odot$ (stage c). Conservation of orbital angular momentum has lengthened the binary period to about 11 days at this stage. After about another half million years the helium star explodes as a supernova, ejecting about 3 $M_\odot$ of gas and leaving a compact remnant of (assumedly) about 2 $M_\odot$, presumably a neutron star (stage d). The orbital period has now increased to nearly 13 days. The 23 $M_\odot$ star continues to evolve and after about another 4 million years it becomes a blue supergiant with a strong stellar wind (stage e). Accretion from this wind and, subsequently, due to beginning Roche-lobe overflow (Savonije 1978, 1979) turns the compact star into a strong X-ray source. This stage lasts for only about 40 000 years, for the supergiant soon expands to overfill its Roche-lobe, and a second stage of mass exchange begins in which the X-ray source is extinguished by the excessive accretion rate (stage f). The subsequent evolution is less certain, but presumably leads to the formation of a binary pulsar or two runaway pulsars (cf. Flannery and van den Heuvel 1975; Taam et al. 1978).

Predicted numbers. The short duration of the strong X-ray stage ($\sim$ 40 000 yrs) combined with an estimated number of some 2400 - 4800 unevolved massive close binaries (with a lifetime of $\sim$ 6 million years) in the galaxy leads in a steady state to a predicted number of some 16 to 32 massive X-ray binaries in the galaxy (cf. van den Heuvel 1978). This seems in fair agreement with the observation of a few dozen of such systems in the galaxy and in M31. Consequently, although many details of the evolutionary picture need further study (such as: details of the accretion processes, spin-down and spin-up behaviour, the effects of tidal interactions, of the supernova explosion and of losses of mass and angular momentum from the systems), the general outline of the evolution of group I sources seems well understood and the presence of these sources fits well into the evolutionary picture of normal massive close binary systems.

4. NATURE OF THE GROUP II SOURCES: THE LOW-MASS BINARY MODEL

Important information on the nature of group II sources is obtained from the fact that several tens of them are X-ray burst sources. The majority of these are field sources belonging to the bulge population, but also five of the globular cluster sources are bursters. This stresses once more the similarity between bulge sources and globular cluster sources and suggests that they form one population class.

Extensive reviews of the present knowledge of X-ray burst sources and globular cluster sources have been given by Lewin

(1979a,b,c), Lewin and Clark (1979a,b,c) and by Joss (1979). We refer the reader to these reviews and restrict ourselves to summarizing their most important conclusions (as to the "bursts", we only consider here the "normal" type I bursts; since the type II bursts from the rapid burster most probably represent a special mode of the steady emission from that source, cf. Lewin 1979a,b,c), which are as follows:

a. The spectral evolution of X-ray bursts suggests that we are observing the black-body emission from the surface of a cooling object which has an average radius of about 7 ± 2 km (van Paradijs 1978), i.e. characteristic of a neutron star.

b. The steady X-ray flux from burst sources is about 10^2 times larger than the flux emitted in the form of bursts. This ratio is about the same as the ratio between the efficiency of energy release due to accretion onto a neutron star and the efficiency of nuclear (helium) burning.

c. Computations by Joss (1978, 1979) show that hydrogen accreted onto a non-magnetized neutron star immediately fuses to helium, but that helium fusion tends to occur in flashes with a time profile strikingly similar to those of X-ray bursts.

d. The optical spectra and colors of optically identified burst sources are strikingly similar to those of Scorpius X-1 (McClintock et al. 1978). The spectrum of the latter consists of a blue continuum with emission lines of HI, HeII, CIII, NIII, and resembles the spectra of accretion disks in cataclysmic variables.

e. Scorpius X-1 is a member of group II, and is known to be a low-mass close binary system ($P = 0.78$). The short binary period and the low optical luminosity of the companion star suggest that it must be a red dwarf, similar to the invisible components in cataclysmic variable binaries (cf. Robinson 1976).

Although all this evidence is circumstantial (no direct evidence of binary motion or eclipses is observed) it suggests a coherent picture in which the group II sources are low-mass close binary systems, consisting of an old (presumably non-magnetized) neutron star which is accreting matter from a Roche-lobe filling low-mass red dwarf companion. In analogy to the cataclysmic variables, the orbital period of such a system is expected to be of the order of a few hours (cf. Faulkner 1971).

This is the low-mass close binary model for group II sources, put forward by Joss and Rappaport (1979) as depicted in figure 2. (In fact, the low-mass main-sequence star may, in this model, be replaced by a lobe-filling degenerate dwarf, which would lead to binary periods of the order of 10 minutes (Li et al. 1980; Rappaport 1980)).

Two further key facts in support of the low-mass close binary model are:

(A). The excessively high incidence of X-ray sources in globular clusters, first noticed by Gursky (1973), i.e. some 10^2

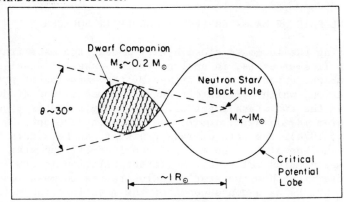

Figure 2. The low-mass close binary model for group II
sources as proposed by Joss and Rappaport (1980).

to 10^3 times higher (per 10^6 solar masses) than in the galaxy as
a whole. This suggests that a special formation mechanism for
X-ray sources is operating in globular clusters, which operates
much less efficiently (or is absent) in other parts of the
galaxy. It seems most plausible that this formation mechanism
is capture, because in the cores of globular clusters the
probability that two stars capture one another and form a close
binary system is much higher than anywhere else in the galaxy
(see below).

(B). The masses of the globular cluster sources, as
determined from their average distances from the cluster centers.
Assuming that the cluster stars have reached equipartition, the
stars with the highest masses will have the lowest space velocities and hence, will reside deep in the potential well close to
the cluster center. Consequently, the average distance to the
cluster core provides a measure of the masses of the sources.
Recent Einstein measurements show the average distance to the
cluster centers to be several arc seconds and indicate a most
probable source mass of about 2 $M_\odot$ (Grindlay 1980). This is just
about the mass expected for a system consisting of a ~ 1.4 $M_\odot$
neutron star and a red (or white) dwarf.

One might consider the absence of X-ray eclipses to
constitute a possible argument against binary models. However, as
pointed out by Milgrom (1978) the presence of an accretion disk –
which is indicated by the optical spectrum – will cause that the
X-ray source is invisible for observers at low inclinations with
respect to the orbital plane, i.e. just at those inclinations
for which the probability for observing eclipses would be highest.
Hence, the observability of the X-ray source and of eclipses are
expected to be mutually exclusive; this might provide a natural
explanation of why for those sources that we observe, eclipses
are absent.

5. THE RATE OF MASS TRANSFER IN GROUP II SOURCES

Adopting the low-mass close binary model, the mass transfer in
group II sources is, in analogy to the case of cataclysmic
variables, expected to be driven by the emission of gravitational
radiation, which carries off angular momentum and gradually
drives the two stars together. The theory for this type of mass
transfer is straightforward and has been developed by Faulkner
(1971, 1974). The rate of mass transfer can be simply calculated
once one knows: (i) the mass of the lobe-filling star and (ii)
its mass-radius relation. The latter relation is crucial, because
it determines how the star's radius reacts to mass loss. If the
star shrinks as a consequence of the mass loss (as main-sequence
dwarfs tend to do), the system will shrink and the stars will
spiral closer and closer together. On the other hand, if the
stellar radius expands as a consequence of the mass loss (which
will be the case if the star is degenerate or fully convective,
e.g. a polytrope of index n = 1.5), the mass transfer rate will
have to adapt itself such that the Roche lobe, and with it the
entire system, expands as a consequence of the mass transfer.
(This expansion is already naturally induced by the fact that
the mass-losing star has the lowest mass of the two, such that in
the absence of gravitational radiation, mass transfer will always
cause the system to expand (cf. Paczynski 1971); taking gravitational radiation losses into account the star will just transfer
mass at such a rate that the expansion of the star goes at the
same rate as the expansion of its Roche lobe.) I.e.: in the case
of a lobe-filling white dwarf, the two stars will spiral away
from each other (cf. Li et al. 1980; Ostriker and Zytkow 1980),
which will cause the mass-transfer rate to decrease in the course
of time. Table 2 lists as an example the binary periods and mass
transfer rates induced by gravitational radiation as a function
of the mass M_2 of the companion for (a) red dwarf companions with
a mass radius relation

$$R_2 = M_2 \qquad\qquad\qquad (1)$$

and (b) helium white dwarfs with a mass-radius relation

$$R_2 = 0.0123 \, M_2^{-1/3} \qquad\qquad\qquad (2)$$

(M_2 and R_2 in solar units). The mass-transfer rates were calculated using the equations given by Faulkner (1971).

The table shows that with red-dwarf companions X-ray
luminosities of the order of 10^{36} ergs/sec ($\dot{M}_2 \simeq 10^{-10}$ $M_\odot$/yr) can
be obtained without much difficulty. However, to reach the highest
observed X-ray luminosities of the bulge sources (> 10^{37} ergs/sec)
some special conditions are required, probably involving some
extra loss of mass and angular momentum from the systems, e.g.
due to a stellar wind from the red dwarf (Ostriker, private
communication). [At the other extreme: degenerate helium dwarfs
are clearly able to provide very high X-ray luminosities, but

Table 2. Binary periods, mass transfer rates (driven by gravitational radiation) and X-ray lifetimes of some representative low-mass close binaries that consist of a 1.4 $M_\odot$ neutron star and a red dwarf or a degenerate helium dwarf, with mass-radius relations given by equations (1) and (2).

M_2 ($M_\odot$)	Red dwarf			Degenerate Helium dwarf		
	P (sec)	$\frac{dM_2}{dt}(M_\odot/yr)$	Lifetime (yrs)	P (sec)	$\frac{dM_2}{dt}(M_\odot/yr)$	Lifetime (yrs)
0.03				1.464×10^3	1.8×10^{-10}	1.7×10^8
0.05				0.845×10^3	3.3×10^{-9}	1.5×10^7
0.07	2.254×10^3	2.06×10^{-10}	3.5×10^8	0.627×10^3	1.2×10^{-8}	$6\ \times10^6$
0.10	$3.22\ \times10^3$	$1.7\ \times10^{-10}$	$6\ \times10^8$	0.439×10^3	5.9×10^{-8}	1.6×10^6
0.20	$6.44\ \times10^3$	1.07×10^{-10}	1.9×10^9	0.220×10^3	1.2×10^{-6}	1.7×10^5
0.30	$9.66\ \times10^3$	0.71×10^{-10}	$4\ \times10^9$			

have the disadvantage of a very short lifetime. It is·not clear at the moment whether such systems could really exist in nature.] Assuming a mass-transfer rate of 10^{-9} $M_\odot$/yr (corresponding to $L_x = 10^{37}$ ergs/sec) and starting from a 0.5 $M_\odot$ companion, the total X-ray lifetime of a low-mass binary will be some 5×10^8 yrs. This is some 10^4 times longer than the lifetime of group I sources, and implies – since both groups have roughly similar numbers of members – that the formation rate of group II sources should be some 10^4 times lower than that of group I sources.

6. THE FORMATION OF THE X-RAY SOURCES IN GLOBULAR CLUSTERS

In the cores of globular clusters the probability that two stars capture one another is much higher than anywhere else in the galaxy. This is due to two facts: (i) the very high star density in the cores of globular clusters (up to 10^5 stars/pc^3) and (ii) the low relative velocities (at maximum a few tens of km/sec) which ensures a large collision cross-section.

A variety of capture mechanisms have been suggested (direct collisions, Sutantyo 1975b; Hills and Day 1976; tidal capture, Fabian et al. 1975; three-body capture and exchange collisions, Hills 1976; Heggie 1977; Shull 1979). We refer here to the review by Hills and Day (1976), who estimate that on the average some 3.3% of all globular cluster stars have experienced a collision during the lifetime of the galaxy. Of course, in order to form an X-ray binary, one of the two stars should be a neutron star.

Using an average X-ray source lifetime of 5×10^8 yrs, an average globular cluster age of 12×10^9 yrs and the fact that 8 out of the about 140 globular clusters in the galaxy contain an X-ray source, one finds that, in order to sustain a steady-state

population of 8 globular cluster sources, one such source should
be formed per cluster every 8.5×10^9 yrs. Using a 3.3% collision
probability per star in 12×10^9 yrs one then finds that on the
average some 40 neutron stars per globular cluster are required
to provide this formation rate. This number is very modest since,
from extrapolation of the cluster mass function beyond the turn-
off point one finds that in an average globular cluster like M3,
originally more than 4000 stars more massive than 8 $M_\odot$ must have
been present. These all are expected to have exploded as supernovae (and to have left neutron stars) during the early life of
the clusters (cf. Sutantyo 1975b). The galactic distribution and
observed space velocities of radio pulsars suggest that pulsars
receive at their birth space velocities of the order of 50 to
100 km/sec (Hanson 1979). As the escape velocities from globular
cluster cores are generally below 20 km/sec one therefore would
expect most neutron stars to have escaped. The above calculation
shows that indeed 99% of the neutron stars that once were formed
in globular clusters must have escaped, since otherwise far more
globular cluster X-ray sources should be observed. The fact that
only about one percent of all neutron stars that were ever formed
in globular clusters, are required to provide the presently
observed incidence of globular cluster X-ray sources seems therefore one more (independent) confirmation that most neutron stars
indeed do receive a high space velocity at their birth.

7. THE FORMATION OF THE SOURCES IN THE BULGE

In analogy to the globular cluster sources, Ostriker (private
communication) has suggested that also the bulge sources were
formed by capture. Of course, the capture probability per neutron
star in the bulge is much lower than in a globular cluster core,
since:
a. the space velocities of the stars are much higher (~ 200
   km/sec), which results in a much lower collision cross-section
   σ, since $\sigma(:)1/v^2$;
b. the star density is much lower.
The total collision probability per neutron star will be
$$F = N\,v\,\sigma\ (:)\ N/v \tag{4}$$
where N is the star density and v is the relative velocity.
Assuming N to be some 10^3 times lower than in average globular
cluster cores, and v to be 10 times higher, one finds that in the
bulge F is 10^4 times lower than in a globular cluster core.
[A 10^3 times lower star density seems reasonable, since in the
bulge of M31 one finds some 10^{10} $M_\odot$ within a 400 pc radius; most
of these stars are probably M-dwarfs (mass ~ 0.4 $M_\odot$; cf. Spinrad
and Taylor 1971), leading to a star density of order $10^2/pc^3$.]
On the other hand, assuming a bulge mass of 10^{10} $M_\odot$, and 2% of
this mass to be present in the form of neutron stars, the total
number of neutron stars in the bulge will be $\sim 2 \times 10^8$, which is

some 3.2×10^4 times larger than the estimated 5600 neutron stars in the 140 globular clusters.

Consequently, one would expect, from this very rough estimate that the collision rate of neutron stars and dwarfs in the bulge is about three times the collision rate in all globular clusters combined. This shows that it is not inconceivable that also the bulge sources were formed by collisions between single neutron stars and dwarfs. An attractive aspect of such a formation mechanism is, that it yields a natural explanation for the stronger central concentration of the bulge sources in M31 as compared to our own galaxy. For a more refined estimate of the collision frequency between neutron stars and field stars in the bulge we refer to Finzi (1978).

An alternative possibility is that all bulge sources were formed in globular clusters and due to three-body encounters have escaped from the clusters. In view of the very favourable conditions for collisions inside globular clusters, also this possibility cannot be discarded.

Although the above estimates suggest that collision and capture seem attractive possibilities for the formation of the bulge sources it is highly desirable that these ideas are worked out in more quantitative detail.

ACKNOWLEDGEMENTS

I am greatly indebted to S. Rappaport, J.P. Ostriker and W. Lewin for stimulating discussion on the character of the group II sources.

REFERENCES

van Beveren, D., de Greve, J.P., van Dessel, E.L. and de Loore, C. 1979, Astron. Astrophys. 73, 19.
Cowley, A.P. and Crampton, D. 1975, Astrophys. J. 201, L65.
Fabian, A.C., Pringle, T.E. and Rees, M. 1975, Mon. Not.Roy.Astr.Soc. 172, 15p.
Finzi, A. 1978, Astron. Astrophys. 62, 149.
Flannery, B.P. and van den Heuvel, E.P.J. 1975, Astron. Astrophys. 39, 61.
Flannery, B.P. and Ulrich, R. 1977, Astrophys. J. 212, 533.
Faulkner, J. 1971, Astrophys. J. 170, L99.
Faulkner, J. 1974, in IAU Symp. No. 66, ed. R.J. Tayler, Reidel, Dordrecht, p. 155.
Giacconi, R. and Gursky, H. (eds.) 1974, "X-ray Astronomy", Reidel, Dordrecht.
Giacconi, R. and Ruffini, R. (eds.) 1978, "Physics and Astrophysics of Neutron Stars and Black Holes", North Holl. Publ. Co., Amsterdam.

Grindlay, J. 1980, Paper Presented at the HEAD meeting,
    Boston (January 1980).
Gursky, H. 1973, Lecture at the Nato Advanced Study Institute
    on Physics of Compact Objects, Cambridge (Engl.) (July 1973).
Hanson, R.B. 1980, Mon.Not.Roy. Astron. Soc. 186, 357.
Heggie, D.C. 1977, Comments on Astrophys. and Space Phys. 7, 43.
van den Heuvel, E.P.J. 1974, in: Proc. 16th Solvay Conf.
    on Physics, Univ. of Brussels Press, p. 119.
van den Heuvel, E.P.J. 1976, in: "Structure and Evolution
    of Close Binary Systems", (P. Eggleton et al., eds.),
    Reidel, Dordrecht, p. 35.
van den Heuvel, E.P.J. 1977, Annals N.Y. Acad. Sci. (M.D.
    Papagiannis, ed.), 302, 14.
van den Heuvel, E.P.J. 1978, in "Physics and Astrophysics
    of Neutron Stars and Black Holes" (R. Giacconi and R.
    Ruffini, eds.), North Holl. Pub. Co. Amsterdam, p. 828.
Hills, J.G. 1976, Mon. Not. Roy. Astr. Soc. 175, 1p.
Hills, J.G. and Day, C.A. 1976, Astrophysical Letters 17, 87.
Jones, C. 1977, Astrophys. J. 214, 856.
Joss, P.C. 1978, Astrophys. J. 225, L 123.
Joss, P.C. 1979, Proc. 9th Texas Symp. on Relativistic
    Astrophysics, Annals N.Y. Acad. Sci. (in press).
Joss, P.C. and Rappaport, S. 1979, Astron. Astrophys. 71, 217.
Kippenhahn, R. and Meyer-Hoffmeister, E. 1977, Astron. Astrophys.
    54, 539.
Lewin, W.H.G. 1979a, Talk at Cosmic Ray Conference, Kyoto, Japan.
Lewin, W.H.G. 1979b, Advances in Space Exploration, 3,
    IAU/COSPAR Symp. on X-ray Astron.,Innsbruck, Austria.
    (eds. W.A. Baity and L.E. Peterson, Pergamon Press, N.Y.).
Lewin, W.H.G. 1979c. In: "Globular Clusters", eds. D. Hanes
    and B. Madore, Cambridge Univ. Press. (in press).
Lewin, W.H.G. and Clark, G.W. 1979a, in: Proc. 9th Texas,
    Symp. on Relativ. Astrophys., Annals N.Y. Acad. Sciences
    (in press).
Lewin, W.H.G. and Clark, G.W. 1979b, in: Proc. Nato Advanced
    Study Institute on Galactic X-ray Sources (P. Sanford, ed.)
    Cambridge Univ. Press (in press).
Lewin, W.H.G. and Clark, G.W. 1979c, Proc. X-ray Symp. Tokyo,
    Japan (in press).
Li, F.K., Joss, P.C., McClintock, J.E., Rappaport, S. and
    Wright, E.L. 1980, Astrophys. J. (in press).
de Loore, C., de Greve, J.P., van den Heuvel, E.P.J. and
    De Cuyper, J.P. 1975, Mem. Soc. Astron. Ital. 45, 893.
Maraschi, L., Treves A. and van den Heuvel, E.P.J. 1977,
    Astrophys. J. 216, 819.
Mc Clintock, J.E., Canizares, C. and Backman, D.E. 1978,
    Astrophys. J. 223, L 75.
Milgrom, M. 1978, Astron. Astrophys. 67, L 25.
Ostriker, J.P. 1977, Annals Acad. Sci. (M.D. Papagiannis, ed.)
    302, 209.

Ostriker, J.P. and Zytkow, A. 1980, Astrophys. J. (in pre-
     paration).
Paczynski, B. 1971, Annual Rev. Astron. Astrophys. 9, 183.
Papagiannis, M.D. (ed.) 1977, "Eighth Texas Symp. on Relativ.
     Astrophys.", Annals New York Acad. Sci, Vol. 302.
van Paradijs, J. 1978, Nature 274, 650.
Rappaport, S. 1980, in: Proc. Nato Adv. Study Institute on
     Galactic X-ray Sources, Cape Sunion, Greece (P. Sanford,
     ed.), Cambridge Univ. Press (in preparation).
Robinson, E.L. 1976, Annual Rev. Astron. Astrophys. 14, 119.
Salpeter, E.E. 1973 in: "X- and Gamma- ray Astronomy", (H. Bradt
     and R. Giacconi, eds.), Reidel, Dordrecht, p. 135.
Savonije, G.J. 1978, Astron. Astrophys. 62, 317.
Savonije, G.J. 1979, Astron. Astrophys. 71, 352.
Shull, J.M. 1979, Astrophys. J. 231, 534.
van Speybroeck, L., Epstein, A., Forman, W., Giacconi, R.,
     Jones. C., Liller, W. and Smarr, L. 1979, Astrophys. J.
     234, L 45.
Spinrad, H. and Taylor, B.J. 1971, Astrophys. J. Suppl. 22, 445.
Sutantyo, W. 1975a. Astron. Astrophys. 41, 47.
Sutantyo, W. 1975b. Astron. Astrophys. 44, 227.
Taam, R., Bodenheimer, P. and Ostriker, J.P. 1978, Astrophys. J.
     222, 269.
Tutukov, A.W. and Yungelson, L.R. 1973, Nautsnie Inform. 27, 58.

STELLAR CORONAE FROM EINSTEIN: OBSERVATIONS AND THEORY

R. Rosner

Harvard-Smithsonian Center for Astrophysics
Cambridge, Massachusetts, U.S.A.

G. S. Vaiana
Harvard-Smithsonian Center for Astrophysics
Cambridge, Massachusetts, U.S.A.   and

Istituto e Osservatorio Astronomico di Palermo
Palermo, Italy

1.   INTRODUCTION

The new EINSTEIN Observatory data have dramatically changed our perspective on the constitution of low-luminosity galactic X-ray sources. Whereas X-ray emitting stars were thought to be the exception, we have now observed soft X-ray emission from stars of all spectral types along the main sequence, as well as from giants, supergiants, white dwarfs, and other exotica of the stellar zoo (Fig. 1). Stellar X-ray emission appears to be the rule rather than the exception. In most cases the levels of observed emission far exceed predictions based upon the "standard" theories of coronal formation and heating, including the most recent modification of such theories (cf. Mewe 1979). We are therefore faced with the tasks of rethinking the problems of coronal formation and reexamining its role in stellar evolution. In order to place these problems in perspective, we will first outline some of the larger issues, and then discuss how the new X-ray observations impinge on the questions that are raised.

The problem of coronal formation is intimately connected with the classical problem of stellar angular momentum loss. The latter problem arises as follows. By balancing gravitation and centrifugal force for "typical" protostellar clouds ($\rho \sim 10^{-24}$ g cm^{-3}), one can estimate the "typical" angular momentum J to be of the order of $\sim 10^{56}$ $(M/M_\odot)^{5/3}$ g cm^2 s^{-1}, if the cloud

R. Giacconi and G. Setti (eds.), X-Ray Astronomy, 129-151.
Copyright © 1980 by D. Reidel Publishing Company.

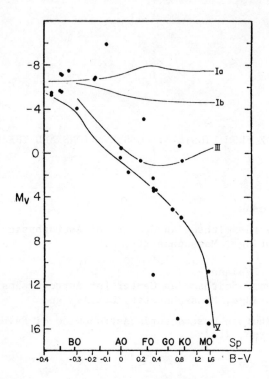

Fig. 1 H-R diagram of optically well-characterized stars detected
to date (6/79) as X-ray sources in the Pointed EINSTEIN/CfA
stellar survey.

acquires its angular momentum from the rotation of the galaxy
(cf. Tassoul 1978). In contrast, very rapidly rotating main-
sequence stars (e.g., B and A stars) typically have $J \sim 10^{51}$-10^{52}
g cm^2 s^{-1}. If we believe that these "typical" cases are physically connected (e.g. that stars evolve from such "typical"
clouds, and not solely from clouds which are particularly slow
rotators), then we must account for the 10^4-10^5 reduction in spin
angular momentum. This problem is exacerbated by the fact that
main sequence stars as a class do not show any uniformity in rotation properties, but rather show distinct and pronounced differences between spectral types. The basic questions to be resolved are therefore:

(1) *When does spin-down occur?* There are three, not mutually
exclusive, possibilities: (a) never; e.g. "slow" rotation is an
initial condition imposed upon initially collapsing proto-stellar
clouds; (b) during pre-main sequence evolution (for example,
during the Hayashi phase or earlier); (c) during evolution on
the main sequence.

(2) *How does spin-down occur?* Again, there are three possibilities: (a) by fragmentation (e.g. by forming multiple components

within the collapsing cloud, so that angular momentum resides in
the component orbits);   (b) by mass shedding;   (c) by torques
exerted by magnetic fields threading the protostar and/or star.

In the following we shall concentrate upon possibilities
(1c) and (2c), which involve the formation of extended stellar
atmospheres; in so doing we face a third crucial question, namely

(3) *How do stellar extended atmospheres form?* This third
question is of particular concern to us, because it leads
directly to the issue of stellar X-ray emission. Our perspective
will in fact be to work backwards, from the data on stellar X-ray
emission to the processes leading to creation of hot plasma and
their role in stellar despinning. We will in fact suggest that
observations of stellar X-ray emission provide a useful new probe
for studying stellar evolution.

First, however, we require a bit of an historical perspective
as the subject of stellar spindown and its relation to coronae,
winds, and magnetic fields constitutes a by now classic subject
in stellar astronomy. An excellent review of the entire subject
of rotating stars can be found in the monograph by J.-L. Tassoul
(1978). The following is a brief abstract of the principal points
relevant to our discussion.

The first crucial fact is that single main-sequence stars
show a sharp break in rotation speed at $\sim$ F5 at roughly the point
at which stellar interior calculations predict the onset of deep
surface zones (Kraft 1967). This effect is evident in data for
$\langle v_e \sin i \rangle$ (Fig. 2a), but is particularly pronounced when one
plots the angular velocity $\Omega (\sec^{-1})$, versus spectral type (thus
taking out the stellar radius; see Fig. 2b). Furthermore, for
fixed spectral type, Skumanich (1972) has shown an age dependence
of rotation speed, with rotation rate scaling like $(\text{age})^{-1/2}$.

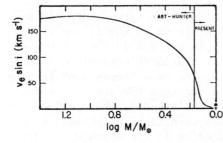

Fig. 2a Mean equatorial rota-
tion speed $\langle v_e \sin i \rangle$ vs. mass
along main sequence (from Kraft
1967).

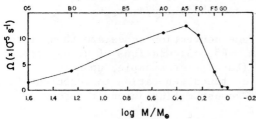

Fig. 2b Mean angular
speed Ω (s^{-1}) along main
sequence (after Allen
1973).

The second key observational point is that Ca II h and k
emission, which is believed to be a good indicator of stellar
surface activity (Wilson 1966), is a prominent feature of stars
of spectral type $\sim$ F5 and later. Because solar Ca II emission
is strongly correlated with surface magnetic fields, it has been
surmised that the observed stellar emission is indicative of
solar-like stellar magnetic field activity; because the level of
Ca II emission also scales as the age of the star [$\propto(age)^{-1/2}$;
Skumanich 1972], and because the onset of such emission again
coincides with the theoretically-predicted onset of deep surface
convection, it has been conjectured that dynamo-generated magnetic fields first occur at spectral type $\sim$ F5, and that they
(and related activity) decay with time according to Skumanich's
(1972) scaling law (see also Durney 1972; see Fig. 3a).

The canonical picture which has evolved, following the
seminal papers of Biermann (1946), Schwarzschild (1948) and
Schatzman (1962), and the observations reported by Wilson (1966),
Kraft (1967), and Skumanich (1972), involves the following
"standard" scenario for single stars (see also Kippenhahn 1973):
During the pre-main sequence evolution, stars lose the bulk of
their angular momentum, placing them on a locus $J(\equiv$ angular
momentum per unit mass) $\propto M^{2/3}$ (see Kraft 1967). Further despinning on the main sequence depends upon the presence of an
outer convection zone (Fig. 3b). If the latter is present, then:

(i)   An extended, hot outer atmosphere and related stellar
wind is generated via the heating of surface layers by shocked
acoustic modes generated by convective turbulence;

(ii)   A convectively driven Babcock/Parker magnetic dynamo
is operative, leading to surface magnetic fields;

(iii) Plasma flows from the hot coronal atmosphere inter-
act with the surface magnetic fields. Corotation enforced by
the coronal magnetic field takes place up to the Alfven radius
R_A [defined by v_{alfven} $(R_A) = U(R_A)$, where U is the coronal wind
speed]; beyond R = R_A, the coronal plasma effectively decouples
from the stellar atmosphere. The effective "lever arm" for angular momentum loss is hence R_A, rather than the stellar radius R_*;
because $R_A \gg R_*$ (at least in the Sun's case), the combined
effect of stellar wind and surface magnetic fields leads to far
more rapid despinning than would occur in the absence of surface
magnetic fields (cf. Weber and Davis 1967).

Because vigorous surface convection zones are thought to
appear on stars later than $\sim$ F5, calculations of atmospheric
heating by acoustic waves (based, for example, upon the
Proudman/Lighthill theory of sound generation in turbulent flows)
lead to predictions of the acoustic flux intensity shown in

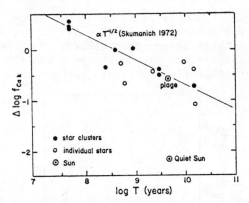

Fig. 3a Ca II k emission fluxes for stellar groups and individual stars as a function of age, normalized to the Hyades main sequence (after Blanco et al. 1974). Note the relative positions of active and quiet portions of the solar atmosphere.

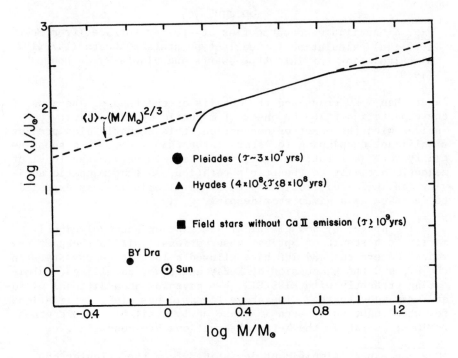

Fig. 3b Angular momentum per unit mass versus mass along the main sequence (after Kraft 1967), under the assumption of solid body rotation. Note that the positions marked are lower bounds if $\partial\Omega/\partial r < 0$.

Fig. 4.* Although the details of this prediction are subject to substantial uncertainties (viz. Jordan 1973), certain aspects will be characteristic of any such theory. For example, the peak in mechanical flux at spectral type F and the rapid decrease in expected mechanical flux for stars earlier than ∿ A5 and

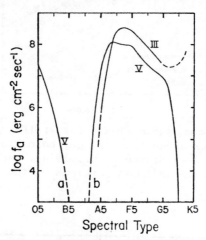

Fig. 4 Available acoustic flux at stellar surface (from Mewe 1979): a) calculation for early-type stars by Hearn (1972,1973); b) calculation for late-type dwarfs and giants by de Loore (1970).

later than ∿ K0 transcend the details of the theory. Because of the expectation that the onset of magnetic dynamo activity coincides with the onset of convection, this scenario also predicts significant despinning for stars later than ∿ F5. On a substantially more qualitative level, dynamo models relate the level of magnetic activity to the star's rotation, so that magnetic activity (as indicated, for example, by Ca II emission) is expected to decrease as a given star despins.

The above, by now classical description has evolved primarily as a result of optical observations. Within the past few years, Copernicus and IUE have allowed a look at nearby stars in the UV, and the succession of X-ray astronomy satellites, culminating presently with EINSTEIN, has given us an additional window at X-ray wavelengths. The above scenario has clear implications for what ought to be seen in these newly available wavelength regimes; as far as the X-ray observations are concerned, we

---

*One can expect significant deviations from the calculations shown if the dominant acoustic mode generation process is, for example, modified by photospheric magnetic fields (viz. Osterbrock 1961).

would expect the level of coronal X-ray emission to follow the
level of acoustic flux generation (Fig. 4). That is, although
the acoustic flux-to-X-ray flux conversion efficiency ε may
dominate the detailed modulation of stellar X-ray emission levels,
predictions of acoustic flux levels place clear upper bounds on
this emission (e.g. ε < 1). In particular, coronal emission
should be negligible for stars earlier than ∿ A5 and later than
∿ K0, and peak at around F2-F5. Furthermore, Parker-type wind
models (upon which stellar despinning models are based; cf.
Weber & Davis 1967) predict that mass loss should vary with the
coronal activity level, e.g. peak at ∿ F0-F5 for main-sequence
stars. In the following, we will examine whether these expectations are in fact fulfilled.

2. STELLAR CORONAE: OBSERVATIONS

In order to best appreciate the impact of the new EINSTEIN
observations, it is useful to consider the historical context.
We shall therefore very briefly recall the status of solar and
pre-EINSTEIN stellar observations before summarizing the
EINSTEIN results.

2.1 The Solar Corona

Following the recent high resolution solar studies, culminating
with Skylab and more recently, OSO-8, our understanding of the
solar corona has been truly revolutionized. Recent reviews can
be found in Withbroe & Noyes (1977), Vaiana and Rosner (1978),
Wentzel (1978), as well as the various symposium proceedings.
The principal results relevant to the present discussion are:

(a) *The solar corona is structured.* Solar X-ray emission domi-
nantly derives from well-defined structures ("loops") whose geometric integrity is provided by coronal magnetic fields. There
is no evidence for a significant contribution to the total solar
coronal X-ray luminosity from volumes other than "loop" structures; regions in which the magnetic field is "open" to the interplanetary medium ("coronal holes") show very low levels of X-ray
surface flux, and contribute negligibly to the integrated coronal
radiative emission. Similarly, the solar mass loss is spatially
extremely inhomogeneous but, in contrast to the X-ray flux, domi-
nantly derives from regions of "open" magnetic field lines; thus,
strong winds (≡ "high speed wind streams") emanate not from
regions of coronal activity (as defined by the surface X-ray
flux), but rather from regions of minimal activity (coronal holes).
Coronal activity and mass loss are hence spatially anti-correlated
and, further, are temporally anti-correlated on time scales short
compared to the solar cycle period. There is as yet insufficient
data to judge whether mass loss and coronal activity are temporally correlated on time scales longer than the solar cycle period.

(b) *Coronal heating is extremely inhomogeneous.* Soft X-ray and
EUV observations provide incontrovertible evidence that coronal
activity levels ($\sim$ intensity of observed emission) correlate with
surface magnetic fields. In addition to the morphological association, quantitative studies show that the temperature of closed
coronal structures scales roughly as $(pL)^{1/3}$ [p the coronal plasma
pressure and L the scale length of the magnetically confined
coronal structure], and that p is positively correlated with the
mean surface magnetic field <B>. If one, in addition, recalls
that thermal conduction is strongly anisotropic under coronal
conditions ($\kappa_{\parallel} \gg \kappa_{\perp}$, where $\kappa_{\parallel}$ and $\kappa_{\perp}$ are the thermal conductivi-
ties along and across the local magnetic field, respectively), it
is evident that local plasma heating must be extremely inhomogeneous (as reflected by the observed structuring); and furthermore
there is a strong implication that this heating process is sensitive to the surface magnetic fields. In contrast there is no
evidence for a correlation between local coronal emission levels
and local surface (photospheric) turbulence levels. This view
had not been universally accepted; but recent OSO-8 observations,
designed   to test for acoustic wave propagation,   have in
fact placed constraints upon the acoustic flux which fall <u>below</u>
levels necessary to heat the corona (viz. Athay and White 1979).

To summarize, the recent solar observations argue strongly
against a coronal model in which heating is simply due to damped
acoustic modes. In contrast, several models of magnetic field-
related heating mechanisms have been proposed which do provide
natural explanations for the observations, in particular inhomogeneity and field-related radiative emission correlations; these
include dissipation of coronal magnetic fields via current-driven
instabilities (Tucker 1973, Rosner et al. 1978) and the damping of
magnetic surface waves (Ionson 1978), Alfven modes (Osterbrock
1961, Wentzel 1974, Hollweg 1979) and fast mode MHD waves (Habbal,
Leer and Holzer 1979). Present data are, however, insufficient to
distinguish between these mechanisms as yet, and further theoretical work will be required for detailed comparison between theory
and data. In all these models the basic idea is to relate
coronal heating to the stressing of coronal magnetic fields by
surface turbulence, and to retrieve the scaling laws obeyed by
the observed coronal structures (viz. Galeev et al. 1979). In
this view, coronal formation requires stellar surface fields as
well as surface turbulence, with magnetic fields providing both
the means of plasma confinement and plasma heating.

2.2   Pre-EINSTEIN Observations of Stellar Coronae

Reviews of relevant stellar observations can be found in
Mewe (1979), Vaiana and Rosner (1978), Linsky (1977), and
Gorenstein and Tucker (1976). The key results which are of
interest to us here are:

(a) Certain types of stars were found to be strong X-ray emitters, including Capella (Catura, Acton & Johnson 1975; Mewe et al. 1975) and the class of RS CVn stars. The latter have surface X-ray fluxes orders of magnitude larger than the Sun's, associated with plasma whose temperature is substantially higher than the Sun's corona, yet have effective surface gravities much lower than the Sun's; clearly, these coronae must be confined (Walter et al. 1979; Holt et al. 1979). For more details on observations of these RS CVn stars, see Walter et al. (1979) and references therein.

(b) Observation of the α Cen A, B system by Nugent and Garmire (1978) confirmed that truly solar-like stars could indeed be observed.

(c) At least one single A dwarf (Vega: A0 V) was observed as an X-ray source (Topka et al. 1979); "standard" theory fails to account for this observation.

(d) A number of stars throughout the H-R diagram have been looked at with Copernicus, BUSS, and IUE, and UV line emission due to ionization states associated in the solar case with the transition region between the chromosphere and corona have been observed. Particularly striking have been: (i) observations of O VI and other transition region line emission from early-type stars (cf. Snow and Morton 1976), suggesting that these stars also may possess coronae; (ii) the possible correlation between UV emission and stellar rotation rate for late-type stars reported by Ayres and Linsky (private communication), suggesting that convective activity levels alone are not the sole determinant of coronal activity; and (iii) indications from the UV that the transition region structures of late-type dwarfs are relatively independent of the stellar effective temperature (Hartmann et al. 1979).

These observations supplement the optical data discussed in §1 and, in and of themselves, suggest that the "standard" scenario outlined above is inadequate. Thus, it is important to remember that the data discussed above, including in particular observations of Ca II emission, already pointed in the direction of a close coupling between coronal and surface magnetic field activity for certain types of stars (the "chromospherically active" stars), but were not generalized as representative of typical coronal activity. A review of the literature published prior to the dissemination of the EINSTEIN results shows that discussions of quiescent (as opposed to flaring) stellar coronal activity were, with few exceptions (for example, see Blanco et al. 1974 and Thomas 1975), phrased in terms of traditional acoustic heating theories (see review by Mewe 1979). If one assumes that the literature reflects the scientific opinion of the day, one is forced to conclude that acoustic coronal heating (in its various theoretical guises) represented the consensus view. A similar situation has characterized solar studies (viz. Athay 1976) with only a minority (whose roots are to be found in the works of T. Gold and H. Alfven) insisting upon the dominance of magnetic field-related heating throughout the

corona. How is one to account for this dichotomy? We conjecture
that in the absence of detailed comparison between theory and observations (which only became feasible for solar coronal data
after the flight of Skylab), alternatives to acoustic heating
models were relatively speculative, so that extending observations
and interpretation of, for example, Ca II emission from "active"
stars to a radically different coronal formation scenario,
involving magnetic fields as the essential element, for (normal)
stars in general would not have been reasonable; simply put,
earlier work did not enjoy the advantage of hindsight.

2.3 EINSTEIN Stellar Survey

The EINSTEIN observations have given us the first opportunity
to directly examine coronal emission from a large sample of
"normal" stars. In consequence, it appears that there is no category of stars to be called "X-ray stars"; that is, our data are
consistent with the hypothesis that all stars are X-ray sources at
some level, so that the two categories of stars and stellar X-ray
sources may be essentially coextensive. In the following we shall
summarize the principal conclusions obtained from the survey, as
abstracted from the detailed report of the preliminary EINSTEIN/
CfA stellar survey results (Vaiana et al. 1980).

The EINSTEIN/CfA stellar survey combines results from:
(1) pointed observations aimed at optically well-characterized
stars; (2) serendipitous detection of stars in fields observed
for other purposes; (3) systematic search for detection of (or
upper limit for) X-ray emission from stars down to visual magnitude V = 8.5 at medium sensitivity; (4) identification of stellar
X-ray sources in the deep survey fields observed for $\sim 10^5$ seconds. This survey has now demonstrated that X-ray emission is
associated with stars throughout the H-R diagram (Fig. 1).

Our typical sensitivity for most fields in the 0.2 to 3 keV
passband ranges from 10^{-13} to 10^{-12} erg s^{-1} cm^{-2}. At these sen-
sitivities we can probe for coronae emitting at $\sim 10^{28}$ erg s^{-1} to
~ 100 pc; for the nearest stars we can typically probe down to
$\sim 10^{26}$ erg s^{-1}. The corresponding count rates for the Imaging
Proportional Counter (IPC) range from $\sim 10^{-3}$ to ~ 1 ct s^{-1} (the
latter for the closest and the brightest objects, such as Sirius
B). Some examples of our detections are shown in Figs. 5-7. Not
only have we observed X-ray emission from stars long known to be
active in other indicators (viz. Ca II emission; π' UMa is shown
in Fig. 5), but also "inactive" stars, such as Vega and Sirius A.
In addition, we have now detected quiescent X-ray emission from
stars as late as dM8 (viz. EQ Peg shown in Fig. 6). This per-
vasiveness of stellar X-ray emission is also well illustrated at
the extreme other end of the main sequence by the EINSTEIN image
of the η Carina field (Fig. 7). Stars as early as O3, and ranging
in luminosity class from V to III, are seen as strong ($\sim 10^{32}$ -
10^{33} erg s^{-1}) soft X-ray sources (Seward et al. 1979); this

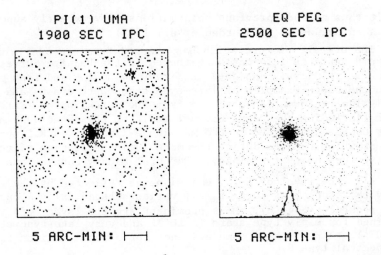

PI(1) UMA
1900 SEC   IPC

EQ PEG
2500 SEC   IPC

5 ARC-MIN: ⊢──┤   5 ARC-MIN: ⊢──┤

Fig. 5   IPC image of π^1 UMa, a nearby G dwarf (G0V, V=5.6)

Fig. 6 IPC image of EQ Peg, a flare star binary (M4Ve + M6Ve, V=10.4/12.4)

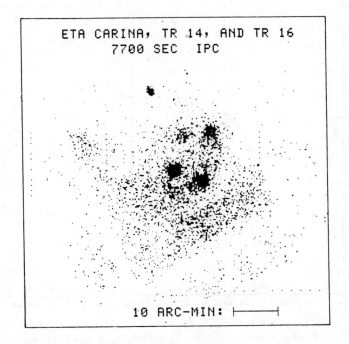

ETA CARINA, TR 14, AND TR 16
7700 SEC   IPC

10 ARC-MIN: ⊢──┤

Fig. 7 IPC image of the η Carina region, showing η Carina itself, a Wolf-Rayet star, and a number of early O stars belonging to the Trumpler 14 and 16 associations.

result thus extends previous observations of very early supergiants in Cygnus OB 2 (Hanrden et al. 1979).

The overall picture which has emerged from our preliminary
analysis reflects information gathered in the Pointed Survey
(which is relatively unbiased in X-ray luminosity and from
which the above examples of late-type stars were drawn) and the
statistical information obtained from the serendipitous detections (including the Deep Surveys) and from the detections and
upper bounds determined in the magnitude-limited $V \leq 8.5$ survey
(all of which tend to define the more luminous portion of the
stellar luminosity functions). Our preliminary conclusions are:

(i) The level of X-ray emission, as reflected by the typical
X-ray to visual flux ratio f_x/f_v, varies substantially along the
main sequence (Fig. 8). Most prominent are the large f_x/f_v values
attained for early-type stars ($\gtrsim 10^{-4}$) and very late-type stars
($\sim 10^{-1}$); there is also some evidence for a local maximum f_x/f_v
at spectral type F.

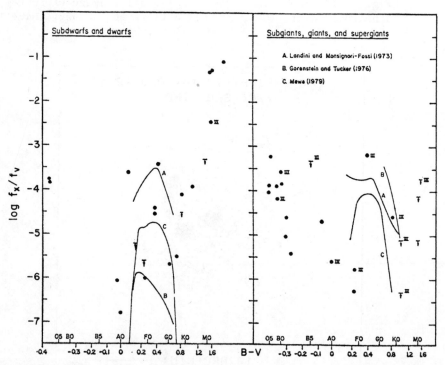

Fig. 8 Ratio of observed X-ray to V-band flux ratio (or upper
bounds) for stars detected by EINSTEIN/CfA stellar survey; the
V-band (rather than bolometric) flux is used provisionally in
order to allow comparison with faint unclassified stars or star-
like objects in serendipitous surveys. Also shown are theoretical
predictions for stellar X-ray emission, converted to f_x/f_v.

(ii) With the possible exception of a subpopulation of the F-G giants (viz. Capella), the level of X-ray emission (as indicated by f_x/f_v) for giants and supergiants declines monotonically as one proceeds to later spectral types (Fig. 8).

(iii) The typical X-ray luminosity L_x along the main sequence, as derived from distance determinations of the nearby sample of stars, shows a decrease from $\sim 10^{33}$ erg s^{-1} for O stars to $\sim 10^{27}$ erg s^{-1} at A0, with a steep increase to $\sim 10^{29}$ erg s^{-1} for F dwarf, a more modest decline to $\sim 10^{28}$ erg s^{-1} for G dwarfs, and a continuation of similar emission levels (at $\sim 10^{28}$ erg s^{-1}) for later-type dwarf to, and including, spectral type M. The corresponding typical surface X-ray flux ϕ_a shows a more dramatic variation with spectral type, principally because the approximate constancy of L_x for stars later than G0 implies an increasing X-ray surface flux as the stellar radius (and mass) is decreased (Fig. 9).

(iv) The range of X-ray emission levels observed for any fixed spectral type cannot as yet be established, although we have some hint that it may be large; for example, the small

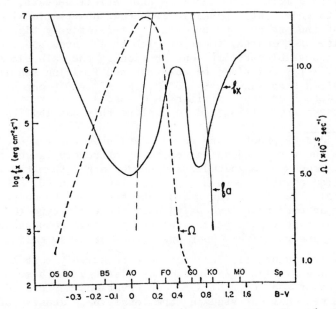

Fig. 9 Comparison of typical X-ray surface flux ϕ_x calculated from preliminary survey data, acoustic surface flux ϕ_a (Mewe 1979), and angular rotation rate Ω (Allen 1973) vs. spectral type for main sequence stars.

sample of G stars detected to date show substantially larger emission levels than the Sun.*

In order to confront available theories for stellar coronal formation with the survey data, we have plotted the predictions of calculations for acoustic flux ($\oint_a$) generation (based upon the Proudman/Lighthill theory) on the same figure as that showing derived X-ray surface fluxes (Fig. 9). The key point is that the curve for $\oint_a$ defines the upper bound for the energy available to heat a corona; acoustic dissipation models define the efficiency of thermalization as well as the dissipation geometry (see Kippenhahn 1973), but cannot change $\oint_a$. The comparison shows, among the several discrepancies for main sequence stars, that the acoustic flux available to heat the corona falls many orders of magnitude short for stars later than K0. More critically, the qualitative behavior of $\oint_a$ with spectral type is precisely oppo-site to that of the observed variation of coronal emission with spectral type. Furthermore, a theory which defines coronal heat-ing levels solely on the basis of convective activity levels will have clear difficulties in accounting for the rather broad range of coronal emission levels seen for G dwarfs. Below we shall address this problem directly, and suggest a possible resolution.

Before delving into the more theoretical aspects, however, we wish to point out a possible important observational conse-quence of the present survey data: the relatively high X-ray luminosity of M dwarfs, together with their large space density, implies a possible stellar contribution to the galactic soft ($\lesssim 1$ keV) X-ray emission; because, at the limiting sensitivity of the EINSTEIN Deep Surveys ($\sim 10^{-14}$ erg cm^{-2} s^{-1}), one should be able to detect dM stars similar to CN Leo down to V $\sim$ 17 mag, the Deep Surveys are well-suited to explore this possibility.[†]

----

* Note added in proof: More recent data show the range of emission levels about the median values to be extremely large, typically 3 to 4 orders of magnitude, and hence larger than the variation in median emission levels for stars later than A0. The X-ray luminosity functions for stars must therefore be quite broad; and it appears that, excepting only the early type (OB) stars, spectral type -- and hence stellar effective temperature or mass -- are relatively poor predictors of X-ray emission levels.

† Note added in proof: Because their relatively soft X-ray spec-trum and very high f_x/f_v ratio may distinguish these stars from other faint sources, the X-ray observations may provide a unique new tool for studying the local space density of the low-mass end of the main sequence. As these stars constitute the bulk of the galactic stellar mass, X-ray observations may place new constraints upon the galactic mass. These possibilities and new tasks must, however, await the future. One of the pri-mary goals of our survey is in fact to better define the X-ray luminosity function of stars, particularly at the low-mass end;

3. SKETCHING A NEW THEORY OF STELLAR CORONAE

The standard theory of stellar coronae, summarized most recently by Mewe (1979), requires the presence of vigorous surface convection. In consequence, the expectation of such a theory is that stellar X-ray emission -- if due to a corona -- should be primarily limited to a subset of stars (viz. those of spectral types F and G), and therefore should be relatively rare. This theory makes detailed predictions about coronal radiative emission which are now subject to test. Thus, Skylab and later observations have supplied us with spatially resolved data of the solar corona, while the succession of high-energy X-ray astronomy satellites, culminating with EINSTEIN, gives us a long-awaited glimpse of stellar X-ray emission throughout the H-R diagram. As we have seen above, there now seems little doubt that the standard scenario of coronal formation is inadequate.

Are there alternative theoretical coronal models? One possibility, which we shall explore further here (cf. Vaiana and Rosner 1978), is that *stellar magnetic fields play the key role in determining the level of coronal emission, and that the modulation of the surface magnetic flux level (by variation of the interior structure of the stellar convection zone and by stellar rotation) and the level of stressing of surface magnetic fields (by surface turbulence) essentially determine the variation of mean coronal activity in the H-R diagram.* The elements in this scenario are thus the assumption that the Sun is indeed representative (not on the quantitative, but rather on the qualitative level) and the theoretical assumptions that magnetic fields provide the means of channeling free energy resident in the stellar surface layers to the overlying extended atmosphere, and that processes exist which can thermalize this energy locally so as to produce a corona.

The pioneering work of Wilson (1966), Kraft (1967), and Skumanich (1972) provided the basic link between stellar rotation, magnetic dynamo activity, and chromospheric activity, a link which has been extended by Ayres, Linsky, and other IUE observers to the chromosphere-corona transition region on the basis of ultraviolet observations, and confirmed by us for stellar coronal X-ray emission in a restricted region of the H-R diagram. That is, as Fig. 9 shows, there is a coincidence between the despinning of rapidly rotating stars and the stars with high X-ray surface

---

it is only then that calculations of, for example, stellar contributions to galactic X-ray emission can be performed with some assurance.

brightness in the F0-F8 main sequence domain. This effect can be understood qualitatively by noting that the Rossby number (= $\tau_{rotation}/\tau_{convection}$), which measures the effect of rotation on convection cells, is large when convection zones first appear on the main sequence, but decreases rapidly from F2 to G0 (Durney & Latour 1978). As Durney and Latour point out, solar-type dynamos are thought to be particularly effective for small Rossby numbers; this notion is in accord with the argument presented by Ayres and Linsky (1979) that for fixed spectral type, rapid rotators should have larger dynamo activity (as their associated Rossby number will be lower). This effect, together with the fact that convection-driven surface turbulence is thought to peak around spectral type F, and that coronal emission is likely to be correlated with both the level of surface turbulence and surface magnetic field activity (Rosner, Tucker, and Vaiana 1978; Vaiana and Rosner 1978), give a qualitative explanation for the intensity of coronal X-ray emission at spectral types F to G. What cannot be understood on this basis alone are, however, the high surface X-ray fluxes for early-type stars and the persistence of high surface X-ray flux levels for stars later than K0. In the following, we shall focus in particular upon the puzzle of late-type stars, and discuss early-type stars only briefly at the end.

3.1 Late-type Stars

If surface magnetic fields are to be the key ingredient for coronal formation, the process whereby such fields are generated must be a primary determinant of coronal activity. We recall the well-known fact that because the observed time scale for changes of the large-scale solar magnetic field is much shorter than the appropriate Ohmic decay time scale, a currently operating cyclic magnetic dynamo must be operative (viz. Cowling 1953); the principal flux generating process in such dynamos is the "ω-effect", e.g. the generation of toroidal magnetic flux by the interaction of differential rotation with poloidal magnetic fields (Parker 1955). Can one say anything substantial about the functioning of the ω-effect for different types of stars?

Recently, we have conducted a study of the spatial and temporal correlation of emerging surface magnetic fields, as manifested by the associated coronal X-ray emission, in collaboration with N.O. Weiss (Golub et al. 1980b). For the present purposes, the result of interest is the observation that the ω-dynamo layer (where toroidal magnetic flux is thought to be generated) must lie deep, close to the bottom of the solar convection zone, and very likely at the interface between the radiative core and the convectively unstable region. A similar conclusion was reached by Parker (1975) on theoretical grounds, based primarily on the efficacy of magnetic buoyancy. However, recent work by Schüssler (1977) and others suggests that the rate of flux rise was

previously overestimated; furthermore, arguments based solely
upon magnetic buoyancy apply properly only to an otherwise
stably stratified atmosphere, which the convection zone is not.
We must therefore rely solely upon the aforementioned argument
based upon emerging surface magnetic fields.

The crucial aspect of the deep ω-effect layer is that it
be stably stratified against thermal convection. In that case
recent theoretical work by Acheson and collaborators (cf.
Acheson 1978, Acheson and Gibbons 1978) becomes relevant; they
have shown that magnetic buoyancy is strongly inhibited in rotating systems by Coriolis forces (see also Gilman 1970) and
that new instabilities enter, having growth rates given by the
expression

$$\tau^{-1} = v_a^2 / 4\omega H^2 \tag{1}$$

where H is the local density scale height, v_a the local Alfven
velocity $[\equiv B/(4\pi\rho)^{1/2}]$, and ω the local angular velocity. The
length of time during which the ω-effect is operative (before
significant toroidal flux eruption occurs) is therefore controlled by: (i) the star's rotation rate, (ii) the structure of
the convection zone, and (iii) the ω-effect layer field
strength; such effects have not, to date, been incorporated into
magnetic dynamo models. The primary modification of the dynamo
equations would occur in the equation for the toroidal mean field
component B_ϕ. For example, one may estimate the effect of flux
loss by writing

$$\left\{ \frac{\partial}{\partial t} - \eta\nabla^2 \right\} B_\phi = \left[B_z \frac{\partial v_\phi}{\partial z} + B_r \left[\frac{\partial v_\phi}{\partial r} - \frac{v_\phi}{r} \right] \right] - \tau^{-1} B_\phi \tag{2}$$

where the last term on the right-hand side simulates the effect of
the buoyant instability. Here we have assumed cylindrical coordinates (r, ϕ, z); η is the effective magnetic diffusivity; the mean
flow is assumed to have the functional form $v = (0, v_\phi(r,z), 0)$.
It is noteworthy that Eq. (2) then resembles the corresponding
equation appearing in Leighton's (1969) phenomenological dynamo
model, in which an ad hoc term describing a threshold value for
surface toroidal flux eruption appears.

In order to appreciate the possible consequences for magnetic
surface activity, we have plotted the variation of τ with stellar
mass $(M/M_\odot)$ along the main sequence (Fig. 10), using the stellar
interior models of Copeland, Jensen, and Jorgensen (1970) and
assuming the ratio of mixing length to pressure scale height to
be unity. The computation assumes the ω-effect layer to lie at
the base of the hydrogen convection zone, and is scaled to values
of ω and B of 10^{-6} sec^{-1} and 10^3 G, respectively; the lower bound
on $M/M_\odot$ for this model is set by the mass at which the star becomes
fully convective (below which the model is no longer meaningful).

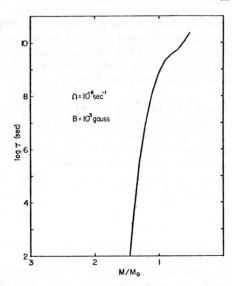

Fig. 10 Time scale variation for magnetic flux eruption
(evaluated near bottom of convection zone) vs. stellar
mass, for constant rotation rate and magnetic field strength.

The computation shows that, for fixed rotation rate and
field strength, the time scale for flux eruption progressively
increases with decreasing stellar mass. Consequently, one
expects relatively larger values of toroidal flux to be generated by the time $\tau^{-1} B_\phi$ (Eq. 2) becomes a significant loss
effect as one goes to lower mass stars of equal rotation rate.
We therefore expect larger values of erupting magnetic flux in
M dwarfs than in G dwarfs (such as the Sun) having equal rotation rates; as there is substantial variation in observed values
of ω for a given spectral type, similar variation in flux production within a given spectral type must also be expected; thus the
above model makes the qualitative prediction that the mean magnetic flux generation follows the variation in mean rotation rate
(viz. Fig. 9), with a scatter determined by the dispersion of
rotation rates about the mean for the various types of stars.
In light of the magnetic-field coupled coronal heating theory
discussed above, this would lead directly to the observed behavi-
or of coronal activity along the main sequence, from dF to dM,
e.g.

   (i)   increasing strong emission-line behavior (associated with
           photospheric field activity) among late K and M stars;

   (ii)   appearance of "spot" stars among M dwarfs, indicating
           progressively larger magnetic fluxes at the stellar
           surface;

   (iii) progressively larger X-ray surface fluxes, indicating a
           systematically larger fraction of the stellar surface
           occupied by (in the solar analogy) confined coronal plasma.

The above suggestion of course requires much fleshing out; not only do we need additional data in order to establish firmer statistical bounds upon the range of emission for various spectral types and luminosity classes, but much theoretical work remains to be done. For example, we are presently revising the scaling laws discussed by Rosner, Tucker, and Vaiana (1978) to take into account the variation of effective surface gravity and surface field activity. Perhaps most exciting is the prospect of using the X-ray data to study the variation of stellar magnetic activity -- and thus the dynamo problem -- over a far wider range of conditions than previously possible.

3.2 Early-type Stars

The above discussion of late-type stellar coronae does not have any obvious relevance to the formation of coronae on early-type stars; many of the basic ingredients, including deep convection zones and the ancillary magnetic dynamo, are thought to be absent (see however Toomre et al. 1976). These very stars are, of course, well-known from radio, optical, and UV observations to show evidence for large mass loss rates ($\dot{M} > 10^{-8}$ $M_\odot$ yr^{-1} for $M_v < -6$, cf. Snow and Morton 1976). The question is then how a corona (in the sense of hot plasma in the vicinity of the stellar surface) is to be energetically maintained in the presumed absence of vigorous, thermally driven convective turbulence.

One possibility, considered by Hearn (1972, 1973) in some detail, is radiation-driven amplification of acoustic modes. Predictions of acoustic flux levels based on such a theory in fact yield values comparable to the X-ray flux levels observed. The difficulty is that the heating efficiency (at $\sim 10^6$-10^7 K) of those modes must be of order unity, a circumstance difficult to envision if the atmosphere is additionally in a state of rapid expansion, as required by UV and other data; a possible remedy is to consider detailed line-driving effects, which may increase the available acoustic flux (Hearn, private communication). We are currently investigating an alternative scenario, in which the strongly heated coronal regions are spatially distinct from the rapidly expanding wind. The idea is to confine a portion of the atmosphere (by magnetic fields), which may then be efficiently heated by a variety of acoustic and magnetosonic instabilities; this X-ray emitting region can then provide both the X-ray photons we observe, as well as the soft X-ray photons necessary for formation of O VI and other observed high ionization states in the wind (as discussed by Cassinelli and Olson 1979). A crucial aspect of spatial separation is that it allows for the possibility that significant numbers of soft (< 1 keV) X-ray photons (see Fig. 11) do not have to traverse the entire wind column density, so that predicted cutoffs at low energies would not be as drastic as calculated for the "thin" coronal models of Cassinelli and coworkers. We are

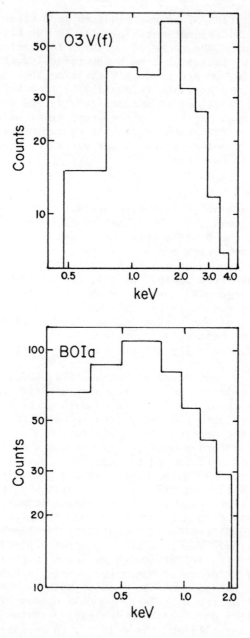

<u>Fig. 11</u> Background-subtracted IPC pulse height spectra for two
early-type stars; note that the two energy scales are different.
The low-energy cut-off of the BOI spectrum is much weaker than
that of the O3V star; preliminary analysis, carried out in collaboration with J. Cassinelli, suggests that the opacity at ¼
keV is lower than expected for this star from standard models.

presently exploring such possibilities in collaboration with
J. Cassinelli; at present it is premature to judge the feasibility of such an approach.

ACKNOWLEDGMENTS

    We would like to extend our appreciation to the many co-
workers who have contributed to the work summarized here,
including G. Fabbiano, L. Golub, P. Gorenstein, F.R. Harnden, Jr.,
C.W. Maxson, J. Pye, and F. Seward, as well as the many EINSTEIN
guest investigators who have permitted us to quote their results.
Particular thanks are extended to Riccardo Giacconi for his
direct participation in the stellar survey and especially for
his prescience in recognizing the possibilities of stellar X-ray
observations before there were any cogent reasons for believing
in them, and for his support of the stellar survey program. This
work was supported by NSG 7176 and the LAP of the Smithsonian.

REFERENCES

Acheson, D.J. 1978, Phil. Trans. Roy. Soc. A289, 459.
Acheson, D.J. and Gibbons, M.P. 1978, J. Fluid Mech., 85, 743.
Allen, C.W. 1973, Astrophysical Quantities (London: Athlone Press).
Athay, R.G. 1976, The Solar Chromosphere and Corona: Quiet Sun
    (Dordrecht: Reidel).
Athay, R.G. and White, O.R. 1979, Ap.J., 229, 1147.
Ayres, T.R. and Linsky, J.L. 1979, preprint.
Biermann, L. 1946, Naturwiss., 33, 118.
Blanco, C., Catalano, S., Marilli, E., and Rodono', M. 1974,
    Astron. Ap. 33, 257.
den Boggende, A.J.F., Mewe, R., Heise, J., Brinkman, A.C.,
    Gronenschild, E.H.B.M., and Schrijver, J. 1978, Astron. Ap,
    67, L29.
Carpenter, K.G. and Wing, R.F. 1979, BAAS, 11, 419.
Cash, W., Bowyer, S., Charles, P.A., Lampton, M., Garmire, G. and
    Riegler, G. 1978, Ap.J.Lett., 223, L21.
Cash, W., Snow, T.P. Jr., and Charles, P. 1979, preprint.
Cassinelli, J.P. and Olsen, G.L. 1979, Ap.J., 229, 304.
Cassinelli, J.P., Waldron, W.L., Vaiana, G.S., and Rosner, R.
    1980, in preparation.
Catura, R.C., Acton, L.W., and Johnson, H.M. 1975, Ap.J.Lett.,
    196, L47.
Copeland, H., Jensen, J.O. and Jorgensen, H.E. 1970, Astron. Ap.,
    5, 12.
Cowling, T.G. 1953, in The Sun, ed. G. Kuiper (Chicago: U. of
    Chicago Press).
Dupree, A.K. 1975, Ap.J.Lett., 200, L27.
Durney, B.R. 1972, Proc. Asilomar Conf. on Solar Wind NASA sp.
Durney, B.R. and Latour, J. 1978, Geophys. Ap. Fluid Dyn.
Evans, R.G., Jordan, C., and Wilson, R. 1975, MNRAS, 172, 585.
Galeev, A.A., Rosner, R., Serio, S., and Vaiana, G.S. 1979, Ap.J.
    (submitted).

Giacconi, R. et al. 1979a, Ap.J., 230, 540.
Giacconi, R. et al. 1979b, Ap.J.Lett., 234, L1.
Gilman, P.A. 1970, Ap.J., 162, 1019.
Golub, L., Maxson, C.W., Rosner, R., Serio, S. and Vaiana, G.S.
    1980a Ap.J. (in press).
Golub, L., Rosner, R., Vaiana, G.S. and Weiss, N.O. 1980b Ap.J.
    (submitted).
Gorenstein, P. and Tucker, W.H. 1976, Ann. Rev. Astron. Ap.,
    14, 373.
Habbal, S., Leer, E. and Holzer, T. 1979, Solar Phys. 64, 287.
Haisch, B.M. and Linsky, J.L. 1976, Ap.J.Lett., 205, L39.
Harnden, F.R., Jr. et al. 1979, Ap.J.Lett., 234, L51.
Hartmann, L., Davis, R., Dupree, A.K., Raymond, J.P.C.,
    Schmidtke, P.C. and Winer, R.F. 1979, Ap.J.Lett. (in press).
Hearn, A.G. 1972, Astron.Ap., 19, 417.
Hearn, A.G. 1973, Astron.Ap., 23, 97.
Hearn, A.G. 1975, Astron.Ap., 40, 355.
Heise, J. et al. 1975, Ap.J.Lett., 202, L73.
Hollweg, J. 1979, preprint.
Holt, S.S., White, N.E., Becker, R.H., Boldt, E.A., Mushotzky, R.F.,
    Serlemitsos, P.J. and Smith, B.W. 1979, Ap.J.Lett. 234, L65.
Ionson, J. 1978, Ap.J., 226, 650.
Jordan, S.D. 1973, in Stellar Chromospheres, ed. S.D. Jordan and
    E.H. Avrett (NASA SP-317), p. 181.
Kippenhahn, R. 1973, in Stellar Chromospheres, ed. S.D. Jordan
    and E.H. Avrett (NASA SP-317), p. 265.
Kraft, R.P. 1967, Ap.J., 160, 551.
Landini, M. and Monsignori-Fossi, B.C. 1973, Astron. Ap.,25, 9.
Leighton, R.B. 1969, Ap.J., 156, 1.
Linsky, J.L. 1977 in The Solar Output and Its Variation, ed. O.R.
    White (Boulder: Colo. Assoc. Univ. Press), 477.
Maxson, C.W. and Vaiana, G.S. 1977, Ap.J., 215, 919.
Mewe, R. 1979, Space Sci. Rev. 24, 101.
Mewe, R., Heise, J., Gronenschild, E.H.B.M., Brinkman, A.C.,
    Schrijver, J., and den Boggende, A.J.F. 1975, Ap.J.Lett., 202,
    L67.
Noci, G. 1973, Solar Phys., 28, 403.
Nugent, J. and Garmire, G. 1978, Ap.J.Lett., 226, L83.
Oster, L., 1975, in Problems in Stellar Atmospheres and Envelopes,
    ed. B. Baschek, W.H. Kegel and G. Traving (NY: Springer), 901.
Osterbrock, D.E. 1961, Ap.J., 134, 347.
Parker, E.N. 1955, Ap.J., 122, 293.
Parker, E.N. 1975, Ap.J., 198, 205.
Pneuman, G.W. 1973, Solar Phys., 28, 247.
Rosner, R., Golub, L., Coppi, B., and Vaiana, G.S. 1978, Ap.J.,
    222, 317.
Rosner, R., Tucker, W.H. and Vaiana, G.S. 1978, Ap.J., 220, 643.
Rosner, R., Giacconi, R., Golub, L., Harnden, F.R. Jr., Topka, K.,
    and Vaiana, G.S. 1980, in preparation.
Schatzman, E. 1962, Ann. d'Ap., 25, 18.

Schüssler, M. 1977, Astron.Ap., 56, 439.
Schwarzschild, M. 1948, Ap.J., 107, 1.
Seward, F.D. et al. 1979, Ap.J.Lett, 234, L55.
Skumanich, A. 1972, Ap.J., 171, 565.
Snow, T. and Morton, D.C. 1976, Ap.J. Suppl., 32, 429.
Stencel, R. 1978, Ap.J.Lett., 223, L37.
Tassoul, J.-L. 1978, Theory of Rotating Stars (Princeton: Princeton Univ. Press).
Thomas, R.N. 1975, Chem. Phys., 32, 259.
Toomre, J., Zahn, J.-P., Latour, J., and Spiegel, E.A. 1976, Ap.J., 207, 545.
Topka, K. 1980, Thesis, Harvard University.
Topka, K., Golub, L., Harnden, F.R., Jr., Gorenstein, P., Rosner, R., and Vaiana, G.S. 1980, in preparation.
Topka, K., Fabricant, D., Harnden, F.R., Jr., Gorenstein, P. and Rosner, R. 1979, Ap.J., 229, 661.
Tucker, W.H. 1973, Ap.J., 186, 285.
Vaiana, G.S. and Rosner, R. 1978, Ann. Rev. Astron. Ap., 16, 393.
Vaiana, G.S. et al. 1980, in preparation.
Walter, F.M., Cash, W., Charles, P.A. and Bowyer, C.S. 1979, Ap.J. (submitted).
Weber, E.J. and Davis, L. Jr. 1967, Ap.J., 148, 217.
Weaver, R., McCray, R., Castor, J., Shapiro, P. and Moore, R. 1977, Ap.J., 218, 377.
Wentzel, D.G. 1974, Solar Phys., 39, 129.
Wentzel, D.G. 1978, Rev. Geophys. Space Phys., 16, 757.
Wilson, O.C. 1966, Science, 151, 1487.
Wilson, O.C. 1978, Ap.J., 226, 379.
Withbroe, G.L. and Noyes, R.W. 1977, Ann. Rev. Astron. Ap., 15, 363.
Zirin, H. 1975, Ap.J.Lett., 199, L63.

THE STRUCTURE AND EVOLUTION OF X-RAY CLUSTERS OF GALAXIES

C. Jones

Harvard-Smithsonian Center for Astrophysics
Cambridge, Massachusetts

Rich clusters are the largest, well-studied aggregates of matter in the Universe. It is a basic observational fact that the distribution of galaxies on the sky shows a high degree of clumpiness. A cluster is essentially an enhancement in the density of galaxies projected on the sky. Abell (1958), who has done much of the basic work of systematizing the concept of cluster, includes in his catalog clusters that have at least 50 galaxies in the magnitude range m_3 to $m_3 + 2$, where m_3 is the magnitude of the third brightest galaxy, and within a 3 Mpc radius of the cluster center. For richer clusters there may be several hundred galaxies within this radius.

Several schemes have been suggested to classify clusters of different morphology. For example, the Coma cluster (at a distance of 138 Mpc) is classified as a so-called "binary" system since it is dominated by a pair of giant galaxies. One of the main types of clusters that figures prominently in the realm of X-ray clusters is called a cD cluster. These clusters are dominated by a single giant cD galaxy. This type of cluster is rather rare; only about 10% of the clusters in Abell's sample are cD clusters. cD's are the largest galaxies known and have the core of a giant elliptical galaxy and an extended envelope of low surface brightness. Many also contain complex nuclei with two or more condensations. Ostriker and his co-workers (Gallagher and Ostriker 1972, Ostriker and Tremaine 1976, and Hausman and Ostriker 1978) have suggested that these galaxies are formed by a process in which massive galaxies in a dense cluster core lose energy by dynamical friction as they move towards equipartition with less massive galaxies, consequently, they spiral in towards the center of the cluster's potential well.

R. Giacconi and G. Setti (eds.), X-Ray Astronomy, 153-170.

There, the cD's grow by cannibalizing other galaxies whose envelopes approach nearby. This theory predicts that cD galaxies will be found at or near the cluster center. The Einstein X-ray observations can be used to determine the cluster center and can even improve on galaxy counts as a means of determining the center. We therefore can compare the X-ray center and the position of the cD galaxy.

The first figure gives an example of a cluster that has quite a different central structure from both Coma and cD clusters. This is the Perseus cluster at a distance of 110 Mpc in which several bright galaxies are spread out along a line. This figure shows a result for the Perseus cluster produced by Forman et al. (197) from the first X-ray satellite, UHURU. Indicated are the contours of an extended radio source associated with the cluster, tailed radio sources around galaxies, as well as the error ellipse and extent for the X-ray source. Observations of this type showed for the first time that X-ray emission from clusters originates in a diffuse region rather than a point-like source. The X-ray core radius of Perseus is 15 arc minutes, or 0.47 Mpc. This ability to distinguish between a point and an extended source was about the limit of UHURU, Ariel V, SAS-C, and HEAO-1 satellites, and then that analysis was only possible for very bright, nearby clusters. However, now the era of imaging allows one to study the detailed structure of clusters.

The discovery of diffuse X-ray emission from clusters and the realization, primarily from the discovery of iron emission by Mitchell et al. (1976) and Serlemitsos et al. (1977), that most of the observed emission was due to thermal bremsstrahlung, brought about

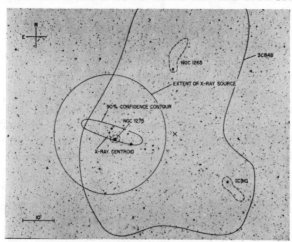

Fig. 1. The extent of the X-ray source, the shape and position of the error ellipse for the cluster are shown. The radio emission from NGC1275, NGC1265, IC310 and the diffuse source are taken from the observations of Ryle and Windram (1968).

the question of the origin of the hot gas. Two basic types of models involved either the infall of primordial gas into the cluster from the intergalactic medium (Gunn and Gott 1972) or the injection (or stripping) of
evolved gas in galaxies to form the intracluster medium (Yahil and
Ostriker 1973, and Cowie and Binney 1977). The iron emission lines
indicated that roughly solar abundances of iron were present and thus
provided support for the models in which the X-ray cluster emission is
produced by gas which has been processed in stars and then subsequently
lost to the intracluster region.

One of the best studied clusters is the rich cluster in Virgo
at a distance of only $\sim$ 20 Mpc. Earlier observations have presented a
picture of the Virgo cluster X-ray emission consisting of two components -- a cool 2 keV gas contained by the gravitational potential of M87,
and a hotter ($\sim$ 10 keV) gas of lower density contained by the gravitational
potential of the entire cluster. The hot material must have a low density
($\sim 10^{-4}$ cm^{-3}) to account for its low surface brightness. The Einstein
observations have added to this picture and show that cool gas surrounds
several of the Virgo galaxies (Forman et al. 1979). Embedded in such
a system, the gas surrounding M86 originated within M86 itself, rather
than from infall of intracluster gas, since the hot intracluster gas could
not cool sufficiently rapidly. Second, our preliminary spectral analysis
of the emission implies that the gas around M86 is similar to that associated with M87 and has a cool temperature of $\sim$ 1 keV, and that is contains line emission, and is therefore evolved.

The Virgo cluster phenomena of gas associated with galaxies
is not unique. We have studied several nearby rich clusters of galaxies
with the Einstein Observatory Imaging Proportional Counter and have
discovered other examples of Virgo-type cluster emission. These observations have shown that clusters display a variety of structures in
their X-ray surface brightness distributions. I will discuss our observations in the context of dynamic evolution in clusters. We have begun
to classify the various clusters we have observed into four groups.
This chapter discusses our observations of nearby clusters and some of
their implications for cluster evolution, as well as the fourth class of
cluster emission - from groups of galaxies - and distant clusters, and
suggests how some of our ideas on cluster and galaxy evolution may need
to be modifed. Much of this work has been done by William Forman,
Joseph Schwarz, Eric Mandel, Pat Henry, Dan Schwartz, and others at
CFA.

The discussion of nearby clusters will begin with other examples of Virgo-type clusters. Three examples are shown in the top row
of Figure 2. This figure shows contour plots of the X-ray emission for
twelve clusters superposed on enlargements from the Palomar Sky Survey. The cluster in the upper left hand corner is A1367. The bright

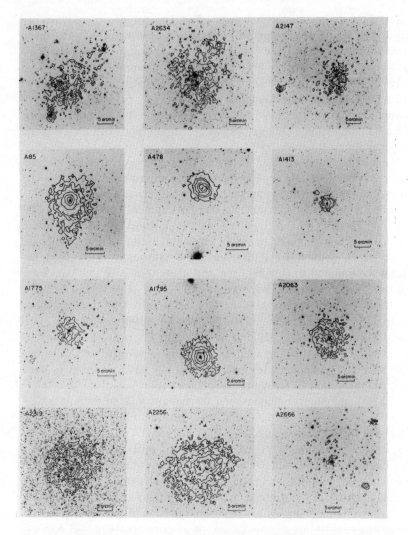

Fig. 2.   The iso-intensity contour plots of the X-ray emission are
         shown superposed on the optical PSS field for each cluster.
         The contour levels are arbitrary.

source at the lower end of the cluster emission is the radio galaxy
3C264. The emission from this galaxy is easily apparent from that of
the cluster even though its luminosity is only a few percent that of the
cluster. The 0.25 to 3.0 keV luminosity of 3C264 is a few (2) x 10^{42}
ergs/sec comparable to the X-ray luminosity of Cen A. The strong
radio emission in this galaxy suggests unusual activity and therefore
a higher X-ray luminosity than for other cluster members. Several

other galaxies in A1367 also are associated with regions of enhanced X-ray emission. The two galaxies, NGC3842 and NGC3841, have X-ray luminosities of several (3.5 and 5.5) x 10^{41} ergs/sec. These luminosities are about one hundred times greater than the X-ray luminosity of our galaxy or Andromeda, but comparable to the luminosity of M86 in the Virgo cluster.

This contour plot shows that the X-ray emission from A1367 is broad and highly clumped. Although some of the X-ray fluctuations correspond to bright galaxies, others do not. Given a relation between the mass of a galaxy and the associated X-ray emission, the enhancements which do not correlate with bright galaxy may indicate the presence of previously undetected mass in the cluster. X-ray measurements can be used to determine the cluster centers. For A1367 the center of the X-ray emission fairly clearly defines the cluster center. Bahcall (1974) counted galaxies in this cluster to determine its center, radial distribution, and spatial core radius. She found the center to be outside the region of X-ray emission shown in this figure.

An extensive Westerbork radio study of this cluster has been made by Gavazzi (1978). He found several weak point sources and an extended area of emission northwest of the X-ray cluster emission. He detected radio emission from the spiral galaxy ZW1141.2+2015, which we do not detect at our present level of sensitivity, and from NGC3842 which we also observe but no radio from the other nearby X-ray emission region around NGC3841. Although our cluster observations are preliminary, we have not in general found good correlations between the X-ray surface brightness distribution and the radio observations.

The next cluster in this figure is A2634. Again, the contour plot shows that the X-ray distribution is not smooth. The two main concentrations correspond to concentrations in the optical galaxy counts made by Baier (1976). What looks from this figure to be one central galaxy is a giant galaxy with two nuclei. Jenner (1974) has spectroscopically measured the redshift of each nuclei and found a projected separation for the two components of 5.34 kpc. The multiple nuclei inside a common envelope suggest that this galaxy is a cD. The Centaurus cluster is another Virgo-type cluster -- that is, a cluster in its early stages of cluster evolution which contains a cD galaxy (NGC4696). Although these observations are not conclusive, they suggest that cD galaxies may form in the early stages of cluster collapse, perhaps during an initial collapse of the cluster core.

The last cluster in this figure (top right) whose emission is broad and associated with individual galaxies is A2147. A2147 is part of the Hercules supercluster and is near A2152 so that optically it is difficult to distinguish the two clusters. Extended cluster emission is clearly present and centered on the ellipticals in the Arp chain. For

this cluster the emission is somewhat more peaked than for A1367.
The other source in the field is a radio galaxy (WE1601+16103) whose
X-ray luminosity is again a few times 10^{42} ergs/sec, similar to that
of 3C264 in A1367.

A very different type of cluster X-ray emission is shown in
the next two rows of Figure 2. Unlike the broad surface brightness
profile for A1367, and the other early type clusters, the X-ray emission
of these clusters is sharply peaked. The peak of the emission coincides
with the dominant central galaxy. The X-ray emission of these clusters
is concentrated toward the cluster center and does not show the clumping that was so apparent for Virgo-type clusters.

For three of the brighter clusters we have measured the
energy spectra as a function of radius from the X-ray center. The
Einstein energy range allows us to determine the low energy absorption
but does not permit a reliable temperature measurement if the temperature is above a few keV. We have measured the cluster spectra in
annuli where the width of each ring is three arc minutes. This width
was chosen based on the energy dependent spatial response of the IPC
and corresponds to the area needed to insure that less than 10% of the
flux was scattered outside each ring. Figure 3 shows the IPC spectra
for each of the clusters A85, A1795, and A1367 in three arc minute
rings in detector counting rates. The background has been subtracted
for each ring. Since the clusters fill a substantial fraction of the IPC
field, the background was determined from the deep surveys in Draco
and Eridanus (Giacconi et al. 1979). Since the deep surveys and the
cluster fields are all at high galactic latitudes, we have assumed that
the deep survey X-ray background is applicable to the cluster fields.
Except for A1367, the background is only a few percent of the source
counts. This figure illustrates the change in the spectrum of A85 at the
center of the cluster. The apparent increase in low energy absorption
at the cluster center also may be interpreted as due to a temperature
gradient and subsequent changes in the line emission by a cool gas contained in the potential of the cD (similar to the M87 observations).
Such emission would mimic an increasing cutoff in the center regions if
the gas temperature was decreasing as one approached the cD galaxy.
Alternatively the apparent cutoff may indicate that there is a contribution to the X-ray emission from the core of the cD galaxy.

Of the clusters which we have studied with the Einstein Observatory, several have shown broad, highly clumped X-ray emission,
while for others the emission is smooth, and centrally peaked. Although time does not permit a detailed discussion of each cluster, we
can use the broad versus peaked features of the surface brightness distribution combined with the core radii to characterize the cluster.
Figure 4 shows the radial distribution of the X-rays for several clusters.

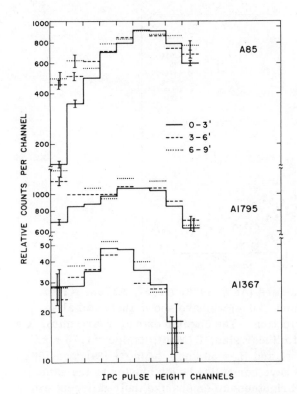

Fig. 3. The spectra for the three clusters A85, A1795, and A1367 are
shown in three arc minute annuli from the cluster center. The
counts above background for each of the IPC pulse height channels are shown for the energy range from 0.25 to 3.0 keV.

The angular dimensions have been converted to linear size with the axis
ending at 1 Mpc. We also have normalized the clusters to the same
peak central intensities to illustrate their different extents. For the
smooth clusters, which contain cD galaxies, the emission is centrally
peaked while for the others the emission is broader. Contour plots of
some of these extended clusters show that the emission is clumped,
while for others the surface brightness distribution is smooth. If we
use the King approximation to an isothermal sphere to characterize the
radial distribution of the X-rays, then the core radii of the cD clusters
is approximately 0.25 Mpc -- about half that of the broad clusters.
        The first two classes of clusters are those that are like
Virgo -- broad with the gas primarily around galaxies -- and those
which are sharply peaked with dominant galaxies at their centers.
There is a third type of cluster -- those which show a broad X-ray

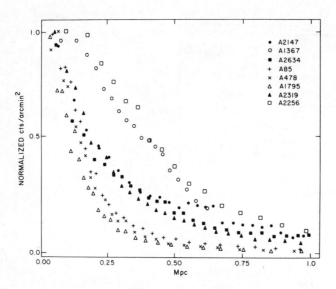

Fig. 4.   For the eight clusters (A85, A478, A1367, A1795, A2147,
A2256, A2319, and A2634) we have shown their radial surface
brightness distribution.   The angular extent, r (arc min), has
been converted to linear size, R (Mpc); R (Mpc) = 1.75 r z /
$(1 + z)^2$ with $q_O = 1$ and $H_O = 50$ km s^{-1} Mpc^{-1} (Sandage 1961).
In this graph we have normalized each cluster to the same peak
central surface brightness to emphasize their different extents.

radial distribution but whose emission is not clumped around individual
galaxies.   Coma is an example of this type of cluster.   Based on UHURU
observations, Forman and colleagues determined the core radius to be
1/2 Mpc (1972).   Based on a rocket experiment with an imaging detector
Gorenstein and his co-workers (1978) confirmed the size of the core and
showed that there were no large fluctuations from a smooth surface
brightness distribution.   A2256 and A2319 also have core radii of ~ 0.5
Mpc and show smooth X-ray distributions.   The bottom center of
Figure 2 shows the iso-intensity contour of the X-ray emission from
the Coma-type cluster A2256 on the PSS print.   This cluster is more
extended in an E-W direction in line with the three large galaxies than
in a N-S direction.

The other example we have observed of a Coma-type cluster
is A2319.   The bottom left portion of Figure 2 shows the X-ray contour
where the cluster is apparently elongated and brighter toward one end.
Faber and Dressler (1977) noted that for A2319 the distribution of
galaxies in both their velocities and positions on the sky suggested that
there were two clusters -- one centered on the brightest galaxy and the

other located 8' to the NW centered on another bright galaxy. If two clusters are present, the velocity dispersion is fairly low -- comparable to that found for the bright galaxies in A1367. This low velocity dispersion would be somewhat out of line with A2319's other cluster properties. In particular, the X-ray temperature measured by Mushtozky et al. (1978) is quite high (12.5 keV +7, -4) and the X-ray surface brightness distribution is quite smooth and is not clumped around individual galaxies. Although the X-ray emission shows an elongation in the direction of the two suggested clusters, it does not resolve two clusters. Therefore, the present X-ray observations favor the existence of only a single cluster in A2319.

These observations of clusters can be interpreted using dynamic cluster evolution. Peebles (1970), White (1976), and Aarseth (1967) have shown, through numerical simulations, that a cluster which begins as a large cloud of galaxies will collapse and finally reach equilibrium with an extended halo around a dense core. Optical studies of clusters have indicated that the "regular" clusters (Abell 1975) which have high central galaxy concentrations and a small fraction of spiral galaxies are dynamically more evolved than "irregular" clusters which are not centrally condensed and have large spiral fractions. The Virgo-type clusters whose X-ray emission is broad and highly clumped could be interpreted as clusters in their early evolutionary stages in which the gas escaping the galaxies is bound more by the gravitational potential of the individual galaxies than by the relatively weak potential field of the cluster. We can use the cluster velocity dispersion as a measure of the cluster potential. Clusters with low velocity dispersions would have weak cluster potentials and the potentials of individual galaxies would cause substantial deviation from a smooth surface brightness distribution. The Virgo cluster and A1367 have low velocity dispersions -- less than half the value measured for Perseus (Yahil and Vidal 1977). Both clusters show substantial clumping of the X-ray luminosity around galaxies. Clusters in this early stage would be expected to have a lower density of hot intracluster gas, and therefore, a higher fraction of spirals than in more evolved clusters in which the ram pressure of the gas would strip the galaxies of their interstellar matter (Gunn and Gott 1972, Gisler 1976) thereby transforming spirals into S0's. Bahcall (1978) has determined the percentage of spirals in A1367 to be 40% which is considerably higher than the ∼ 10% spirals she found in the dynamically evolved Coma and Perseus clusters.

As the cluster evolves, a high density core or sub-cluster is formed, thereby enhancing the chances for building a cD galaxy. The clusters in our rather limited sample which show strongly peaked X-ray radiation are all Bautz-Morgan type I clusters. These cD clusters have greater central surface brightnesses than do the less evolved clusters. This increased surface brightness is probably due to the increased gas density at the cluster center as the gas becomes bound by the cluster potential.

When the cluster has reached equilibrium, the X-ray emission should follow the cluster potential as traced by the galaxies. Since the gas distribution is dominated by the cluster potential, it should be smooth, and the percentage of spiral galaxies remains small. Both the clusters with dominant galaxies (e.g. A85) and the Coma-type clusters (e.g. A2256) are clusters in their final relaxed stage.

In summary, we have found from the first observations of the structure of clusters with the Einstein Observatory that the nature of the X-ray emission is complex and varies from broad and highly clumped, to smooth and centrally peaked. The clusters whose emission is clumped tend to be rich in spirals and to have X-ray temperatures in the few kilovolt range or two-component spectra and low velocity dispersions. For many of the clusters the emission is irregular and cannot be described by the simple, spherically symmetric models for a hot isothermal or adiabatic cluster gas. For these clusters the low density intracluster gas is strongly influenced by the potential of individual bright galaxies.

The smooth, centrally peaked clusters are spiral poor and have higher X-ray temperatures and larger velocity dispersions. These evolved clusters may be divided into two classes -- one whose X-ray emission is sharply peak around a dominant cD type galaxy and the other in which the emission is less peaked. The clusters in the first group have isothermal core radii of ~ 0.25 Mpc while those in the second, including Coma, have core radii of ~ 0.5 Mpc. The other properties of the evolved clusters are similar. In general, their X-ray temperatures are high (~ 10 keV) and their surface brightness distribution is smooth.

Perrenod's (1978) discussion of models for the evolution of clusters includes predictions for changes with time of the X-ray properties. In particular, he has suggested that as clusters evolve their X-ray luminosity and temperature should increase as the cluster potential deepens. These new Einstein observations of clusters, combined with the good correlation between decreasing spiral fraction and increasing X-ray luminosity found by Bahcall (1978), support the predicted evolutionary changes.

An additional consequence of these observations is that the previous extended X-ray emission reported by Forman et al. (1978) for the two clusters A1367 and A2634 no longer need be interpreted as evidence for a massive halo. Instead the extent can be understood as broad, highly clumped emission produced by a relatively small mass of gas.

Tne observed X-ray structure combined witn optical properties sucn as tne spiral fraction and central density complement tne
theoretical model for dynamic evolution in clusters. Many of the X-ray
properties of clusters, including tneir structure and luminosity, appear
to be strongly influenced by tneir dynamical evolution. However, since
clusters do not all evolve at the same rate due to differences in cluster
properties sucn as richness, we are able to observe clusters at tne
same epoch but in a variety of evolutionary pnases.

Although X-ray emission is a common characteristic of rich
clusters of galaxies, less rich condensations of galaxies also have been
associated with X-ray sources. But unlike Abell clusters, the concept
of groups has not been well defined, nor have groups been well cataloged.
Turner and Gott (1976) have cataloged statistically significant groupings
over a small fraction of the sky. Morgan and his co-workers have surveyed fields in the Palomar Sky Survey to search for cD galaxies in
poor clusters (Morgan, Kayser, and White 1975; Albert, White, and
Morgan 1977).

One should not be surprised to see cD galaxies or, in fact,
X-ray emission from poor clusters. One of the parameters which
determines the relaxation time of a cluster or group is the number of
galaxies in the group (T_{relax} time $\propto V^3_{RMS}/(G^2 M_{gal} \rho_{cl}) \propto N_{gal}^{5/4}$)
(Cavaliere 1979). Therefore, poor clusters should evolve faster than
richer ones.

In their first sample of poor clusters Morgan, Kayser, and
White (1975) selected 50 galaxy clusterings which they felt had a high
probability of physical association. Of these 50, 7 contained cD galaxies and 6 contained D galaxies. The number of cD galaxies in this
group is comparable to the ~ 10% cD clusters found among the Abell
rich clusters. Morgan and his colleagues noted one important difference between the rich clusters and the groups. They found that the cD
galaxies were often not located at the center of the poor cluster, which
is not the situation for rich clusters and also not what one would expect
if cD galaxies are produced by galactic cannibalism.

Since evolved clusters are generally somewhat brighter X-
ray emitters than unevolved clusters, one might expect that evolved
groups would also emit X-rays. In fact, in the Ariel V catalog of high
latitude X-ray sources, several sources were suggested to be associated with groups of galaxies. Observations of groups of galaxies also
have been made with the HEAO-1 scanning modulation collimator
(Schwartz, Schwarz, and Tucker 1979).

At the lower right of Figure 2 is an example of X-ray emission from a group of galaxies. Although this group, A2666, is contained in Abell's catalog of rich clusters, Butcher and Oemler (1978)

have noted that this cluster does not contain sufficient numbers of galaxies to have met Abell's minimum of 50 galaxies. From Figure 2 it
is apparent that the X-ray emission is located around a cD galaxy. However, the X-ray source is quite weak so it is difficult to determine if
there is an extent associated with it. We have put an upper limit on the
extent of 1 1/2' which at the distance of this group corresponds to a size
of about 65 kpc. The intrinsic luminosity of the source is quite low,
$\sim 2 \times 10^{42}$ ergs/sec -- ten times weaker than any of the rich clusters
shown in this figure.

        Optical studies of the colors and spectra of cD galaxies in
groups and rich clusters have shown that these properties of cD galaxies
did not change substantially with their different environments (Schild
and Davis 1979). However, the extent and magnitude of the cD's in
groups are generally substantially less than for the cD's in rich clusters (Schild and Weekes 1979).

        Hausman and Ostriker (1978) have discussed the formation of
cD galaxies by galactic cannibalism in clusters with different richnesses.
Optical studies done by Sandage (1976) showed that the optical magnitude
of cD galaxies was not well correlated with the cluster richness.
Hausman and Ostriker suggested that although the total luminosity of
the cD galaxy should increase as more galaxies merge, that the optical
magnitude measured through a small aperture would at first increase as
the galaxy grows to fill the aperture but would then remain constant or
might even decrease in magnitude as the cD galaxy grows an extended
envelope and much of the light from the galaxy falls outside the optical
aperture. The optical luminosity of the core of the cD does not increase substantially with time. The Einstein observation of A2666 suggests that there is probably not a large contribution to the X-ray luminosity of clusters from the cD galaxies themselves.

        Table 1 shows the properties of several groups which have
been observed as X-ray sources (Schwartz, Schwarz, and Tucker 1979).
The properties of the rich cD cluster A85 have been added for comparison. Four of these groups have luminosities comparable to those which
have been observed for rich clusters. Their isothermal core radii are
comparable to or somewhat less than those of rich cD clusters. For
A2666 the size and luminosity are considerably less than for the rich
clusters or for the other groups. From the X-ray size and luminosity
a gas density can be inferred. If we describe the X-ray luminosity by
thermal bremsstrahlung in an isothermal sphere, the computed densities are comparable to those found in clusters. From these observations one can see that the properties of X-ray emission from groups of
galaxies containing cD galaxies are often not that different in character from those of rich cD clusters.

TABLE 1

X-ray Source	Cluster Type	L_x	Core Radius	Density
2A0335+096	cD	6×10^{44} erg/sec	140 kpc	$27 \times 10^{-3}/cm^3$
MKW11 = 4U1326+11	D	4×10^{43}	<180 kpc	$>5 \times 10^{-3}$
AWM7 = 2A0251+413	cD	1×10^{44}	220 kpc	6×10^{-3}
A2666	cD	2×10^{42}	<65 kpc	$>6 \times 10^{-3}$
A85	cD	5×10^{44}	260 kpc	13×10^{-3}

We are planning to survey poor galaxy clusterings with the Imaging Proportional Counter on the Einstein Observatory to determine if groups which do not contain dC galaxies also show cluster type emission.

In addition to nearby clusters, X-ray observations have been made of clusters at cosmological distances (Henry et al. 1979). The Einstein observations of distant clusters were undertaken primarily to search for changes in the total X-ray luminosity due to evolution of the galaxies or of the cluster as a whole.

The nature of the X-ray luminosity evolution is determined primarily by the efficiency of galaxy formation in the protocluster and by the evolution of the cluster gravitational potential. If galaxy formation is very efficient, so that no intracluster gas remains, then the X-ray luminosity is low until the interstellar gas starts being stripped by galaxy collisions or galactic winds. Thus, initially there would be little intracluster gas to produce the X-ray emission so that the X-ray luminosity will remain low for a substantial fraction of the cluster life. On the other hand, if galaxy formation is very inefficient, a large fraction of the cluster mass is left as gas so that the X-ray luminosity would be high at the time the cluster forms. For moderately efficient galaxy formation, leaving about 10% of the mass as gas, the luminosity evolution then depends primarily on the evolution of the cluster potential.

So far, 11 clusters of galaxies with redshifts greater than 0.1 have been observed with the Einstein Imaging Proportional Counter. Significant fluxes have been detected from 8 of these. Prior to Einstein, the most distant X-ray cluster had a redshift of $\lesssim 0.1$. The great distance of these clusters reduces their flux and apparent size so that it is difficult to measure precisely the core radii. Henry et al. (1979) found the core radii of these clusters to be about 0.5 Mpc which is within the range of cluster sizes for nearby clusters. Although the sample of distant clusters is still small, their X-ray luminosities range from 10^{43} to 10^{45} ergs/sec. Again, this is the same range as was found for nearby clusters.

Figure 5 shows the suggested evolutionary tracks for various cluster models. These have been normalized at $z = 0.1$ to the average luminosity of a cD cluster which is about 7×10^{44} ergs/sec. The curve labelled 0.0 is what one would expect if the luminosity is not changed as we move earlier in time and if $q_o = 0$. If we change the value of q_o to 0.5 this curve moves slightly. The curve M2 is one of Perrenod's (1978) models which assumes that there is mass injection from the galaxies into the cluster and that the cluster potential evolves. The curves labelled "all gas" refer to the model for very inefficient production of galaxies in the initial cluster formation, while the "no gas" curve is efficient galaxy production. These curves are plotted for the detected spectral density at 2 keV as a function of redshift.

The various types of clusters are shown with different symbols. Clusters have been classified into five different Bautz-Morgan classes based on the magnitude differences between the brightest galaxy and the next brightest one. Those clusters with cD galaxies are generally B-M type I or I-II, and are shown with filled symbols.

Optical studies have shown that as cD galaxies evolve, other galaxies are deleted from the high end of the luminosity function, that is, the large bright galaxies seem to be the first to merge to form the cD. Therefore, in general, those clusters with the largest difference in the magnitude between the first and second brightest galaxies will be the most evolved. And those galaxies with approximately equally bright galaxies will be the least evolved. These type of clusters are classified as later type Bautz-Morgan types -- types II or III. They are shown with open circles in this figure. As Hausman and Ostriker have noted, the Bautz-Morgan classification can be considered as indicative of a cluster's evolutionary state. However, since clusters do not all have the same number of galaxies, and since they do not all evolve along the same path, there is not a unique correspondence between the Bautz-Morgan class and the evolution of the cluster. In particular, all clusters do not form cD galaxies. For example, a cluster like A2256 or Coma have all the standard X-ray and optical properties of evolved clusters, except they do not contain cD galaxies.

In general, from Figure 5, one sees that the less evolved clusters are less luminous than the evolved cD clusters (the underluminous cD is the group A2666). This spread in luminosity for the different types of clusters makes it very difficult to determine if the

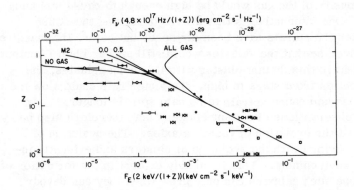

Fig. 5.   Received spectral density at an emitted energy of 2 keV as a
function of redshift for cluster X-ray sources. The boxes are
OSO-8 data and the circles are Einstein data; closed symbols
refer to Bautz-Morgan types I and I-II; open symbols, types II,
II-III, and III; the cross refers to an unknown type. The
curves are for a source of 7.1×10^{44} erg s^{-1} at low redshift
which undergoes no evolution (with a q_0 of 0 and 0.5) or which
evolves according to the Perrenod (1978) model M2, the
Sarazin (1979) model with no intracluster gas (No Gas), or the
Silk (1976) model where at formation most of the mass of the
cluster is in gas (All Gas).

formation and evolution of clusters has changed from one epoch to
another. The present observations are consistent with the model of
Perrenod or those in which the evolution in the past is the same as in
the present. They are less consistent with the extreme models. To
make further progress a larger sample of clusters at high redshift is
needed. Then one can examine the distribution in luminosity of distant
clusters and compare that distribution to that which has been found for
nearby clusters.

     One of the distant clusters is of special interest. This is the
cluster associated with 3C295 which has a redshift of 0.46. Butcher
and Oemler have studied the color distribution of the galaxies in this
cluster and have found a high fraction of blue galaxies which would suggest that it is spiral rich. However, its morphology is that of a Bautz-
Morgan type I which would imply that it should be spiral poor. One
explanation had been that we were observing this cluster before the
galaxies have been stripped so that there would be no intergalactic
medium. However, the X-ray luminosity of 10^{45} ergs/sec and 0.5
Mpc core radius determined from the IPC observation suggest that if
this emission is produced by thermal bremsstrahlung of intracluster

gas, that the density of the gas would be high enough to cause stripping of the spirals. One alternative is that we are observing the 3C295 cluster just after the spirals have been stripped, but before the massive stars have evolved so that the galaxies would still appear blue. Another possibility is that in this distant cluster with its young galaxies, the rate of gas injection from stars is high. This high rate could slow the stripping process and cause spirals to remain spirals longer.

The observations of distant clusters may therefore also be teaching us about the evolution of young galaxies. The evolution of galaxies is intertwined with the evolution of clusters and evidently influenced by its environment. This is clearly evident in the formation of cD galaxies since such galaxies can only grow when they can devour nearby galaxies from a group or cluster.

In summary, the Einstein observations of nearby X-ray clusters have shown a surprising complexity and variety in their surface brightness distribution. Much of this variety can be ordered by associating the different types of cluster morphology with various stages. in their dynamic evolution. The observations of groups of galaxies extend the parameter ranges of cluster properties and should allow one to determine better the influence of these parameters on the X-ray emission. The Einstein observations of distant clusters will determine if their average X-ray luminosities and extents are different in the past than in the present epoch and therefore better define the evolutionary processes of clusters and galaxies.

Acknowledgements

I am grateful to the many people who have been essential to the success of the Einstein Observatory. In particular, thanks is due Riccardo Giacconi, Harvey Tananbaum, and Leon VanSpeybroeck for their contributions. In addition, I acknowledge W. Forman, P. Henry, D. Schwartz, J. Schwarz, and their colleagues at CFA whose scientific analyses of X-ray observations of clusters contributed greatly to the results presented here. I thank M. Twomey for her assistance in the preparation of this manuscript. This research was sponsored under NASA contract NAS8-30751.

References

Aarseth, S.J. 1966. M.N.R.A.S., 132, 35.
Abell, G.O. 1958. Ap. J. Suppl., 3, 211.
Abell, G.O. 1975. In Galaxies and the Universe; eds. A. Sandage,
        M. Sandage, J. Kristian. Chicago: Univ. Chicago Press.
Albert, C.E., White, R., and Morgan, W.W. 1977. Ap. J., 211, 309.
Bahcall, N. 1974. Ap. J., 198, 249.
Bahcall, N. 1977. Ap. J. (Letters), 218, L93.

Baier, F.W. 1976. Astron. Nachr., 297, 6.

Butcher, H.R., and Oemler, A. 1978. Ap. J., 226, 559.

Cavaliere, A. 1979. To be published in the Proceedings of the International School of Astrophysics, Erice, Italy, July.

Cowie, L.L., and Binney, J. 1977. Ap. J., 215, 723.

Faber, S., and Dressler, A. 1977. A.J., 82, 187.

Forman, W., Jones, C., Murray, S., and Giacconi, R. 1978. Ap. J. (Letters), 225, L1.

Forman, W., Kellogg, E., Gursky, H., Tananbaum, H., and Giacconi, R. 1972. Ap. J., 178, 309.

Forman, W., Schwarz, J., Jones, C., Liller, W., and Fabian, A.C. 1979. Ap. J. (Letters), 234.

Gallagher, J., and Ostriker, J. 1972. A. J., 77, 288.

Gavazzi, G. 1978. Astr. Ap., 69, 355.

Giacconi, R., Bechtold, J., Branduardi, G., Forman, W., Henry, P., Jones, C., Kellogg, E., van der Laan, H., Liller, W., Marshall, H., Murray, S.S., Pye, J.P., Sargent, W.L.W., Seward, F., and Tananbaum, H. 1979. Ap. J. (Letters), 234.

Gisler, G. 1976. Astr. Ap., 51, 137.

Gorenstein, P., Fabricant, D., Topka, K., Harnden, F.R. Jr., and Tucker, W.H. 1978. Ap. J., 224, 718.

Gunn, J.E., and Gott, J.R. 1972. Ap. J., 176, 1.

Hausman, M.A., and Ostriker, J.P. 1978. Ap. J., 224, 320.

Henry, J.P., Branduardi, G., Briel, U., Fabricant, D., Feigelson, E., Murray, S.S., Soltan, A., and Tananbaum, H. 1979. Ap. J. (Letters), 234.

Jenner, D. 1974. Ap. J., 191, 55.

Kellogg, E., and Murray, S. 1974. Ap. J. (Letters), 193, L57.

Mitchell, R.J., Culhane, J.L., Davison, P.J.N., and Ives, J.C. 1976. M.N.R.A.S., 176, 29p.

Morgan, W.W., Kayser, S., and White, R. 1975. Ap. J., 199, 545.

Mushotzky, R.F., Serlemitsos, P.J., Smith, B.W., Boldt, E.A., and Holt, S.S. 1978. Ap. J., 225, 21.

Ostriker, J., and Tremaine, S.D. 1976. Ap. J. (Letters), 202, L113.

Peebles, P.J.E. 1970. A. J., 75, 13.

Perrenod, S. 1978. Ap. J., 226, 566.

Ryle, M., and Windram, M.D. 1968. M.N.R.A.S., 138, 1.

Sandage, A. 1961. Ap. J., 133, 355.

Sandage, A. 1976. Ap. J., 205, 6.

Sarazin, C.L. 1979. Preprint.

Schild, R., and Davis, M. 1979. A. J., 84, 311.

Schild, R., and Weekes, T.C. 1979. Preprint.

Schwartz, D., Schwarz, J., and Tucker, W. 1979. Preprint.

Serlemitsos, P.J., Smith, B.W., Boldt, E.A., Holt, S.S., and
        Swank, J.H. 1977. Ap. J. (Letters), 211, L63.
Silk, J. 1976. Ap. J., 208, 646.
Turner, E., and Gott, J.R. Ap. J. Suppl., 32, 409.
White, S. 1976. M.N.R.A.S., 177, 717.
Yahil, A., and Ostriker, J.P. 1973. Ap. J., 185, 787.
Yahil, A., and Vidal, N.V. 1977. Ap. J., 214, 347.

THE X-RAY SPECTRA OF CLUSTERS OF GALAXIES

Richard Mushotzky

Laboratory for High Energy Astrophysics
NASA/Goddard Space Flight Center
Greenbelt, Maryland 20771 U.S.A.

1.   INTRODUCTION

Clusters of galaxies were established as a class of X-ray sources
by observers on the Uhuru satellite (Gursky et al. 1971).   These
early results indicated that clusters were a powerful (L_x in the
2-6 keV band > 10^{44} erg/sec) and numerous ($\gtrsim$ 9) class of extragalactic X-ray sources.   These early results also indicated that
the sources were extended (Kellogg 1972; Kellogg and Murray 1974)
with a characteristic size $\gtrsim$ .25 Mpc.   Early spectral data
(Kellogg, Baldwin and Koch 1974; Catura et al. 1972; Gorenstein
et al. 1973) showed that the spectra were not strongly cutoff at
low energies and could be fit equally well by power laws of energy
index $\sim$ 1 or by the emission from optically thin hot gas (henceforth referred to as thermal bremsstrahlung and abbreviated as
T-B) with a temperature of $\sim 10^{8o}$K.
      These early results suggested several models for the origin
of the X-ray emission.   Among these were (1) the integrated emission from single galaxies (2) thermal bremsstrahlung from a hot
intergalactic gas (3) inverse Compton scattering off the 3^oK background.   Models (2) and (3) provide definite spectral predictions.
The inverse Compton model predicted power law X-ray spectra with
no features while the T-B model predicted a thermal X-ray spectrum
with no features since it was expected that the gas was primordial
and had no heavy elements in it.

2.   PRE-HEAO OBSERVATIONAL RESULTS

The most important result of the first generation X-ray spectrometers flown on OSO-8 and Ariel-5 was the detection of Fe emission
from the Virgo, Perseus and Coma clusters of galaxies (Mitchell et

171

R. Giacconi and G. Setti (eds.), X-Ray Astronomy, 171-179.

al. 1976; Serlemitsos et al. 1978). This feature was interpreted
as emission from iron in collisional equilibrium with a hot gas.
Thus it was dramatic proof of the T-B origin of the X-ray emission
and demonstrated that the intergalactic medium in clusters was
not primordial but, since it had Fe in it, had been processed
through stars.

Further observations with OSO-8 (Mushotzky et al. 1978) and
Ariel-5 (Culhane 1979) detected Fe lines in 5 more clusters and
groups of galaxies (3C129, NGC1129 group, SC0626-52 and Centaurus)
established that the Fe line feature was a common property of
clusters and indicated that the Fe abundance in the intercluster
gas was roughly constant from cluster to cluster at Fe/H $\sim$ 1.5 x
10^{-5} (approximately half solar). These observations implied that
clusters and their member galaxies have undergone substantial
evolution (DeYoung 1978) with material being ejected from stars .
into the intergalactic medium and mixing with primordial material.

In addition to the Fe emission line results the large number
of spectra observed with reasonable statistics by OSO-8 and
Ariel-5 (Mushotzky et al. 1978; Mitchell et al. 1979) showed that
the shape of the continuum emission favored a T-B model. These
results also showed that clusters possessed a range of temperatures
from 2 x 10^{7}°K to 1.2 x 10^{8}°K. The temperature results enable one
to calculate the emission integral, $<n_e^2>$ V, and bolometric
luminosities of the clusters. These results indicate that the hot
gas has $\sim$ 10% of the virial mass of the cluster and a density of
$\sim$ 10^{-3} cm^{-3}. Thus the prime observational quantities determined
for clusters were their iron abundance, temperature, emission
integral and bolometric luminosities. All of these were calculated
for the assumption that cluster spectra (with the exception of the
Virgo cluster) could be well fit by isothermal bremsstrahlung.

3.   PRE-HEAO CORRELATIONS AND INTERPRETATION

The large body of homogenous spectral data enabled one, for the
first time, to correlate the detailed X-ray properties of clusters
with their optical and radio properties. (The following discussion
has been drawn from Mushotzky et al. 1978; Smith, Mushotzky and
Serlemitsos 1979; and Mitchell et al. 1979).

In a thermal bremsstrahlung model the temperature of the X-ray
gas is a measure of the cluster potential. A similar measure is
the optical velocity dispersion of the galaxies. It was found
that, for non-CD clusters, the X-ray temperature was proportional
to the velocity dispersion squared. Treating the cluster sample
together evidence was also found from this correlation that many
clusters are nearly isothermal. This relationship between velocity
dispersion and X-ray temperature strengthens the argument that
clusters truly possess their virial mass; however the X-ray gas,
with a mass similar to the optically luminous material, does not
provide the missing mass.

Another strong correlation was found between the galaxy density in the core of the cluster, N_O, and the cluster temperature and emission integral. In fact it was found that KT was proportional to N_O which is expected if $KT \propto \Delta V^2$ and N_O is proportional to the true virial mass density. The correlation with emission integral along with a luminosity-temperature correlation indicated that high temperature clusters are, on the average, more centrally condensed or that they contain proportionally more intergalactic gas.

Correlations with the optical morphology of clusters and their member galaxies are also significant. (1) There appears to be a strong inverse correlation between the percentage of spirals in a cluster and the ram pressure experienced by an average galaxy in the cluster. This suggests that the remaining spirals in a cluster survive because they are too massive to be stripped by ram pressure. (2) There is a strong correlation between cluster spectral properties and their degree of central condensation as measured by Rood-Sastry or Bautz-Morgan type. The more central condensed clusters have, on the average, higher X-ray temperatures and higher emission integrals and therefore higher X-ray luminosites. This suggests that centrally optically condensed clusters have different X-ray properties (as has been recently shown for their X-ray morphology by Einstein results (see contribution by C. Jones this proceedings)).

4.   HEAO 1 AND 2 SPECTRAL RESULTS

In this section I shall discuss the spectral results derived by the GSFC experiments on HEAO-1 and HEAO-2.

A.   HEAO-1

HEAO-1 has observed $\sim$ 20 clusters in the 2-60 keV band with high counting statistics (see Fig. 1) and has detected Fe line emission in 18 of them. We now are beginning to see that there are small variations in the Fe abundance from cluster to cluster in the framework of an isothermal model. However one of the new discoveries from HEAO-1 is that many low luminosity clusters are not well discribed by an isothermal model. Before HEAO-1 we had only seen that for the Virgo cluster, now we have evidence (see Table 1) for $\sim$ 6 clusters that there probably exist at least 2 thermal components to their spectra.

The fact that this two-temperature structure is only seen in low L_x clusters by HEAO-1 may be related to the fact that only low L_x clusters show a complicated X-ray morphology (see C. Jones this volume). Unfortunately since we do not, with HEAO-1, have spatially resolved spectroscopy we cannot assign the Fe line to one component or the other thus making the Fe abundance, but not the equivalent widths, somewhat uncertain.

The vastly increased statistics available with HEAO-1 has also enabled us to detect the KB line of Fe. This transition

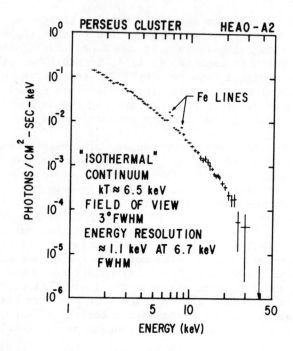

Fig. 1. The X-ray spectrum of the Perseus cluster as observed by HEAO-1 A-2. The line emission from the Kα and Kβ transitions of iron at 6.7 and 7.9 keV is quite prominent. Note also the unresolved line emission from 1.8 to 4 keV due to lines of Si, S, Ca and A.

proves that the line emission is not fluorescent but is truly due to recombination.

The extension of the number of very high quality spectra combined with the increase in the certainty of the cluster identification makes many of the correlations first described by OSO-8 and Ariel-5 data more certain. In Table 1 we show a summary of available spectral data for clusters. The HEAO-1 data are preliminary and the exact values should be used with caution.

B.   HEAO-2

The solid state spectrometer (SSS) on HEAO-2 (Einstein) has an energy resolution of ∿ 160 eV (see contribution by S. Holt this volume) and a bandwidth from 0.8 to 4.0 keV. It is therefore capable of seeing recombination lines from the K shells of the lighter elements (such as Si, S and Mg) and from the L shell of Fe. However it cannot see the Kα and Kβ lines of Fe.

The SSS has detected lines due to Fe, Mg, Si and S in the X-ray spectrum of M87 (Fig. 2). For Si and S both the Hydrogenic and Helium lines have been seen. The line ratios are very sensitive to temperature, assuming collisional equilibrium, and a

TABLE 1

NAME	kT_1	kT_2	6.7 keV E.W.	Fe/H	7.9 LINE?	OTHER LINES?	L_x 44	OBSERVATIONS[*]
A85	$6.8 \pm .5$	No	300	1.2	?	?	6.7	4
A119	$5.4 \pm .5$	?	300	0.95	?	?	3.7	1,4
A262	$2.4 \pm .8$	?	?	?	?	?	0.38	1,2,4
NGC 1129 group	> 10	$4.1 \pm .3$	630	1.8	Yes	?	2.0	2,4
A401	7.6 ± 1.5	No	300	1.3	?	?	21.9	2,4
A426	$6.4 \pm .4$	Very Weak	400	1.4	Yes	Si,S,Fe	12.4	1,2,3,4,5
0340-538	6.5 ± 1.0	?	800	2.9	?	?	4.3	4
A478	7.3 ± 1.0	No	360	1.5	?	?	22	2,3,4
Ser 40/6	$9.0 \pm .5$	?	150	0.77	?	?	7.6	3,4
3C 129 cluster	5.4 ± 1.0	?	330	1.0	?	?	2.7	2
0626-54	6.3 ± 3	?	800	2.9	?	?	6.9	2,3
A754	11.0 ± 3.0	No	< 150	< 1.0	?	?	10.4	4
A1060	> 10	2.0	800	2.5	?	?	0.17	2,4

Table 1 (continued)

A1367	2.8 ± 1.0	?	?	?	?	?	0.49	2
Virgo	> 10	2.2	1100	3.2	Yes	Mg,Si,S,Fe	0.29	1,2,4,5
Centaurus	8 ± 2	$2.4 + .3$	600	> 1.4	Yes	?	0.71	1,2,3,4
A1656	$7.9 \pm .3$	No	200	0.88	Yes	No ?	8.47	1,2,3,4,5
1326-311	$8.2^{+7.3}_{-3.0}$	?	?	?	?	?	12.0	2
A1795	5.8 ± 1.0	No	360	1.2	?	?	10.1	3.4
A2029	$6.2^{+2.6}_{-1.6}$	?	?	?	?	?	19.5	2
A2142	10.9 ± 1.0	No?	200	1.3	?	?	16.1	2,3,4
A2147	> 10	< 2	600	> 1.0	?	?	2.7	2,4
A2199	> 9	1.8	1000	> 3.3	?	?	2.8	2,4
A2256	$7, +3, -2$	?	?	?	?	?	8.2	2
A2319	12.5^{+7}_{-4}	?	?	?	?	?	12.7	2

*1 = Uhuru, 2 = OSO-8, 3 = Ariel 5, 4 = HEAO-1 A-2, 5 = HEAO-2 SSS

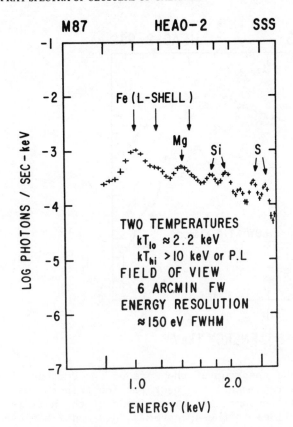

Fig. 2. The SSS spectrum of M87. Note the resolved line emission from both the hydrogenic and helium like lines of Si and S. The spectrum also displays line features due to Mg and the L shell transitions of Fe.

KT ∿ 1.8 keV is derived. The abundances of Si, S and Mg are approximately solar. We thus now know that as expected, elements other than Fe exist in the intragalactic cluster gas.

In the Perseus cluster the situation is more complicated. Here (Fig. 3) we also see lines due to both the Hydrogenic and Helium like lines of Si and S but the continuum gives a temperature such that one should only have the hydrogen-like lines. We interpret this to mean that there exists a very small amount of lower temperature gas (an amount small enough to escape detection by HEAO-1) in the core of the Perseus cluster and that this is evidence for cooling (as predicted by Cowie and Binney 1978) in the core of the cluster. The formal values derived from our observations give ∿10^{12} M$_\odot$ of material to have cooled in a Hubble time and a cooling time of the low temperature material is less than 2 x 10^9 yrs. Therefore, a substantial fraction of the mass of

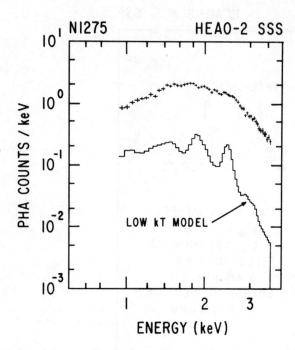

Fig. 3. The SSS spectrum of the core of the Perseus cluster.
The solid line represents the best fit low temperature
model which can account for the presence of the helium-
like S and Si lines in the spectrum. This model has been
correctly normalized to the total flux and shows the extreme
sensitivity of a spectrometer to the presence of low
temperature gas.

NGC 1275 is made up of material that has cooled and collapsed onto
it from the intergalactic gas.

5.   SUMMARY AND FUTURE

Spectral observations of clusters of galaxies have demonstrated
that the X-ray emission is due to a hot evolved intracluster gas
with ∿ 10% of the virial mass of the cluster. In low luminosity
clusters there is an indication that there is thermal structure
that may be related to the X-ray spatial structure. In the Perseus
cluster there is evidence for cooling at the core and therefore
for evolution of the cluster gas.
     The X-ray gas is direct evidence for the existence of a
virial mass in the cluster and responds to the cluster potential.
This gas interacts with the galaxies in the cluster in a dynamic
way and may change their morphology.
     However, there are still many questions left. For example,

what is the temperature structure of the gas? Does there exist
gradients in the elemental abundances across the cluster? What
is the evolution of the elemental abundance of the gas with the
age of the cluster? When and how does this gas arise?

Future observations, especially with spatially resolved
spectroscopy and more sensitive X-ray telescopes may answer these
questions.

ACKNOWLEDGMENTS

I thank B. Smith, P. Serlemitsos, E. Boldt and S. Holt for
extensive discussion, encouragement, and their large contributions
that have made this work possible.

REFERENCES

Catura, R.C., Fisher, P.G., Johnson, M.M., and Meyerott, A.J.
1972, Ap. J. 177, L1.
Cowie, L. and Binney, J. 1977, Ap. J. 215, 723.
Culhane, J.L. 1979 in IAU Symposium, ed. M.S. Longair and J.
Einasto, Reidel (Boston).
DeYoung, D.S. 1978, Ap. J. 223, 47.
Gorenstein, P., Bjorkholm, P., Harris, B., and Harnden, F. 1973,
Ap. J. 183, L57.
Gursky, H., Kellogg, E., Leong, C., Tananbaum, H., and Giacconi,
R. 1971, Ap. J. 167, L81.
Kellogg, E. 1973, X-Ray and γ-Ray Astronomy IAU Symposium No. 55,
ed. H. Bradt and R. Giacconi, Reidel (Boston).
Kellogg, E., Baldwin, J.R. and Koch, D. 1975, Ap. J. 199, 299.
Kellogg, E. and Murray, S. 1974, Ap. J. 193, L57.
Mitchell, R.J., Dickens, R.J., Bell, Burnell, S.S. and Culhane,
J.L. 1979, M.N.R.A.S., in press.
Mushotzky, R.F., Serlemitsos, P.J., Smith, B.W., Boldt, E.A., and
Holt, S.S. 1978, Ap. J. 225, 21.
Smith, B.W., Mushotzky, R.F., and Serlemitsos, P.J. 1979, Ap. J.
227, 37.

EINSTEIN OBSERVATIONS OF THE VIRGO CLUSTER

W. Forman

Harvard-Smithsonian Center for Astrophysics
Cambridge, Massachusetts

The Virgo cluster lies at a distance of only 20 Mpc and is therefore the nearest rich cluster of galaxies. This relatively small distance allows us to study both the individual galaxies and the cluster as a whole using the Einstein X-ray Observatory. With Einstein's tremendous increase in sensitivity over previous X-ray experiments, we can study in some detail the indivudual galaxies in the cluster.

Review of Earlier Work

X-ray emission from the region of sky centered around M87 was first observed in 1966 and 1967 by groups at the Naval Research Laboratory and the Massachusetts Institute of Technology (Byram, Chubb, and Friedman 1966, Bradt et al. 1967). These early observations represented one of the first detections of an extragalactic X-ray source.

In the early days before there was any indication that clusters of galaxies were luminous X-ray sources, the X-ray emission was believed to be associated only with M87 because of its unusual properties -- strong radio emission and the optical jet.

In the intervening years our picture of the Virgo cluster and M87 has become increasingly complex, keeping pace with increasingly sophisticated observations. First it was found that the X-ray emission was extended and centered on M87. The spectral observations showed that this gas was cool compared to that observed in other clusters -- 3 keV rather than the 8 to 10 keV temperature found for Coma and Perseus. These observations, along with the low luminosity of only a few times 10^{43} ergs/second compared to 10^{44} ergs/second and above

R. Giacconi and G. Setti (eds.), X-Ray Astronomy, 181-195.
Copyright © 1980 by D. Reidel Publishing Company.

for other rich clusters, led various workers, including, for example, Bahcall and Sarazin (1978) to propose that the X-ray emission was associated with M87 and not with the cluster as a whole.

The spectral results of Serlemitsos et al. (1977) who observed emission lines from iron at about 6.8 keV showed that the X-ray emission from clusters in general, and from the M87 vicinity in particular, was produced by thermal bremsstrahlung from a hot gas. Also, their studies, as well as those of Mushotzky et al. (1978), showed that the iron abundances in this hot intergalactic gas were roughly solar.

In the last few years various workers, including Malina, Lampton, and Bowyer (1976) and Gorenstein et al. (1977), had begun obtaining structural information on the M87 source with a resolution of several arc minutes. These observations were used, notably by Mathews (1978), to attempt to derive a mass for M87. Fabricant et al. (1979) also showed that the iron abundance and temperature were most likely uniform throughout most of the X-ray source centered on M87.

Most of the detailed work mentioned above has dealt with the emission centrally concentrated on M87. However, a number of papers have shown various indications of a second component of lower surface brightness which is more extended and hotter than the M87 component. Lawrence (1978) and Davison (1978) show that there is a harder component with a temperature of at least 7 or 8 keV having a radius of roughly 1° which is centered to the northwest of M87, and closer to the cluster center. The UHURU observations also showed evidence for this second component. Mushotsky et al. (1978) in a detailed analysis of the spectrum of the M87-Virgo X-ray radiation confirmed the presence of a second hotter component.

While this diffuse, large component has been hard to study because of its low surface brightness and the complicating effects of M87, its presence is clearly indicated in the Einstein observations. This second component plays a very important role in our understanding the Virgo cluster.

The early X-ray observations of M87 and the Virgo cluster described above led to the following picture:

1) Extended source centered on M87 produced by thermal radiation from a gas with solar abundances;

2) A hotter, more extended component associated with the cluster as a whole. This component is analogous to that observed in other clusters and is also probably of thermal origin. However, unlike the other cluster X-ray sources, this emission is only 10^{43} ergs/second -- more than 10 times weaker than that observed in the Coma cluster indicating a lower density.

Overview of Virgo Observations from the Einstein X-ray Observatory

Figure 1 is a reproduction from the catalog of Zwicky, Herzog, and Wild (1961) of the Palomar Sky Survey field containing M87 and the center of the Virgo cluster. The galaxies are indicated by the squares, circles, and triangles. Clusters of galaxies beyond Virgo are outlined with solid lines.

The square fields of which you see five indicate the five fields already observed with the Einstein Observatory. The detector used for this study is the Imaging Proportional Counter (IPC) which has a field of view of $1^{o} \times 1^{o}$ with a spatial resolution of ~ 1.5 minutes of arc. Also there is pulse height information which permits crude spectral analyses.

Let us discuss first the emission from M87 and then proceed to discuss the individual galaxies which we observe and their interaction with the diffuse cluster gas.

M87

M87 is not properly a giant elliptical galaxy and is not a cD. It is neither much brighter than the other nearby galaxies in Virgo-M86

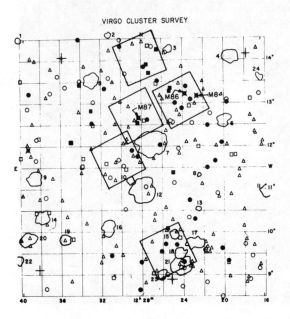

VIRGO CLUSTER SURVEY

Fig. 1.   The fields surveyed with the IPC (squares) are shown superimposed on the catalog of Zwicky, Herzog, and Wild (1961).
Galaxies are indicated by the small symbols and background
clusters are outlined with solid lines.

or NGC4472, nor does it have a large, extended optical envelope. It does, however, have one of the highest internal velocity dispersions of the Virgo galaxies (Faber and Jackson 1976) indicating a larger mass than its neighbors. Also it has a recession velocity which is almost precisely that of the Virgo cluster mean. This puts it in the unique position of being nearly at rest, at least in the line of sight, and near the center of the cluster. Thus it is in an ideal position to grow through the cannibalism of smaller galaxies by larger ones as described by Ostriker and Tremaine (1976) and Hausman and Ostriker (1978).

The X-ray image of M87 is shown in Figure 2. The optical galaxy is fairly small with a diameter of only a few arc minutes, so most of the X-ray emission is beyond the optically bright portion of the galaxy. The emission is remarkably symmetrical. There is no evidence for any differences from a purely radial distribution.

Dan Fabricant, Paul Gorenstein, and Mike Lecar have used the X-ray observations of M87 to determine the distribution of the gas

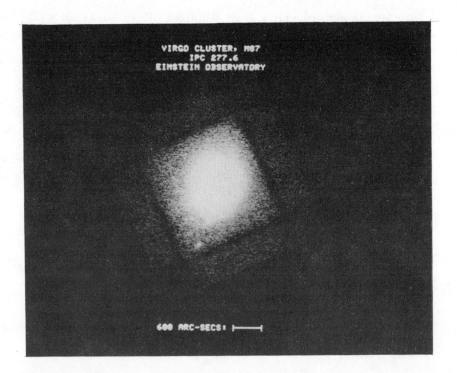

Fig. 2.   The X-ray image of M87 is shown as seen by the Einstein Observatory moderate resolution detector, the IPC. There is no evidence of a point source in this image which has about 1 arc minute resolution.

and to determine the potential in which it is located and the mass
needed to define this potential. The following discussion is based on
their work.

Figure 3 shows the radial distribution of X-ray surface
brightness derived by Fabricant, Gorenstein, and Lecar. It is made
in the energy range 0.7 to 3.0 keV. The dynamic range covered is almost 3 orders of magnitude.

Fabricant, Gorenstein, and Lecar also investigated the spectral behavior of the diffuse emission as a function of radius. This analysis was complicated by the existence of the hard cluster component
which contributed a portion to the background which needed to be subtracted to derive the component associated with M87. In spite of this
difficulty it was found that there was no marked change in spectrum
from the central regions to as far as 50 arc minutes from M87.

The next step involves an assumption -- hydrostatic equilibrium. First, any large outflow of gas would require substantial re-
plenishment of gas which is difficult. The isothermal nature argues

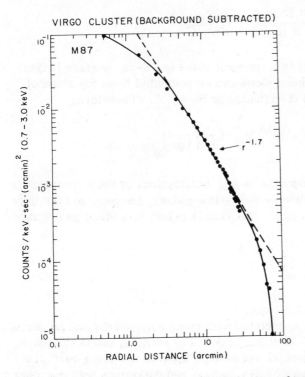

Fig. 3. The surface brightness distribution of M87 in the energy range
0.7 to 3.0 keV is shown. The dynamic range is almost 3 orders of magnitude

that there is not substantial inflow which would probably produce a temperature gradient. Furthermore, free fall or free expansion would tend to produce a steeper density profile than that observed.

Let us then assume hydrostatic equilibrium:

$$\frac{d\,P_{gas}}{dr} = -\frac{G\,M(r)\,\rho_{gas}}{r^2} \tag{1}$$

where P_{gas} and ρ_{gas} are the pressure and density of the X-ray emitting gas. The gas is sitting in a potential well defined by matter having a radial distribution $M(r)$. Then substituting into equation (1)

$$P_{gas} = \frac{\rho_{gas}}{\mu\,M_H}\,k\,T_{gas} \tag{2}$$

and since $T_{gas} \approx$ constant we can rearrange equation (1) as follows:

$$M(r) = -\frac{d\,\ln \rho_{gas}}{d\,\ln r}\,\frac{k\,T_{gas}}{\mu\,M_H\,G}\,r$$

But $d\,(\ln \rho_{gas})/d\,(\ln r)$ is a constant since the X-ray surface brightness is a power law whose slope can be computed from the slope of the surface brightness distribution to be ~ 1.3. Therefore,

$$M(r) = 1.0 \times 10^{13}\,M_\odot\,\left(\frac{r}{100\text{ kpc}}\right)$$

Figure 4 shows the radial distributions of the 3 components of M87 we have described -- the visible galaxy, the gas, and the dark halo needed to bind the gas. The visible galaxy was taken as a King model

$$\rho(r) \propto \left(1 + \frac{r^2}{a^2}\right)^{-3/2}$$

with a core radius $a = 0.84$ kpc.

The masses associated with these components can be derived. The visible galaxy is $\sim 10^{12}\,M_\odot$, the X-ray emitting gas is the same, while the dark halo material has a mass between $1 - 7.5 \times 10^{13}\,M_\odot$ (depending on the choice of outer radius) and dominates both the other components.

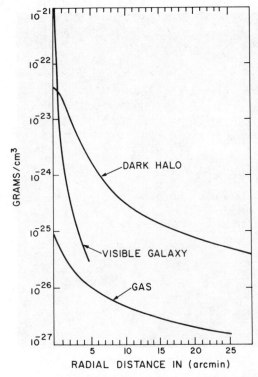

Fig. 4.   The radial distribution of the visible material, the X-ray gas,
          and the derived distribution of the dark material inferred from
          the X-ray observations.

        One can consider this massive halo in at least two ways:
first, if one generalizes to assume it is present in all galaxies, but only
observable when gas is present to permit its observation, then it represents much of the missing mass needed to bind clusters. Also, it is the
halo material needed to stabilize disks which Ostriker and Peebles
(1973) have suggested is required.
        Alternatively, one could interpret this halo material as being
unique to M87. Its origin has to do with the unique situation of M87 at
rest in the center of a cluster where it is growing by the accretion and
cannibalism of other galaxies. As discussed in the chapter by C. Jones,
the Virgo cluster is not very evolved dynamically. Therefore, M87
might be an early stage of a cD galaxy in the process of developing its
outer halo.

Galaxies in the Virgo Cluster

        Figure 5 shows the X-ray image of the field containing M86
and M84. Five X-ray sources are visible. All but the source at the

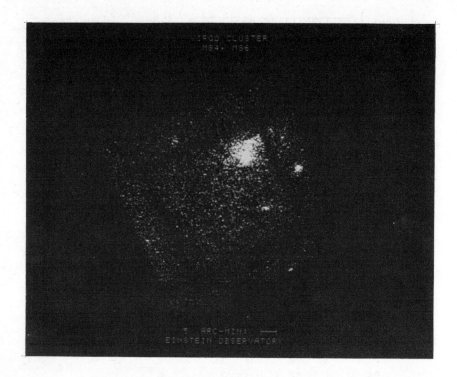

Fig. 5.   The X-ray image of the field containing M86, M84, and other
          bright galaxies. Of the five sources in the field, all but the
          one at the bottom right are galaxies in the Virgo cluster.

bottom right are galaxies in the Virgo cluster. The galaxies detected
in this field are listed in Table 1. As Table 1 shows each of these
galaxies exhibits some peculiarity which could account for its higher
than expected X-ray luminosity (M31 and our galaxy have an X-ray
luminosity of $\sim 2 \times 10^{39}$ ergs/sec in the energy band 0.5 to 3.0 keV).
However, it may be that these various peculiarities play little or no
role in explaining the enhanced X-ray luminosity as is noted later.

        Figure 6 is a superposition of the X-ray contours on the op-
tical photograph made with the Kitt Peak 4 meter telescope showing
pictorially the identifications of the X-ray sources with the optical
galaxies. All the galaxies we have observed, with the exception of M86,
are consistent with point sources -- sizes less than 1 arc minute. We
have planned high resolution observations of M84 to see whether or not
the emission is all from the nucleus or whether it too is slightly extended.

TABLE 1

Galaxy	L_x (.5 - 3.0 keV) 10^{39} ergs/sec	m (1)	Galaxy Type (1)	Radial Velocity (km/sec) (1)
M86 [c]	70.8 $\pm$ 3.8 [a]	10.1	E 3	-419
M84 [d]	26.0 $\pm$ 2.9	10.4	E 1	854
NGC4388 [e]	16.9 $\pm$ 2.5	12.0	SA	2535
NGC4438 [f]	4.5 $\pm$ 1.4	11.0	SA	182
NGC4477	11.1 $\pm$ 1.3	11.4	SBO	1190
NGC4459	11.5 $\pm$ 2.4	11.6	SAO	1039
NGC4473	4.6 $\pm$ 1.4 [b]	11.2	E 5	2205
NGC4424	2.5 $\pm$ 0.4	12.3	SB	358
IC3510	7.0 $\pm$ 2.1	15.2 [(2)]		

a   Value in a 2.5 arc minute square region. The total luminosity is 2×10^{41} ergs/sec.

b   Source near rib structure. Observed flux is therefore only a lower limit.

c   Extended X-ray source.

d   Radio source 3C272.1.

e   Strong, narrow emission lines.

f   Interacting galaxy, Arp 120.

References: (1) de Vaucouleurs and de Vaucouleurs (1963)
                   (2) Zwicky, Herzog, and Wild (1961)

_____

Figure 7 shows an optical spectrum of NGC4388 obtained by W. Liller. It has an optical spectrum very much like that of other narrow emission line X-ray galaxies. Most prominent are the forbidden [O III] lines. Also present but weaker are $H\alpha$, [N II], and [S II]. Ward et al. (1978) had suggested that narrow emission line galaxies were a bridge between normal galaxies like M31 and the luminous type I Seyferts with luminosities of 10^{43} and above. This galaxy, if it is a member of this class, has the smallest luminosity -- 1.6×10^{40} ergs/sec -- compared to NGC2992 whose luminosity is in excess of 10^{43} ergs/sec.

Much of the remaining discussion will concentrate on the observation of the extended X-ray emission surrounding M86. Figure 8 shows the extended nature of this emission which you saw on the earlier reproductions of the X-ray images. We fitted this distribution to an isothermal sphere (which fits the observations adequately). The core radius we determine for the density distribution is $\sim$ 3 arc minutes.

The spectrum of the emission is very similar to that of M87, although it appears somewhat cooler. We do see a strong contribution from the Fe XVII-XXIV line complex which is similar to that seen in

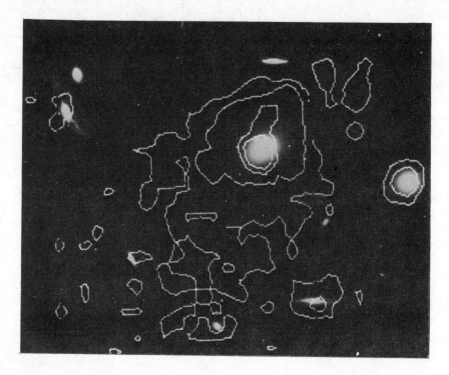

Fig. 6. The X-ray contours derived from the observation of Fig. 5
are shown superposed on the optical photograph.

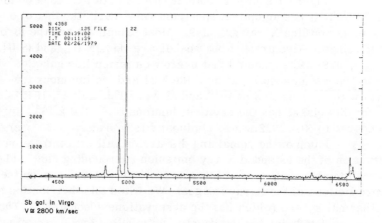

Fig. 7. The optical spectrum of NGC4388 obtained by W. Liller
shows strong emission lines characteristic of objects such
as NGC2992.

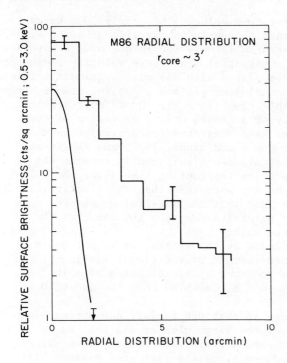

Fig. 8.   The radial distribution of M86 is shown along with that of a known point source.

M87 and indicates the gas is by no means primordial.

With this basic information and the total observed luminosity of 2×10^{41} ergs/sec we find the central particle density to be about $4 \times 10^{-3} cm^{-3}$. The core mass is $2 \times 10^{9} M_{\odot}$ and the total mass within 60 kpc is $\sim 6 \times 10^{9} M_{\odot}$. Alternative radial distributions, e.g. a galactic wind with a slower radial falloff, could change the total mass by a factor of 2.

Before continuing the discussion of M86, I would like to mention briefly the hot cluster gas. In the same field as M86 we find the X-ray surface brightness to be much higher than in fields outside the cluster. Taking a core radius for the cluster of 500 kpc, we find that this surface brightness is equivalent to a density of $\sim 6 \times 10^{-4} cm^{-3}$. This is the same value one gets by subtracting the M87 contribution from the total emission seen by UHURU in the $5^{\circ} \times 5^{\circ}$ collimator.

We now consider the evolution of M86 in such an environment. This work has been done in collaboration with Christine Jones, Andy Fabian, and Joseph Schwarz.

The notable peculiarity of M86 is its very high velocity of approach of almost 500 km/sec (see Table 1). Combining this velocity of approach with the 1000 km/sec recession velocity of the Virgo cluster we conclude that M86 is moving toward us

through the Virgo cluster with a velocity of nearly 1500 km/sec.
The direction of motion must be nearly in our line of sight since
the cluster velocity dispersion is only 666 km/sec and therefore
the galaxy cannot have a substantial transverse velocity component.

The gas we observe associated with M86 must originate within
the galaxy itself. The hot cluster gas has a cooling time substan-
tially in excess of a Hubble time ($\tau_c = 2 \times 10^3 \ T^{\frac{1}{2}}/n$ yrs) which
eliminates the possibility of pressure driven accretion of the
hot gas as various authors have suggested for accretion in much
denser cluster media (see Cowie and Binney 1977, and Fabian and
Nulsen 1977). Nor can the galaxies significantly enhance the
local density, thereby reducing the cooling time since their internal velocity dispersions are much less than the cluster velocity
dispersion. Finally, the fact that the material we observe contains iron makes the most natural assumption the one that the
material is produced by the galaxy.

While the gas is produced by the galaxy, it is not bound by
the visible material we see there. Gravitational binding of the
outer portions of the gas requires a galactic mass of $\sim 10^{13} \ M_\odot$.
A more reasonable mass of $10^{12} \ M_\odot$ can bind only the inner 10 to
15 kpc.

The picture we propose is that M86 is just now approaching
the far side of the core of the Virgo cluster and that as it
crosses the core where it sees a relatively dense intracluster
medium the gas it has produced since its last core passage will
be stripped by ram pressure.

We can verify that such a scenario is possible. First, we
compute the radius r_{max} to which M86 will rise against the cluster
potential. To do this we assume that the galaxies follow an isothermal sphere distribution and that their velocity dispersion
can be used to estimate the cluster mass. With these assumptions
we find that M86 can rise to 3 cluster core radii or about 1800
kpc from the cluster center.

Next, we computed the time for M86 to traverse the cluster
from the center to r_{max} and back. This is about 5×10^9 years.
If we take M86's mass as $10^{12} \ M_\odot$ and a specific mass loss rate
of $10^{-12} \ M_\odot \ yr^{-1}$ then during the time when M86 is in a very low
density medium the mass built up would be about 1 $M_\odot$/year or
$5 \times 10^9 \ M_\odot$. This gas then is stripped in the next 5×10^8 years
which is the time it takes M86 to cross the cluster core. The
stripping occurs in the denser core region since here the ram
pressure exceeds the force holding the gas in the galaxy. Furthermore, the time scale for stripping is such that the gas is
stripped during a core crossing time.

While the bulk of the material will be stripped, the central
core having the highest density could remain bound and co-moving
with the galaxy. This remaining core may be what we are observing in the other galaxies. The spectrum of M84, the radio galaxy observed in this same field, has an X-ray spectrum very
similar to that of M86. Future observations will allow us to

Fig. 9.   An optical photograph of the Centaurus cluster (taken
          from a photograph made with the 4 meter CTIO telescope).
          This cluster has properties intermediate between those
          of Virgo and the rich evolved clusters like Coma.

tell if the emission centered on this other galaxy is only the
remains of a larger M86-type system, or is a true point source
with enhanced nuclear X-ray emission.
        C. Jones has described the X-ray observations of a wide
variety of X-ray clusters made with the Einstein X-ray Observatory
(another paper in this volume). These observations fit in well
with the dynamical evolutionary scenario described by Peebles
(1970), White (1978), and Perrenod (1978). In this dynamical
type of picture the Virgo cluster would be at a relatively early
phase of evolution. The evidence for this includes its low
velocity dispersion (666 km/sec compared to 900 for Coma), its
irregular shape, and its high fraction of spiral galaxies. The
X-ray characteristics of the early evolutionary stages include
lower gas temperatures, emission clumped around individual galaxies, low X-ray luminosities, and asymmetrical gas distributions.
These properties also characterize the Virgo cluster. As the
cluster evolves and relaxes dynamically its central potential
will grow as will the density of gas in the core. This gas will
then increase its ability to strip any gas associated with
galaxies.

Figure 9 is a reproduction of a photograph of the Centaurus cluster. It too is probably in an early evolutionary phase having a lower velocity dispersion, a cooler X-ray temperature, and an irregular shape as determined from optical galaxy counts. However, the dominant galaxy NGC4696 has been classified as a true giant elliptical galaxy. Observations in the near future of this cluster should help us place the time of the development of a cD galaxy in relation to the other evolutionary indicators.

We must still resolve the question as to whether the large mass of M87 results from its unique position in the cluster and is an indication that it too will eventually become a bonafide cD galaxy, or whether the extra dark mass is characteristic of all galaxies. The future observations should help us begin to resolve some of these new questions.

Acknowledgements: I would like to thank Drs. Dan Fabricant, Paul Gorenstein, and Mike Lecar for discussions of their results on M87 and for permitting me to present them in advance of publication. I would also like to acknowledge the contributions of Drs. Christine Jones, Joseph Schwarz, William Liller, and Andy Fabian to the paper upon which the discussion of the Virgo cluster galaxies is based. I would like to thank Mary Twomey for her careful work in preparing this manuscript.

REFERENCES

Bahcall, J. and Sarazin, C.L., 1978, Astrophys.J., 219, 781.
Bradt, H., Mayer, W., Naranan, S., Rappaport, S., and Spada, G., 1967, Astrophys.J.(Letters),150, L199.
Byram, E.T., Chubb, T.A., and Friedman, H., 1966, Science,152, 66.
Cowie, L.L., and Binney, J., 1977, Astrophys.J.,215, 723.
Davison, P.J.N., 1978, Mon.Not.Roy.Astr.Soc.,183, 39p.
Faber, S.M., and Jackson, E.R., 1976, Astrophys.J.,204, 668.
Fabian, A.C., and Nulsen, P.E.J., 1977, Mon.Not.Roy.Astr.Soc.,180, 479.
Fabricant, D., Topka, K., Harnden, F.R., Jr., and Gorenstein, P., 1978, Astrophys.J.(Letters),226, L107.
Gorenstein, P., Fabricant, D., Topka, K., Tucker, W., and Harnden, F.R., Jr., 1977, Astrophys.J.(Letters),216, L95.
Hausman, M., and Ostriker, J., 1978, Astrophys.J.,224, 320.
Lawrence, A., 1978, Mon.Not.Roy.Astr.Soc.,185, 423.
Malina, R., Lampton, M., and Bowyer, S., 1976, Astrophys.J.,209, 678.
Mathews, W.G., 1978, Astrophys.J.,219, 413.
Mushotzky, R.F., Serlemitsos, P.J., Smith, B.W., Boldt, E.A., and Holt, S.S., 1978, Astrophys.J.,225, 21.
Ostriker, J.P., and Tremaine, S.D., 1975, Astrophys.J.(Letters), 202, L113.
Ostriker, J.P., and Peebles, P.J.E., 1973, Astrophys.J.,186, 467.

Peebles, P.J.E., 1970, Astron.J., 75, 13.
Perrenod, S.C., 1978, Astrophys.J., 226, 566.
Serlemitsos, P.J., Smith, B.W., Boldt, E.A., Holt, S.S., and
Swank, J.H., 1977, Astrophys.J.(Letters), 211, L63.
de Vaucouleurs, G. and de Vaucouleurs, A., 1963, Astron.J., 68, 96.
Ward, M.J., Wolson, A.S., Penston, M.V., Elvis, M., Maccacaro,
T., and Tritton, K.P., 1978, Astrophys.J., 223, 788.
White, R.A., 1978, Astrophys.J., 226, 591.
Zwicky, F., Herzog, E., and Wild, P., 1961, Catalogue of Galaxies
and of Clusters of Galaxies, (California Institute of Technology),
Vol. 1.

CLUSTER PARAMETERS[1]

Guido Chincarini

Department of Physics and Astronomy
University of Oklahoma
Norman, Oklahoma 73019 USA

Introduction

        Structural and morphological parameters of clusters of
galaxies should reflect the dynamical and evolutive status of a
cluster. The case is, however, that in answering such simple
questions as: What is the shape of a cluster and what is the
size of a cluster?, we find technical and intrinsic difficulties.
I am not sure they will be easily overcome unless we look at the
problem in a less classical or model dependent way. The technical problems are easily overcome and with the most recent digitized microdensitometers and data analysis it is already possible
to have high quality data.
        Structural parameters are obtained from counts of
galaxies. The published counts can be divided into three categories:[2]
A. Counts obtained on prints of the Palomar Sky Survey or photo-
        graphic plate. Such counts, while quite useful, have gen-
erally one limitation and one bias at the very beginning: in the
data. (1) The counts are often made to some not very well de-

_____

        [1]Lecture given in Erice, July 1979, International
School of Astrophysics "Ettore Majorana" Centre for Scientific
Culture.
        [2]I refer only to counts done in clusters of galaxies.
Counts over large areas of the sky are generally used for other
purposes either because they lack the resolution needed to determine the cluster profile [Shane and Wirtanen, 1967], or because they do not go very faint [Zwicky et al., 1963], or because
they are limited to a small region of the sky [Rudnicki et al.,
1973].

R. Giacconi and G. Setti (eds.), X-Ray Astronomy, 197-216.

fined limiting magnitude (the ideal would be to have the number
of galaxies at different limiting magnitudes), and (2) after a
cluster center has been found (maximum number density on strip
counts for instance) the galaxies are counted using concentric
rings. Symmetry is forced in the system.
B. On well calibrated plates counts are done together with the
    photometry of cluster galaxies. The only limitation here is
that the counts generally do not extend to the faintest objects
and are limited to a fairly small area.
C. Automatic counts. By this I refer to the most recent tech-
    niques where the plate is digitized and the galaxies recog-
nized directly by computer. In such a way it is possible to have
to the plate limit magnitudes and coordinates for each object.
The data are still virgin.
        Of course the above sequence is a sequence in time and
technology. Since a rather large amount of data exist under
category A, we will try to discuss some of these data in view of
their relevance to the x-ray observations of the intergalactic
gas.
        The lecture is organized as follows: In Section I, I
will discuss the cluster environment; in Section II the parameters which can be derived using the isothermal, or King, profile.
In Section III I will give structural parameters derived from
some of the published data. The Appendix will deal with the
richness parameter.

I. The environment of clusters of galaxies

        It is customary to correct counts of galaxies for
"background". The procedure is generally to choose a region
"far enough" from the cluster to estimate the "homogeneous" background. The operation is fair and meaningful in first approximation, however it could be very incorrect (physically it is) because clusters are not islands in a sea, as stars against the
sky background. Evidence is that clusters are in superclusters
and, so far, no sharp boundary has been found. The space distribution of visible matter is beautifully contained in the spatial
pair correlation function (Peebles and Hauser, 1974),

$$\delta_p(r) \equiv n^2 [1 + \xi(r)] \delta v_1 \delta v_2$$

where $\xi(r) \propto (r/r*)^{-1.8}$.

        The distribution of galaxies is represented in statisti-
cal terms by a covariance function in a stationary random process
and therefore it is an invariant with respect to changes in the
origin. In other words, the autocorrelation function not only is
a consequence of the clustered distribution of galaxies, but contains in it the assumption that the universe is homogeneous and
isotropic. (For an excellent review on the subject, see Fall,

1979.)   What we observe in the Coma/A1367 supercluster is, in
spite of important and large asymmetries, a number density distribution

$$\nu(r) \propto r^{-\alpha} \; ; \quad \alpha = 2.3 \pm 0.2$$

[Chincarini and Rood, 1976], see Fig. 1.

Theories and models, therefore, should take such a distribution into account rather than idealize a model "isolated
cluster-background". Note that we really have no evidence of
discontinuity on the number density distribution. The cluster
slowly fades in the supercluster. This may raise questions: Why
at large distances from the cluster center the radial dependence
still simulate, even within large observational uncertainties,
an isothermal distribution. How would this compare with the density distribution of a non-bound density enhancement?

In addition to the immediate environment, the supercluster, counts of cluster galaxies may be affected by the presence
of a background cluster projected. Such probability can be computed by the relation:

$$P = 1 - e^{-h^* \Omega}$$

where h* is the cumulative number of clusters per unit solid angle with the 10th brightest galaxy brighter than m, the limiting
magnitude of the sample. The center of the background cluster is
supposed to fall in the solid angle Ω. Assuming similar population for object and background cluster the resulting luminosity
function is represented in Fig. 2. Obviously structural parameters may be affected in various ways.

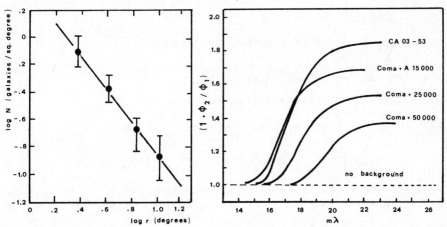

Fig. 1. Number density distribution in the outskirt of the
Coma cluster; $\nu(r) = r^{-2.3 \pm .2}$.
Note however that the distribution of galaxies is asymmetric.

Fig. 2. Variations of the luminosity function due to the
presence of a background cluster.

These and other effects will distort the derivation of parameters which are based on symmetric models so that, while we may use them until we find better ways of analysis, we must be aware of their intrinsic limitations. Nevertheless, idealized models may give some indications on the structure of clusters and, in what follows, I derive parameters which are related to an isothermal, or similar, cluster. The theory of the isothermal sphere is well described, for instance, by Chandrasekhar [1942], and the application to cluster of galaxies in Zwicky [1957]. An excellent review has been published by Bahcall [1977] and a comprehensive application to the Coma cluster can be found in the analysis by Rood et al. [1972].

II. Structural parameters and mass-to-luminosity ratio

Let's consider a spherically symmetric distribution of mass in equilibrium. Using Poisson's equation (only the radial coordinate remains) we have:

$$\nabla^2 \Phi = \frac{1}{\xi^2} \frac{d}{d\xi} (\xi^2 \frac{d\Phi}{d\xi}) = -4\pi G\rho \tag{1}$$

The potential is determined by the mass distribution. Assuming hydrostatic equilibrium

$$\frac{dP}{d\xi} = -g\rho \tag{2}$$

and the equation of gas

$$P = \rho \frac{kT}{\mu} \tag{3}$$

we derive, in terms of density, the equation

$$\frac{d}{d\xi} (\frac{\xi^2}{\rho} \frac{d\rho}{d\xi}) + 4\pi G \frac{\mu}{kT} \xi^2\rho = 0 \tag{4}$$

or, using dimensionless quantities:

$$\xi' = \frac{\xi}{\beta} ; \quad \rho_1(\xi) = \frac{\rho(\xi)}{\rho(0)} ; \quad \frac{4\pi G\mu}{kT} \beta^2\rho_0 = 1 \tag{5}$$

$$\frac{d}{d\xi'} [\xi'^2 \frac{d}{d\xi'} \ln \rho_1(\xi')] + \xi'^2 \rho_1(\xi') = 0 \tag{6}$$

From (6) and from the relation between temperature of the gas (a gas of galaxies in this case) and velocities of the particles:

$$\frac{1}{3} <v^2> = \frac{kT}{\mu}$$

we derive for the central density the equation:

$$\rho_0^* = \frac{<v^2>}{12\pi G\beta^2} \tag{7}$$

In other words from the knowledge of the velocity dispersion $<v^2>$ and of the structural length, β, two observational parameters (of course assuming clusters fit such a model), we can derive the central density in grams cm^{-3}.

Tables exist for the numerical solution of eq. (6) and its projection on the celestial sphere. For small radii, however King has shown that a good approximation is given by the relation

$$\rho = \rho_0 [1 + (\frac{\xi}{R_c})^2]^{-3/2} \tag{8}$$

where $R_c = 3\beta$.

Equation (8) at large radii is much steeper than the isothermal. The use of equation (8) eases the visualization of the projection on the celestial sphere

$$r^2 = x^2 + y^2$$
$$\xi^2 = r^2 + z^2$$

$$\sigma(r) = 2\rho_0 \int_0^\infty (1 + \frac{r^2+z^2}{R_c^2})^{-3/2} dz \simeq 2.0\rho_0 R_c (1 + \frac{r^2}{R_c^2})^{-1} \tag{9}$$

where σ is the projected density, from which we derive

$$\frac{\sigma(r=R_c)}{\sigma(r=0)} = \frac{1}{2}$$

and R_c is called the core radius $\equiv$ the distance from the center at which the projected density is 50% of the central density. When we use either the isothermal or King's formula to fit the observational data (these are generally in the form of counts per square degree as a function of the distance from the cluster center in minutes of arc) we have some free parameters which need to be optimized. In particular we can write the density as:

$$\sigma = \alpha q(r/\beta) + \gamma \tag{10}$$

with $q(r/\beta)$ the non-dimensional projected density (tabulated), α a normalization constant which depends on the population of the cluster, and γ a quantity which is a function of the "background" and eventual cut-off radius. Once α, β and γ are determined (for instance minimizing the χ^2 function) we have, after conversion of units, σ_0 in galaxies per Mpc^2 and from (9)

$$\rho_0 (galaxies/Mpc^3) = \frac{\sigma_0}{6.06\beta}$$

and from (7)

$$\rho_0^* = \frac{3<v^2>}{4\pi GR_c^2} = \frac{9}{4} \frac{<V_{||}^2>}{\pi GR_c^2} \;\; ; \quad \text{if } <V^2> = 3<V_{||}^2> \; .$$

Note that such relations are valid for a variety of profiles and not only for the isothermal sphere. The luminosity distribution of a cluster may be approximated in a similar way. Expressing the projected profile as (Rood et al., 1972)

$$I = I_0 [1 + (R/R_c)^2]^{-1} \; .$$

The space luminosity profile must have the form

$$\lambda = \lambda_0 [1 + (\xi/R_c)^2]^{-3/2}$$

and projecting as in (9) we derive

$$\lambda_0 = I_0/2R_c \; .$$

The observational parameter generally used is:

$$L(corr) = \int_0^{R_c} \frac{I}{[1 + r^2/R_c^2]} 2\pi r dr = \pi I_0 R_c^2 \, \ell n \, 2$$

from which we derive the relation:

$$\lambda_0 = I_0/2R_c = L(corr)/2\pi R_c^3 \, \ell n \, 2$$

and the mass to luminosity ratio

$$(M/L)_{r=0} = \rho_0^*/\lambda_0 \; .$$

Of course the determination of the luminosity of the single gala-xies is not a trivial process, especially if we plan to measure to rather faint magnitudes. Furthermore if we determine λ_0 from L(core) rather than from a luminosity profile we may have some uncertainties due to the small sample or the presence of unusual-ly bright galaxies. In a statistical sense we may use of counts of galaxies and of the cluster luminosity function if the latter is cosmological. At the present we suspect variations from clus-ter to cluster [Dressler, 1978], and between clusters and back-ground [Tarenghi et al., 1980]. Such differences, however, are not yet well established and are, in any case, much finer than the first approximation with which we are dealing here.
    If $\Phi(M)$ is the cluster luminosity function, then

(11) $\qquad n_0(\leq M) = \int_{M_{min}}^{M_{sample}} A \, \Phi(M) dM$ $\qquad\qquad$ (11)

where A is a normalization constant depending on the population of the cluster. In (11) $n_0(\leq M)$ is determined by fitting the number density profile so that we can determine A. We have then

$$I_0 = \int A\Phi(M)L(M)dM .\tag{12}$$

As an example for the Coma cluster we estimate from the figure in Rood et al. (1972), $n_0(\leq M)$ = 2512 down to m $\sim$ 17.9 and we derive, using the method above, λ_0 = 2.46x10^5 L$_\odot$/pc^3 which compares very well with the value $\lambda_0 \doteq$ 2.74x10^5 L$_\odot$/pc^3 derived by the above authors using photometry of galaxies in the cluster core, L(core). Of course the application of the above procedure on a large sample requires at least a fairly homogeneous set of data and the normalization to a standard limiting absolute magnitude, M sample. This seems to be our almost impossible task due to the inhomogeneity of the data in the literature. In some cases the limiting magnitude is poorly known. Recently however good homogeneous samples have been published by Oemler (1973), Dressler (1978). Other high quality data are being produced, see for instance Carter and Goldwin (1979). A useful and large quantity of material, counts, have been produced by Bahcall (1977) and Baier (1978). Such data may give information on a structural parameter, β, which has been often suggested to be a standard meter: Zwicky (1957), Bahcall (1977).

III. The cluster profiles and the core radius

As shown in Fig. 3, the isothermal profile is very well approximated by King's equation in the region near the cluster center. At distances larger than r/β $\simeq$ 10 the projected density

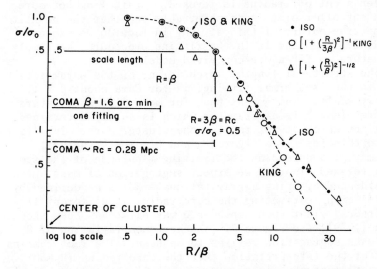

Fig. 3. Comparison between various model profiles. Coma parameters have been also indicated.

is proportional to r^{-1}. I do
not believe we can answer the
question which curve fits better
clusters in general. The χ^2
value, in fact, may reflect more
the distribution of points in
the density-radius plane than
reality. It is of some interest, for instance, to leave in
the equation $(1 + r^2/\beta^2)^n$, n as
a free parameter.

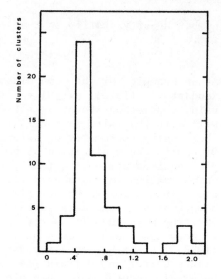

The distribution of
n, Fig. 4, shows a sharp peak
at n=.5 and a minor peak or
tail at larger n. For n=0.5
the isothermal at large distance from the center predom-
inate and for n=1 the King's
approximation near the center.
While we will look into this
carefully it seems that the
value of n depends mainly on
the distribution of the observational points. Perhaps

Fig. 4. Distribution of the
exponent n fitting the cluster
profile by the model:
$$\alpha[1 + (R/\beta)^2]^{-n} + \gamma .$$

data should be homogenized in metric distances before any solution is attempted. I should add that in some cases the minimization of four parameters is unstable. For one thing background
subtraction may become a source of uncertainty for the reasons
described in Section I. Furthermore to limit the mass of the
cluster often a cut-off radius is invoked. Again I am not sure
that such a cut-off radius is an observable and that the quantity
reflects the true nature of things. Why in fact do we expect a
cut-off radius? By using it in the fitting procedure, we simply
may force nature to fit a pre-conceived model.

The structural length, which is one of the parameters
determined in the isothermal fitting, is for Coma about $\beta = 1.6$
arc min, very small. This means that for more distant clusters
the structural length is a parameter which is solely determined
by observations which do not resolve the cluster core. This fact
may add uncertainties. (See Figures 5 and 6.) The structural
length, β, may also depend on the limiting magnitude of the sample. For a relaxed cluster, we expect segregation of mass which
could be evidenced by using a gravitational radius as defined by
Oemler [1973] or by estimating the core radius at different limiting magnitudes. That such dependence may exist has been shown
by Quintana [1979], see Figure 7. Whether such variation of β
are due to mass segregation remains to be seen. The fact remains
independent of the interpretation that the core radius ($R_c=3\beta$)
may be a function of the limiting magnitude of the sample and, if
due to mass segregation, a function of the dynamical evolution of

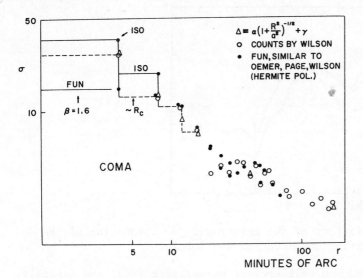

Fig. 5. Fitting of the Coma cluster profile. Observations (*) mark the right end of the bin size.

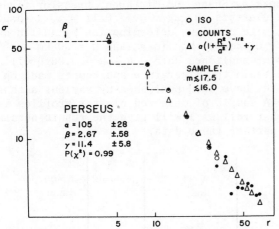

Fig. 6. Fitting of the Perseus cluster profile (counts by Bahcall (1977)).

the cluster. The evidence, however, that a cluster is relaxed through two-pair interaction (energy equipartition) is, to say the least, very meagre. Some evidence exists probably in A194 [Chincarini and Rood, 1977]. However, in Coma the effect is limited to the two brightest galaxies. Segregation in luminosity has been observed by Quintana, Rood and other authors.

Bahcall [1977] who, after Zwicky [1957], drew interest to the structural length, β, finds R_{core} = 0.25 ± 0.04 M_{pc}. The small standard deviation, however, may be due to the fitting

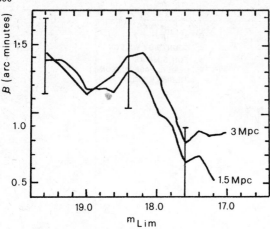

Fig. 7.   Variations of the core radius as a function of the
          limiting magnitude of the sample.

procedure used and be therefore unrealistic [Avni and Bahcall,
1976].

        Using plates obtained at the Lick Crossley reflector
(small field) and counts by Shane and Wirtanen, Dressler finds,
for a set of about 12 clusters, R_{core} = 0.47±0.11 M_{pc}, a result
which is in agreement with previous determination by Austin and
Peach who find R_{core} = 0.38±0.11 M_{pc} ($q_0 = \frac{1}{2}$ and H_o = 50 km sec^{-1}
M_{pc}^{-1}; their analysis of published data: $R_{core} \cong$ 0.26±0.07).   A
large set of data exists in the literature and counts made on
Palomar prints or plates have been published by various authors,
especially by Baier.   A sample of observed cluster profiles is
given in Fig. 8; in what follows I will give the result obtained
by analyzing a large part of these data.

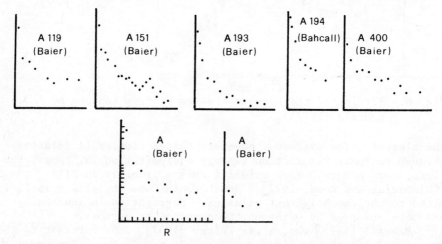

Fig. 8.   A subsample of the data used for this study.

To test our minimization procedure Materne and I determined the isothermal parameters for a set of clusters studied by
Bahcall and shown in Table I. The agreement is generally good,

Table I

Cluster	β (B) Minutes of Arc	β (C-M) *(1)* Minutes of Arc	Error (C-M) Minutes of Arc
A 194	2.3	0.30 *(2)*	0.06
A 401	0.72	0.54	0.15
A 426	2.68	2.67	0.59
A 1132	0.37	0.39	0.01
A 1367	2.70	0.09 *(2)*	0.05
Coma	2.0	1.86 *(3)*	0.29
A 1775	0.73	0.54	0.38
A 1795	0.80	0.76	0.27
A 1904	0.70	0.74	0.44
A 2029	1.0	0.75	0.49
A 2052	1.6	2.89	0.65
A 2065	1.0	1.14	0.29
A 2256	0.7	0.56	0.28
A 2319	0.80	0.42	0.19

(1) Background was left as a free parameter during the fitting.
(2) It is not clear why we obtain such a small value, may be due
to background.
(3) Counts by Zwicky; counts by Wilson give β = 1.66, error = 0.05;
Bahcall's value determined using her own counts.
B ≡ Bahcall; C-M ≡ Chincarini and Materne.

Table II

Cluster	R_c (D) M_{pc}	R_c (C-M) M_{pc}
A 154	.44 (.19)	.11 *(1)*
A 168	.55	.24 *(1)*
A 401	.40	.19 *(2)*
A 1413	.57	.44 *(3)*
A 2029	.68 (.35)	.27 *(2)*
A 2256	.49	.15 *(2)*
A 2670	.31	.28 *(1)*

(1) Counts by Baier.
(2) Counts by Bahcall.
(3) Counts by Noonan.
D ≡ Dressler, counts by Dressler.
C-M ≡ Chincarini and Materne

especially if we take into account that we left all our parameters free. The error on β is, however, fairly large. A set of
clusters had counts both by Dressler and Baier. Generally cluster parameters using counts by various authors are in agreement,
see Table II. The core radius we determined using the counts by
Baier is in fair agreement with the core radius determined by
Dressler. Our results [Chincarini and Materne, 1980] are preliminary since we have not yet reduced all the clusters we found in
the literature. In some cases, we found no acceptable solution.
The present sample is, nevertheless, sizable and the results are
summarized in Fig. 9, where I give the number of clusters per interval Δβ=0.1. It is a fairly spread distribution. If we determine the core radius for clusters at various z, Fig. 10, the lack
of cluster at large z is evident. Of course the measurements become more difficult and we lose resolution toward the cluster
center. Such counts are, nevertheless, needed. A core radius
defined in a somewhat different way has been determined by Bruz-
val and Spinrad [1978] by fitting the equation

$$\sigma(r) = \sigma_0 (1 + r/a)^{-2} + f_0 .$$

Again the core radius seems to be fairly constant from cluster to
cluster. It is however doubtful that, using the data presently

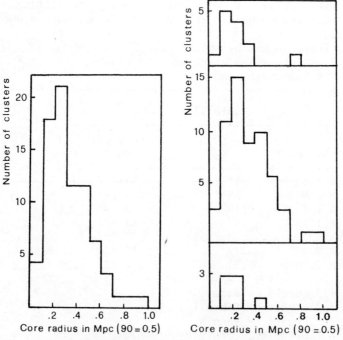

Fig. 9. Distribution of the core radius for the clusters in the
sample studied.

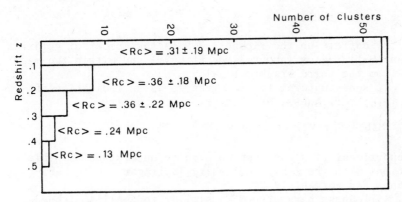

Fig. 10.   Distribution of the sample cluster as a function of
           redshift.   The mean core radius is also marked.

available, we can determine it to an accuracy higher than .15 Mpc.
        The core radius is therefore a fair standard size for
clusters of galaxies.   Whether, however, the core radius as determined for a single cluster is an accurate parameter to be used
in connection with X-ray observations it is uncertain.   We may
have to take into account asymmetries (a fact mentioned also by
Branduardi and Binney at this school) and perhaps find parameters
less sensitive to the observational uncertainties or better reflecting the true nature of a cluster.   In this respect the dynamics itself may have to be taken into account both to determine
evolutionary effects and to account for the interaction cluster-
supercluster.
        A new approach has been proposed by Hickson [1977] who
suggests the use of an harmonic radius

$$(13) \qquad \theta_c = [\frac{2}{N(N-1)} \sum_{i>j} \frac{1}{r_{ij}}]^{-1} \; ; \quad \lambda_c = \frac{cz}{H} \theta_c \qquad (13)$$

which give $\langle \lambda_c \rangle = 1.20$ Mpc.
        Strubble [1978], using computer models prepared by Wie-
len, studies the variation of various parameters as a function of
dynamical evolution.   He shows that the core size so defined,
which is essentially the harmonic radius, is time invariant.
Perhaps the understanding of the structure of clusters of galaxies relies heavily on computer simulations and, to this aim,
the approach by Binney [1979] and the recent possibility to fully
analyze a digitized plate by computer, see for instance Carter
and Goldwin [1979] and Kron [1979], will give new insights to the
problem.   The role played by dynamical evolution in clusters is
discussed, among others, by Hausman and Ostriker [1978].

Appendix.   Abell's Richness

By definition the richness $\equiv$ R is the number of galaxies within an Abell radius (about 3 M_{pc} at H=50) and within two magnitudes from the third brightest galaxy.

It seems natural to use richness as an indication of mass, $M_{virial} \propto$ richness, or, better, of luminosity

$$M/L \propto R_{vir}^2 \cdot V_{vir}^2 / \text{richness} .$$

Such proportionalities can then be used to understand possible correlations with the X-ray luminosity [Solinger and Tucker, 1972; Jones and Forman, 1978].

A richness parameter N_C^{48} similar to Abell's richness has been defined by Sandage and Hardy [1973]. The various richness classes are very well defined so that "richness" is a well defined and determined parameter. The question is: Can we prove that it gives an indication of the cluster mass or luminosity? Is it related to the central density as determined, for instance, using an isothermal fit?

Following the definition, if $\Phi(M)$ is the cluster luminosity function, then

$$(14) \qquad R = \int_{M_3}^{M_3+2} \Phi(M)\,dM \qquad\qquad (14)$$

and the luminosity $L = \int L\Phi(M)\,dM$.

Figure 11 shows that when we measure the richness parameter we are testing only the bright part of the luminosity function; our estimate is based on at most 40% of the total luminosity. We judge therefore the cluster population by looking at the

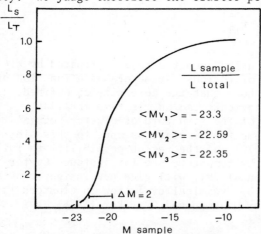

Fig. 11.   Fraction of the sampled luminosity function versus absolute limiting magnitude of the sample.

tip of the iceberg. Any variation of the luminosity function
could cause large differences between the real and estimated
counts.

Sandage and Hardy [1973] studied the variation of the
first, second and third ranked galaxy in clusters as a function
of the Bautz and Morgan class and as a function of the Abell's
richness. They find that the third brightest galaxy is a strong
function of the Bautz and Morgan class ($\Delta M \approx 0.8$, between Class I
and Class III). After a correction is applied for such correlation, the magnitude of the third brightest galaxy still correlate
with the richness class, between richness class 0 and richness
class 4 the effect is somewhat larger than 1 magnitude, see Fig.
12. The latter effect works in the sense that for rich clusters,
class 4, the third brighter galaxy becomes brighter. If this is
a statistical fluctuation of the bright end of the luminosity
curve we would underestimate the real population of the cluster,
see equation (14). The effect therefore should eventually emphasize any correlation with total mass or luminosity. The point,
however, I want to stress is that the third brightest galaxy is
a function of the morphology and other parameters so that the
definition of richness is cluster dependent.

Finally variations in the luminosity function from cluster to cluster as suggested by Dressler [1978] may confuse the
physical significance of the parameter richness. Such variations
are visible in Fig. 12. More work is needed in this direction on
a fairly homogeneous sample of clusters. The evidence is not yet

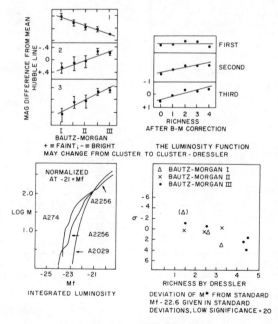

Figure 12.

conclusive one way or the other and we still have disagreement
among various authors. Of course the problem is of basic importance, also in relation to the formation and evolution of clusters.

Oemler's sample [1973] shows a correlation between
Abell's richness class and the cluster luminosity. To a large
extent such correlation is based on the luminosity of the cluster
A 665, the only cluster of richness class 5. Such correlation
does not exist in Dressler's sample; in particular the cluster
A 665 is not as luminous as measured by Oemler. The four clusters
in common between Oemler and Dressler are shown in Fig. 13. The
two different estimates for A 665 are critical in this, and any
other, correlation. As pointed out by Dressler, Abell's richness
does not correlate with the central number density either, see
Fig. 14. The correlation with virial mass could be hidden by the
inhomogeneity of the published data. The virial mass is given by

$$M_{VT} = \frac{1}{G} V^2_{vir} R_{vir}$$

where

$$V^2_{vir} = \frac{3}{ML}\left\{\Sigma m_i (V_{oi} - V_{cm})^2 - \Sigma m_i \sigma^2_{err_i}\right\}^2$$

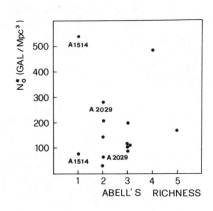

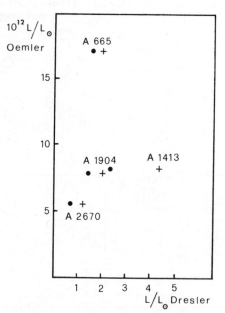

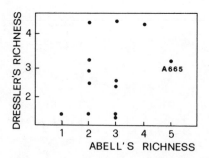

Fig. 13. Oemler's versus
Dressler's total luminosity
• $\dot{L}_{core}/10^{12}$; + $L_{tot}/10^{13}$ (Dressler).

Fig. 14. Abell's richness
class versus central number
density.

is the square of the velocity dispersion, V_{cm} the center of mass corrected for dispersion due to the errors in redshifts, σ_{err}.

For the Coma cluster we have a large amount of redshifts within 3 degrees from the center. The velocity dispersion decreases up to a distance of about 1.5 M_{pc} to increase again at about 6 M_{pc} from the center, Fig. 15, a consequence probably of the influence by supercluster galaxies. The virial radius is a harmonic radius weighted by mass (or by luminosity). It is a function, therefore, of the distribution of galaxies and of the area over which it is computed. In general, observations of clusters refer to different areas and in some cases, due to limitations on observing time, the sample is not complete. In general, however, the mean velocity dispersion is determined fairly well and the correlation with Abell's richness is shown in Fig. 16. Similar results have been discussed by Danese, De Zotti and Di Tullio. It is unfortunate we have no observations for clusters of richness class 4 and only a few for richness class 3 since we

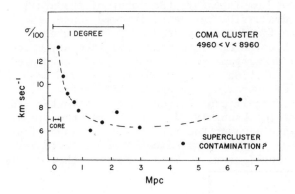

Fig. 15. Velocity dispersion versus distance from the center of the Coma cluster of galaxies.

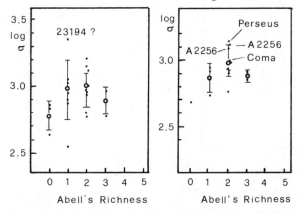

Fig. 16. Velocity dispersion versus richness class.

can not really say whether we reach a maximum at richness class 2 or the low dispersion observed in richness class 3 is a statistical fluctuation due to the small sample.

A rather good indication on the virial radius may be obtained from the sample by Hickson who measured the harmonic radius, equation (13), of about 50 clusters.

Within the measured error, Fig. 17, the harmonic radius is constant. We conclude, therefore, that if richness correlates with the virial mass, such correlation is similar to the correlation between Abell's richness and velocity dispersion. Due to the importance and frequent use of the richness parameter, it is of primary importance to understand its physical meaning, if any, and to extend and improve, therefore, the available observations.

Conclusions

Aim of the above discussion was to evidence limitations in the classical determination of cluster structural parameters and make, even if in a not complete way, the point of the available observations. The discussion reflects, in some sense, the rapid improvements in the field and the need of using state of the art reduction facilities in order to reach the accuracy which is required for a better understanding of clusters and the use of structural parameters in relation to X-ray observations. The very recent computer model approaches [Binney, 1979; Strubble, 1979] seem to be the most significant in conjunction with the use of new cluster parameters [Hickson, 1977]. In general the cluster models should take into account their environment, superclusters, and the dynamics and kinematics studied in relation to it. Richness, as defined by Abell, seems to be a parameter that we

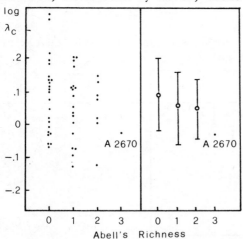

$< \lambda_c > = 1.17$ Mpc $($ Ho $= 55$ Km sec^{-1} Mpc$^{-1})$

Fig. 17. Hickson's harmonic radius versus richness.

use quite often, we know what it should mean and yet we do not fully understand it observationally. Whether this is due to lack of resolution in our observations, lack of accurate and homogeneous data or to intrinsic factors is, at the present, hard to determine. We badly need to know the velocity dispersion of a sizable group of clusters of richness class 3 and possibly of a sample of clusters of richness class 4. Fortunately a fair number of clusters is under analysis by various scientists and the photometric results from digitized plates will be soon available.

Acknowledgements

This work was prepared during my stay at ESO, Geneva. It is a pleasure to thank Dr. L. Woltjer for computing facilities and discussions.

References

Austin, T.B., and Peach, J.V., 1974, M.N.R.A.S. 167, 437.
Avni, Y., and Bahcall, N., 1976, Ap.J. 209, 16.
Bahcall, N., 1977, Ann. Rev. Astron. Astrophys. 15, 505.
Baier, F.W., 1978, Astron. Nachr., Bd. 299, 311, and references therein.
Bautz, L.P., and Morgan, W.W., 1978, Ap.J. 222, 23.
Binney, J., 1979, this volume.
Bruzual, G., and Spinrad, H., 1978, Ap.J. 220, 1.
Carter, D., and Goldwin, J.G., 1979, M.N.R.A.S. 187, 711.
Chandrasekhar, S., 1942, Principles of Stellar Dynamics, New York, Dover.
Chincarini, G., and Materne, J., 1979, in preparation.
Chincarini, G., and Rood, H.J., 1972, Ap.J. 206, 30.
Chincarini, G., and Rood, H.J., 1977, Ap.J. 214, 351.
Dressler, A., 1978, Ap.J. 223, 765.
Fall, M.S., 1979, Rev. Mod. Phys. 51, 21.
Hausman, M.A., and Ostriker, J.P., 1978, Ap.J. 224, 3.
Hickson, P., 1977, Ap.J. 217, 16.
Jones, C., and Forman, W., 1978, Ap.J. 224, 1.
King, I.R., 1972, Ap.J. 174, L123.
Kron, R., 1979, preprint.
Oemler, A., 1973, Ph.D. Thesis, California Institute of Technology.
Peebles, P.J.E., 1974, Astron. Astrophys. 32, 197.
Peebles, P.J.E., and Hauser, M.G., 1974, Ap.J. Suppl. 28, 19.
Quintana, H., 1979, A.J. 84, 16.
Rood, H.J., Page, T.L., Kintner, E.C., King, I.R., 1972, Ap.J. 175, 627.
Rudnicki, K., Dworak, T.Z., Flin, P., Baranowski, and Sendrakowski, A., 1973, Acta Cosmologica 1, 7.
Sandage, A.R., and Hardy, E., 1973, Ap.J. 183, 743.

Shane, C.D., and Wirtanen, C.A., 1967, Pub. Lick Obs., Vol. 22,
        Part I.
Solinger, A., and Tucker, W., 1972, Ap.J. 175, L107.
Strubble, M.F., 1979, A.J. 84, 27.
Tarenghi, M., Chincarini, G., Rood, H.J., and Thompson, L., 1980,
        Ap.J., January 15.
Zwicky, F., 1957, Morphological Astronomy, Berlin, Springer-
        Verlag.
Zwicky, F., and Herzog, F., 1963, Catalogue of Galaxies and
        Clusters of Galaxies, Pasadena, California Institute of
        Technology.

MODELS OF X-RAY EMISSION FROM CLUSTERS OF GALAXIES

A. Cavaliere

University of Rome
Laboratorio Astrofisica Spaziale CNR
Frascati, Italy

1. PHYSICAL PROCESSES

The dominant X-ray emission from galaxy clusters is very likely
to be thermal Bremsstrahlung from a diffuse, hot intracluster medium (ICM) with maximum density $n \simeq 10^{-3}$ cm^{-3} , temperature
$T \simeq 10^8$ °K and near-solar composition. The main evidence is based
on integrated spectra: shape and intensity of the continuum, and
strength of very high excitation lines up to Fe XXV and XXVI in
several clusters (Mushotzky et al. 1978 and references therein).In
dependent evidence of a dense hot ICM extending out to several core
radii r_c has been provided by the discovery and the interpretation of "head-tail" radio galaxies (TRG, reviewed by Harris 1977).
Direct evidence of extended X-ray emission had been obtained from
coarse scannings by the UHURU satellite (cf. Gursky and Schwartz
1977) and from imaging observations of Perseus and Coma Clusters
(Gorenstein et al. 1978, 1979).
    The Einstein Observatory is revealing in many clusters coherent
morphologies extending over scales of several r_c , coupled with
rich, sometimes dominant, structure down to galactic scales. Two
main morphological classes may be discerned in these data (cf.
Jones, these proceedings):
"Early" structures: very clumpy and sparse emission, with galaxies
standing out; some correlation with lower ICM temperatures. "Evol
ved" morphologies: relatively smooth, compact emission. In the
observations there is scope for distinguishing the subclasses:
A) centrally enhanced; B) centrally peaked; C) poor groups, with
emission concentrated around the central cD galaxy.
    The notion implied is that the morphology of the ICM with its
Bremsstrahlung emissivity $j_B \sim n^2 T^{\frac{1}{2}}$ exhibits and marks the clus
ter evolutionary stage. The latter is the result of many, inter-

217

R. Giacconi and G. Setti (eds.), X-Ray Astronomy, 217-237.
Copyright © 1980 by D. Reidel Publishing Company.

playing and often superposed processes, and my guide-line will be
to single out three main evolutionary components:
i) the underlying gravitational evolution of the cluster as a whole,
through the stages of dynamical collapse out of the Hubble flow
(time-scale t_d , see Appendix), violent relaxation and virializa-
tion ($\Delta t \simeq 3t_d$) , collisional relaxation (basic scale t_r) causing
segregation of the massive galaxies and a long term contraction of
the core (cf. Bahcall 1977, Ostriker 1978). Within this standard
picture, I would like to stress the following points. Since the
relative lengths of the collisional and the dynamical timescales
depend on the member number N ,

$$\frac{t_r}{t_d} \simeq 10^{-1} \frac{N}{\ln N/4} \quad ,$$

(1.1)

collisional relaxation effects mix in at $\Delta t \simeq$ few t_d in numerical N-body computations like the often referred to Peebles 1970
(N = 300), and may be responsible for a core-halo differentiation
not necessarily representative of the current conditions in truly
rich clusters. At the other extreme, in poor groups collisional
effects are indeed fast, especially because the presence of a dis
tribution of masses causes central segregation of the largest gal
axies which speeds up even more central relaxation and dynamical
friction ($t_r \sim M_G^{-2}$, $t_f \sim M_G^{-1} \rho_G^{-1}$) : the morphology Evolved - C
most likely reflects this state of affairs. In addition, Ostriker
1978 notes that in statistical samples of groups selected at equal
contrast, $t_r \sim N^{5/4}$ holds. If the clustering is hierarchical
(cf. discussion by Rees 1978), the specific binding energy increases
but weakly with the clustered mass, $< v^2 > \sim M^{2/3-a}$, so that rem
nants of initial subclustering persist for a few t_d : large scale,
soft X-ray clumpiness should mark such initial cluster stages.
ii) individual evolution of the member galaxies in terms of stellar populations and related recycling of their interstellar medium.
iii) action of the galaxies on the ICM by injection of mass, enrichment and heating; back-reaction of the ICM on galaxies by ram-
-pressure stripping that should transform many spirals into SO's,
and by accretion associated with a cooling phase that may feed ac
tivity in the central galaxies and perhaps add substantial mass.

2. ICM EVOLUTION

The behaviour of the ICM may be described with the conservations
laws for a single, ideal fluid:

$$\frac{d\rho}{dt} + \rho \Delta \cdot \underline{v} = \dot{m}$$

$$\rho \frac{d\underline{v}}{dt} = - \Delta p + \underline{g}\rho$$

$$\rho T \frac{dS}{dt} = \Delta \cdot \underline{Q} + j$$

(2.1)

where $S = c_v \ln(p/p^{5/3})$ is the change in specific entropy, Q is
the conductive heat flux and j is the heat input or loss (for
Brems. radiative losses, $j_B \simeq - 2 \ 10^{-27} \ n^2 \ T^{\frac{1}{2}}$). The time-scales
associated with various terms of Eqs.(2.1) are listed in the Ap-
pendix. Viscosity terms, not needed here, are given by Perrenod
1978, who also summarizes the values of the transport coefficients
to be adopted when the macroscopic time scales are very long and
when they become comparable with the microscopic mean free times.
The validity of the single fluid description rests on the condition that the energetic coupling of electrons with ions (equipartition time $t_{eq} \simeq 10^3 \ t_e$) is still fast on the macroscopic time
scale; when ions are preferentially and impulsively heated (as for
instance in shocks) and collective plasma interactions are not
operative, different electron and ion temperatures must be allowed
for.

Evolutionary histories of the ICM corresponding to Eqs.(2.1)
(with viscosity and $T_e \neq T_i$ when needed) have been computed nu-
merically,most thoroughly by Perrenod 1978. These time-dependent
models are based on the gravitational evolution $g = g(t)$ taken
over from a 700-body computation by White 1976 that includes all
the stages mentioned in Sect.1 (i). The mass source for the ICM
is provided by infall during the protocluster condensation of cool
material left over from galaxy formation (fraction of cluster
virial mass $= \varepsilon$) , and/or by early continuous injection into the
cluster of metal-enriched gas lost by stars within the member ga-
laxies (rate $\dot{M}/M = \alpha_0 \ t^{-1}$, implying $\Delta M_{tot}/M \simeq 5\alpha_0$). The results
indicate that in the models the energy is partly supplied by·
infall to the core during the dynamical stage, and by subsequent
compression associated with the deepening of the potential well
on the time-scale t_r ; but Field 1978 points out that a comparable role is apparently played by the conversion of the gas kinetic
energy associated with the orbital motion of the parent galaxy (a
process more important, I would add, in the early evolutionary
stages,and if the injection is actually delayed as described in
Sect.6).From $z \simeq 1$ to $z \simeq 0$ the overall L_x increases by
$\gtrsim 10$ as the central density raises; in fact, in infall models
limits are required to the amount of cool primordial gas, hence
to ε and Ω_0 , to avoid excessive luminosity: e.g., $\varepsilon = 0.1$
implies $\Omega_0 \lesssim 0.2$ whereas $\varepsilon = 0.5$ implies $\Omega_0 \ll 0.2$. With
10-20% mass loss from galaxies, injection models reproduce within
factors 2 - 3 the integrated quantities of Coma Cluster taken as
the standard source.

It is important to note that in all these models, over a substantial (> 5 Mpc at $z \simeq 1$) and increasing (> 10 Mpc by the
present) region of the cluster, the ICM reaches an approximately
static configuration wich these time - dependent models can describe only coarsely because of inevitable finite grid-size effects. Quasi-static evolution had been discussed and computed by
Takahara et al. 1976.

Given this framework, it is my task to concentrate on the $z \simeq 0$ configuration of the ICM and on models that try to derive information by testing a space - resolved, time - independent description of the gas.

Three categories of such models have been considered in detail. Hydrostatic models (Lea 1975, Gull and Northover 1975, Cavaliere and Fusco - Femiano 1976, 1978, Bahcall and Sarazin 1977) assume no dynamics, $(t_g \ll t_s)$ in the absence of driving terms $\dot{M}$ and j ; in particular, they imply negligible cooling $(t_{cl} \ll t_c)$. Cooling - accretion models (Cowie and Binney 1977, Fabian and Nulsen 1977, Mathews and Bregman 1978) envisage a flow onto cooling regions $(t_c < t_{cl}$ locally) which is steady owing to a continuous mass injection $(M/\dot{M} \simeq t_c)$. Wind models (Yahil and Ostriker 1973) are often dismissed on account of the large amount of power required, some 10^{46} erg s^{-1} (corresponding, if produced by Supernova explosions, to a rate $\simeq 1500$ yr^{-1} per rich cluster); while this is clearly excessive at present, cluster winds may have blown once, driven by the early star generations needed to produce the chemical abundances required by the observations and the mass injection implied by the previous models (cf. de Young 1978 vs. Norman and Silk 1979). To be sure, the random orientation of the TRG's (Harris 1977) speaks against large systematic velocities in the ICM at present.

The pressing alternatives concerning ICM that models should help to resolve, center on the relative distributions of gas and galaxies in cluster central regions, in particular on the effects of cooling; on the large-scale structure of ICM, presence of any halos and merging with any substantial intercluster medium; on the heating mechanisms. I would like to discuss critically the models in this perspective: as it is fair to say that only partial conclusions have been drawn so far from detailed model fittings (which anyway make sense only for evolved morphologies of the type A and B), it is appropriate to ask first whether we can do away with models after all.

3. A MODEL - INDEPENDENT APPROACH

The old hope may be revived here of retrieving 3-dimensional information in a model-independent way, by deprojection of 2-dimensional brightness distributions on the sky plane: the continuous nature of the X-ray emissivity suggests better prospects here than in the classical case of the galaxy counts.

A projection integral may be treated as an Abel integral equation; in spherical geometry, its direct inversion yields the basic relationship linking any volume distribution f(r) with its projected counterpart, the surface distribution F(s) (s being the radial coordinate in the sky plane)

$$f(r) = \frac{1}{\pi r} \frac{d}{dr} \int_R^r \frac{ds \; s \; F(s)}{\sqrt{s^2 - r^2}} = \frac{1}{\pi} \int_R^r \frac{ds}{\sqrt{s^2 - r^2}} \frac{dF}{ds} + \frac{F(R)}{\pi \sqrt{R^2 - r^2}}$$

$$(3.1)$$

In particular, $f(r)$ may be identified with the measured Brems. emissivity, $j_b(r) \sim n^2(r) \; T^{\frac{1}{2}}(r) \; W(E/kT(r))$, W representing the effect of the instrumental spectral response; and $F(s)$ with the corresponding brightness distribution $L(s)$ (details: Cavaliere and Fusco-Femiano 1976).

The problem with Eq.(3.1) is obviously constituted by the sensitivity of the solution to statistical fluctuations in the input data, which are amplified by the differential part of the operator; two viable solutions are known. Either an analytical (but arbitrary) model of $L(s)$ is fitted to the data in the first place: an often' used, 2-parameter representation is constituted by

$$\frac{L(s)}{Lo} = \left[1 + \frac{s^2}{a^2} \right]^{-b} \rightarrow j(r) \sim \left[1 + \frac{r^2}{a^2} \right]^{-b - \frac{1}{2}} \qquad (3.2)$$

core radius (HWHM): $s_c = a \left[2^{1/b} - 1 \right]^{\frac{1}{2}}$

$$r_c = s_c \left[2^{1/(b + \frac{1}{2})} - 1 \right]^{\frac{1}{2}} \left[2^{1/b} - 1 \right]^{-\frac{1}{2}}$$

The parameter a is easily pinned down to within 20% say, from the central brightness peak; the parameter b enters weakly the determination of a, but sensitively rules the outwards fall-off of the brightness distribution, and its precise determination again runs against difficulties with the data statistics. As a result, even when $L(s)$ converges strongly (e.g. for $b = 5/2$ and $T(r) = $ const, so that $n(r) \sim r^{-3}$) the gas mass $M(r)$ may formally diverge ($M(r) \sim \ln r$ in this example); in other words, it remains essentially undetermined.

Or else, when the volume distribution is apparently complex, the original purpose of minimizing the assumptions must be pursued with more direct deprojections. But then delicate techniques of data filtering (suppressing the noisy small wave-length components, yet retaining a maximum of real structural detail) must be used; such techniques are discussed, with extension to spheroidal distributions, by Binney in these proceedings. Alternatively, iterative procedures (Lucy 1974) can keep in check the instability of the solution of the projection integral equation.

In any case, since $j_b(r) \sim n^2(r) \; T^{\frac{1}{2}}(r) \; W(E/kT(r))$, the X-ray brightness is by itself insufficient to determine even the gas distribution $n(r)$ alone: independent information on $T(r)$ is also needed. Compromising on the objectivity of the final result, one may adopt at this point a polytropic assumption $T \sim n^{\gamma-1}$ (discussed in next Sect.); this mixed approach looks promising for a most efficient use of the current quality of the X-ray data, but it is still insufficiently tested so far. In principle, a direct information could be provided by space-resolved X-ray spectroscopy of the continuum or rather of the emission lines; or else by resolved measurements of the Sunyaev-Zeldovich effect, the apparent diminution ΔT of the temperature of the cosmic microwave radiation T_b towards clouds of hot gas whose electrons Compton scatter the photons to higher energies (Sunyaev and Zeldovich 1972). However, each way has its own uncertainties (cf. Smith et al. 1979 and Cavaliere et al. 1979), and either one runs into the usual, or even into harder, problems connected with high space re solution. As for the Sunyaev-Zeldovich effect, it yields $\Delta T/T_b =$ $= - 2.2 \ 10^{-34} \ \int dl \ nT$, and Eq.(3.1) in which $f(r)$ is identified with $n(r) \ T(r)$ may be used to derive

$$ n(r) \ T(r) \sim \frac{1}{\pi r} \frac{d}{dr} \int_R^r \frac{ds \ s}{\sqrt{s^2 - r^2}} \frac{\Delta T(s)}{T_b} \tag{3.3} $$

(cf. also Silk and White 1979); however, the resolution achieved to date in the microwave band (Birkinshaw et al. 1979) is still far from that required for a meaningful direct use of Eq.(3.3), and it may be constrained anyway by the presence of cluster radio sources (Tarter 1978, Schallwich and Wielebinski 1978). It appears that at present and in the near future the fully direct approach can provide only coarse, if very useful, checks or constraints on otherwise devised distributions. Sect.7 tackles one such case.

4. A BASIC STATIC MODEL

The simplest model, and a useful comparison term, is that characte rized by the conditions:

$$ t_d \ , \ t_s \ \ll \ t_{cl} \ \ll \ t_c \qquad , \ t_h \quad ; \tag{4.1} $$

with the addition of spherical symmetry and of the condition that the gravitational potential difference associated with the single member galaxies (Fig.1) be negligibly shallow:

$$ < v^2 > \ \gg \ < v_G^2 > \tag{4.2} $$

To such conditions tend for $z \lesssim 0.5$ the evolutionary tracks com puted by Perrenod 1978, while the observations by HEAO-2 suggest

that actually they may be met only by the morphological types
Evolved A and B.

When Eqs.(4.1) (4.2) are satisfied, the gas simply conforms to
the general gravitational field $\underline{g}$ prevailing in the cluster. The
disposition of the gas relative to that of the galaxies is control
led by the ratio of the respective scale heights, $H = kT/\mu mg$
(μ = mean molecular weight; for cosmic abundances: $\mu = 0.6$) and
$H_G = v^2/g$ (v henceforth denotes the line $-$ of $-$ sight velocity dispersion $< v_{//}^2 >^{\frac{1}{2}}$). To a first approximation, it is useful to
single out: a) the core behaviour, ruled by the central value

$$\frac{H_0}{H_{Go}} \equiv \tau = \frac{\mu m \, v_0^2}{kT_0} = 0.7 \, \frac{v_8^2}{T_8} \qquad (4.3)$$

so that the core radius of the gas is $r_{cg} \simeq r_c \, \tau^{-1}$; and b) the
outwards fall-off of the density, ruled by

$$\frac{H(r)}{H_G(r)} \sim \frac{T(r)}{v^2(r)} \qquad . \qquad (4.4)$$

Possible dispositions range between the very extended marked by
$\tau < 1$ and by $T(r)/v^2(r) = const$, and the compact ones marked
by $\tau > 1$ and by a decreasing ratio $T(r)/v^2(r)$.

A finer description requires the full use of the hydrostatic
equilibrium for the gas (cf. Eqs.(2.1) with $d/dt = 0$); this may
be directly related to the equilibrium condition for the galaxies
to obtain:

$$\frac{1}{\rho}\frac{dp}{dr} = -g = \frac{1}{\rho_G}\frac{dp_G}{dr} \qquad (4.5)$$

which holds in isotropic conditions, but independently of assumptions concerning the form taken by the cluster's missing mass
(provided only that the luminous component is in equilibrium with
the full potential). The essential ingredients are constituted by
some knowledge (or by an apt parametrization) of the thermal state
of the gas to be inserted into $p = \rho kT/\mu$, and by some knowledge about the velocity dispersion of the galaxies to be inser-
ted into $p_G = \rho_G v^2$. For example, a polytropic gas ($T \sim n^{\gamma-1}$)
and an "isothermal" galaxy velocity distribution ($v^2 = const$)
entail the explicit and simple link of $n(r)$ with $\rho_G(r)$ given
by

$$\frac{n}{n_0} = (\frac{\rho_G}{\rho_{G_0}})^\tau \qquad (\gamma = 1)$$

$$(\frac{n}{n_0})^{\gamma-1} = \frac{T}{T_0} = 1 + \frac{\gamma-1}{\gamma} \, \tau \, \ln \frac{\rho_G}{\rho_{G_0}} \qquad ,$$

$$(4.6)$$

As for the gas, the widely used polytropic index γ should parametrize with a single number the results of the complex thermal history of the ICM. A constant γ corresponds to a distribution of the mean energy per particle $3kT/2 \sim n^{\gamma-1} \sim p^{1-\gamma}$, and to a distribution of the specific entropy $S \sim \ln n^{\gamma-5/3} \sim \ln p^{1-5/3\,\gamma}$: both quantities are thus constant on isopotential surfaces as long as $dp/dr = -g\rho$ holds. The inequality $\gamma \geq 1$ corresponds to $dT/dr \leq 0$, and the limit $\gamma = 1$ to fast conduction $(t_D \ll t_{cl})$. The inequality $\gamma \leq 5/3$ corresponds to $dS/dr \geq 0$, that is, to stability against convection (onset of convective instability would reset γ to 5/3); and the value $\gamma = 5/3$ corresponds to the adiabatic distribution. The weakness of the polytropic assumption is that values of γ other than 1 and 5/3 are not uniquely associated to a specific physical condition.

As for $\rho_G(r)$, we may use just empirical representations of the counts (taken from an analysis based on relationships formally similar to Eqs.(3.1), (3.2)); a simple, yet remarkably successful representation is given by (Rood et al. 1972)

$$\frac{\rho_G(r)}{\rho_{G_0}} = \left[1 + \frac{r^2}{r_c^2} \right]^{-3/2} \rightarrow 1 - \frac{3}{2} \frac{r^2}{r_c^2} \; (r \lesssim r_c) \; , \; \sim r^{-3} (r \gg r_c) \tag{4.7}$$

with a core radius $s_c = r_c$, cf. Eq.(3.2), and a weak divergence of the mass $M_G(r) \sim \ln r$.

Such a direct link between the X-ray and the optical data as given by Eqs.(4.6) and (4.7) is worth some detailed discussion. The galaxy distribution is characterized by the observables r_c and v^2; an upper bound ρ_{v_0} to the normalization ρ_{G_0} is given by the Virial Theorem applied to the core, which yields $r_c = 3(v_0^2/4\pi G\rho_{v_0})^{\frac{1}{2}}$, this being of course the missing mass problem. The X-ray distribution contains directly only r_c and v_0^2, and depends on three gas parameters: n_0, T_0 and γ. The normalized $L_x(s)$ is governed by $\tau \sim v_0^2/T_0$ (see Eq.(4.3)) and by γ; the total luminosity $L_x \sim n_0^2 T_0^{\frac{1}{2}} r_x^3$ is very sensitive to n_0, less sensitive to T_0 and still less to γ. Its core radius r_X is given by

$$r_X \simeq r_c \left[2^{2/(6\tau-1)} - 1 \right]^{\frac{1}{2}} \qquad (\gamma = 1, \; \tau > 0.5) \tag{4.8}$$

independently of the energy; for $\gamma \neq 1$ it is energy dependent, and for instance (Sarazin and Bahcall 1977)

$$r_{1keV} \simeq 1.4 \, r_c \simeq 1.14 \, r_{10keV} \; (\gamma = 5/3, \, T_0 = 2 \; 10^8 \; °K).$$

Fits to the observed brightness distributions have been repeat-
edly performed using the limited amount of structural data provided by the first-generation satellites (Lea 1975, Cavaliere and
Fusco-Femiano 1976, 1978), by softer experiments (Malina et al.
1978) and by the first focusing telescopes (Gorenstein et al.
1978, 1979). For Coma Cluster, the fits essentially converge to
require $\tau \simeq 0.5$ $(1 \pm 20\%)$ and $1.1 \leq \gamma \leq 1.4$ (90% confidence),
using the optical data $r_c \simeq 0.25$ Mpc (Rood et al. 1972): Table 1.

The combined picture may be checked for self-consistency as
indicated by Gorenstein et al. 1979: τ may be evaluated directly
from the optically measured velocity dispersion $v = 1060$
$(1 \pm 20\%)$ km s^{-1} (Rood et al. 1972), and from T_0 derived on the
basis of the X-ray spectra taken by Mushotzky et al. 1978, emission – weighting the temperature: $T_0 \simeq 1.4 \ 10^8$ °K for $\gamma = 1.1$
and $T_0 \simeq 2.1 \ 10^8$°K for $\gamma = 1.4$. The results, $\tau = 0.6$ and
$\tau = 0.4$ respectively, are consistent with the results of the
fits considering the respective uncertainties. Strikingly different is the situation found by Gorenstein et al. 1978 for the
Perseus Cluster with a similar computation (after subtracting the
central $< 4'$ source at NGC 1275): using $r_c = 8'$ (Bahcall 1974)
they obtain $\tau = 0.6$ from the brightness fit of a $\gamma = 1$ model,
against $\tau = 1.8$ as evaluated from the spectral temperature
$T_0 = 7 \times 10^7$ °K and $v = 1420$ km s^{-1} (Chincarini and Rood 1971).
The inconsistency is well beyond any statistical uncertainty and
possible variation of γ ; it might indicate that the velocity
dispersion is very anisotropic with $< v_\perp^2 > \ << 2 < v_{//}^2 >$, or
alternatively that the mass-bearing objects have a distribution
different (and wider) than that of the bright galaxies (cases of
segregation, not necessarily dynamically relevant,are discussed
by Quintana and Havlen and references therein). While either possibility is puzzling on dynamical grounds, it may well be expected
that in 'the absence of cooling the ICM, with its short relaxation
times, provides a closer and more true representation of the present gravitational status of the cluster than the galaxies themselves. In such a situation the mixed approach of Sect.3 may prove very valuable.

Fig.1. The gravitational
potential associated with
a rich cluster including
large member galaxies: ΔV_G
$/\Delta V_{cl} = < v_G^2 >/< v^2 > \simeq$
$\simeq (500/1800)^2$. The ICM
fills a given potential
well to a hight determined
by the combined parameter
$\tau(\gamma-1)/\gamma$. (Adapted from
Longair, M.S., 1975 ESTEC
Conference).

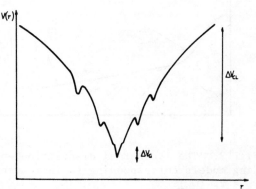

Table 1

Static models of the Coma Cluster

	n_0 $10^{-3} cm^{-3}$	T_0 $10^8 \,°K$	M_0 $10^{15} M_\odot$	M_{Gc}	$M(5Mpc)$	$M_G(5Mpc)$
$\gamma = 1.1$	2.0	1.4	$3.7 \ 10^{-3}$	$1.1 \ 10^{-1}$	0.51	1.7
$\gamma = 1.4$	1.2	2.1	$2.3 \ 10^{-3}$	$1.1 \ 10^{-1}$	0.69	1.7

Notes: Here, as throughout the text, n = number density of protons. $\rho_{V_0} = \rho_{G_0} = 2.1 \ 10^{-25} g \ cm^{-3}$.

Table 2

Values of r^*/r_c from Eq.(5.1); for $\gamma = 1$, $r^*/r_c = \infty$

$\tau \diagdown \gamma$	1.1	1.2	4/3	5/3
0.4	9573	148.4	28.7	8.0
0.5	1530	54.5	14.4	5.2
1.0	39.1	7.3	3.7	2.1
1.5	11.5	3.7	2.2	1.4

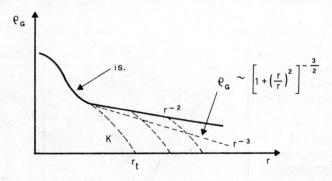

Fig. 2. The isothermal disposition envelopes the sequence of King models corresponding to increasing values of the cut-off $r_t = r_t \ (w_0)$.

5. LARGE-SCALE STRUCTURE

After the model in Sect.4, the ICM terminates $(n = T = 0)$ at the radius r^* given by

$$\frac{\rho_G(r^*)}{\rho_{G_0}} = e^{-\frac{\gamma}{\tau(\gamma-1)}}, \quad r^* = r_c \left[e^{\frac{2\gamma}{3\tau(\gamma-1)}} - 1 \right]^{\frac{1}{2}} \tag{5.1}$$

Table 2 gives values of $r^*(\tau,\gamma)$; if gaseous "halos" are defined by $r^* > 10\, r_c$, they require $\tau(\gamma-1)/\gamma < 0.14$. At such distances, as noted by Gorenstein et al. 1979, the brightness inferred from a $\tau \simeq 0.5$ and $\gamma \simeq 1$ model would still be observable with an instrumentation like that on board HEAO-2.

On the other hand, finite-mass rather than divergent models are fit to answer questions on large scale structures, or to attempt a continuous description of both the IGM and its clumpiness associated with the cluster potential wells. Such problems require undertaking a self-consistent formulation of the cluster potential, in spite of the foreseeable difficulties: a) The use of optical information rests on the condition that bright galaxies are fair markers of the virial mass distribution. b) At variance with seemingly similar structures such as globular clusters and E galaxies, in galaxy clusters the optical information vanishes rapidly outside a few $10\, r_c$: at surface density levels $\sigma/\sigma_0 \simeq 10^{-2}$ the counts become unreliable because of severe difficulties with the "field" galaxies (van der Bergh 1977) c) Large scale condensations are likely to be relaxed to some equilibrium state only in their central region; even rather regular structures like Coma Cluster may be still dynamically active in their outskirts (Oemler 1974, White 1976). However such an analysis does indicate a number of model-independent points and can provide upper limits to the energy density of the IGM; eventually, the X-ray observations themselves will help towards a reasonably complete picture.

The gas with its short relaxation times is amenable to a fully local description

$$\frac{n}{n_0} = e^{-\tau \frac{V(r)-V(0)}{v^2}} \qquad (\gamma = 1)$$

$$\left(\frac{n}{n_0}\right)^{\gamma-1} = \frac{T}{T_0} = 1 - \frac{\gamma-1}{\gamma} \tau \frac{V(r)-V(0)}{v_0^2} \tag{5.2}$$

in terms of the potential:

$$\frac{V(r)}{v_0^2} = \frac{G}{v_0^2} \int_{\infty}^{r} dr \; \frac{M(r)}{r^2} \equiv \frac{4\pi G \rho_{v_0} r_c^2}{v_0^2} \; \psi \left(\frac{r}{r_c}\right) \tag{5.3}$$

normalized with its natural units. The Virial Theorem applied to
the central core yields $4\pi G \rho_{v_0} r_c^2 / v_0^2 \simeq 9$, while $\psi(0) \simeq 1$
should hold for a finite well.

The description of self-gravitating galaxies (as the mass-bearing
objects) consistent with $v^2 = const$ is local and simple: it is
the classical "isothermal" disposition (cf. Chandrasekhar 1967),
a 2-parameter function $\rho_G(r/r_c)/\rho_{Go}$, with the asymptotic limits

$$\frac{\rho_G}{\rho_{Go}} \rightarrow e^{-\frac{3r^2}{2r_c^2}} \;\; (r \lesssim a) \; , \;\; \sim r^{-2} \;\; (r \gg a) \tag{5.4}$$

and with $r_c = 3(v_0^2/4\pi G \rho_{vo})^{\frac{1}{2}}$. It is held to approximate well the
counts out to several r_c in many clusters (cf. Bahcall 1977).
However, the associated mass diverges badly, $M_G(r) \sim r$: for a finite system in equilibrium it is required that $v^2(r)$ should
eventually become smaller than $|V(r)|$; moreover, v^2 must
vanish at some finite r_t if $t_{cr} < t_{cl}$ is to hold. Thus r_t is
expected to enter explicitly the description.

King's model (originally devised for globular clusters and
E-galaxies, but often invoked in the present context following
Rood et al. 1972) is based on a simple and drastic "closure" con-
dition, namely that all galaxian orbits should be closed:
$E_{max} = V(r_t) < 0$. The constraint is realized choosing as velocity
distribution a truncated Maxwellian

$$f(v,r) = e^{-\frac{V(r)-V_0}{v_M^2}} \left[e^{-\frac{v^2}{2v_M^2}} - e^{-\frac{V(r_t)-V(r)}{v_M^2}} \right] \tag{5.5}$$

(v_M , the dispersion of the full Maxwellian, in practice is very
close to v_0 , see Cavaliere and Fusco-Femiano 1978 for details).
The rationale (Lynden-Bell 1967) is that during the short dynamical collapse, violent relaxation should give rise to a Maxwellian
population of the lowest energy states, whereas the high-energy,
long-period orbits cannot acquire their full population quota;
the high velocity tail of $f(v)$ is subject to a diffusion process,
stationary for the form (5.5).

It is easy to compute density and velocity in terms of the potential difference from the boundary, $w = (V_t - V)/v^2$; it turns
out that density and v decrease as given by:

$$\rho_G = 4\pi \int dv \ v^2 \ f(v) = \rho_{Go} \ e^{w-w_0} \ \frac{I_{3/2}(w)}{I_{3/2}(w_0)} \sim w^{5/2} \quad (w \to 0)$$

$$\overline{v^2} = \frac{4\pi}{3} \int dv \ v^4 \ f(v) = v_0^2 \ \frac{2}{5} \ \frac{I_{5/2}(w)}{I_{3/2}(w)} \sim w \quad (w \to 0)$$

$$(5.6)$$

where

$$I_n(w) \equiv \int_0^w dx \ e^{-x} \ x^n \quad .$$

Galaxy and gas density (if relevant: note from Eqs.(4.6) that the ratio $\rho(r)/\rho_G(r)$ tends to increase when $\tau < 1$ and $\gamma \to 1$) once expressed in terms of the potential, may be inserted into the Poisson equation to obtain $w(r)$ integrating from the center with the usual boundary conditions $w = w_0$, $(dw/dr)_0 = 0$. For negligible gas density the result is the pure King model, Fig.2, a sequence of finite galaxy dispositions $\rho_G/\rho_{Go} \ (r/r_c, \ w_0)$ which tend to the isothermal limit non-uniformly for $w \gg 1$: that is, they tend to the isothermal limit at the center for any $w_0 > 1$, and are cut-off at a radius r_t increasing with w_0.

After Eq.(5.3), we expect $w_0 \simeq 9$ to hold when the core is approximately isothermal, and in fact King uses $w_0 = 9.32$ to model Coma Cluster; the overall depth of the potential well is $w_0 \ v_0^2 - GM/r_t = 9.70 \ v_0^2$. King 1972 and the references therein give further numerical details: here it is worth recalling that the empirical law Eq.(4.7) constitutes a fair approximation to the full solution; the latter however, is finite, and thus must contain explicity a third parameter, the cut-off w_0 . The parametrization of the X-ray emission is as in Sec.4, even when the ICM mass is non negligible; Cavaliere and Fusco-Femiano 1978 compute also cases where this condition prevails and correspondingly the value of r_t is appreciably reduced.

In fact, one drawback of the pure King model is that it places r_t at a distance implying crossing times $t_{cr} \gtrsim t_H$ such that the equilibrium assumption becomes dubious. But beyond all the details of this particular model, we stress that its main qualitative features must be shared by any finite model. In such cases, a fit in the central region $r \lesssim 10 \ r_c$ determines much as in Sect.4 the values of τ , γ , n_0 . When extrapolated to the galaxy "boundary" (marked by a normalized potential difference relative to the center $w_0 \simeq 9$) the ICM pressure reads, model-independently

$$n_t \ T_t = n_0 \ T_0 \ \left[1 - \frac{\gamma-1}{\gamma} \ \tau \ w_0 \right]^{\gamma/(\gamma-1)} \qquad (5.7)$$

It is $\neq 0$ if $\gamma < (1-1/\tau w_0)^{-1}$ is satisfied, and then it must
be balanced by the pressure of the external medium. Thus the value of $3n_t kT_t$ should constitute an upper limit to the energy den
sity of the IGM considering that outside r_t , $V(r) \to 0$ holds
implying $(nT)_{IGM} \lesssim n_t T_t$; and that we expect, if anything, T_t
to be smaller that the simple extrapolation $T \sim n^{\gamma-1}$. The limit
$\gamma \gtrsim 1.1$ for Coma Cluster implies $n_t T_t < 300$ °K cm^{-3} (Cavaliere
and Fusco 1978); the values $n_{IGM} = n_{critical}$ and $T_{IGM} = 3\ 10^8$°K,
the reference terms for a diffuse thermal Bremsstrahlung interpre
tation of the 2-50 keV background (see Field and Perrenod 1977),
imply $(nT)_{IGM} = 900$ °K cm^{-3} . If $\Omega_0 < 1$ holds as several lines
of evidence suggest, the above is essentially an upper limit to
the amount of hot IGM.

6. HEATING

As the cooling time t_c is generally long, it is the time integral of the various heating processes over the cluster life-time
that determines the current thermal state of ICM. In addition,
the steady state temperature is regulated by the cluster potential to the equilibrium value $kT \simeq \mu m\ v^2$: hotter gas escapes,
cooler gas collapses (Lea and Holman 1978). Finally, the current
thermal state itself is poorly known. Thus, direct observational
constraints still cannot discriminate between the various heating
mechanisms proposed since the first discovery of clusters as
X-ray sources.

The following general constraints, however, should be considered: at the beginning of the ICM's thermal history, at temperatures around 10^5 °K , the cooling time was comparatively short
due to abundant line emission; to overcome this "thermal barrier",
impulsive heating (e.g. by shocks) or gradual mass injection together with energy input had to take place. On the other hand,
the iron lines detected in several cluster spectra (Mushotzky et
al. 1978) constantly indicate a near-solar abundance, evidence
that most of the emitting ICM has been processed within stars.
The Butcher-Oemler (1978) effect indicates that little ram-pres
sure ablation took place before $z \simeq 0.5$, evidence against a pre-
existing, high density ICM.

The simplest evolutionary history (Field 1978, De Young 1978,
Norman and Silk 1979) that explains mass and energy production
consistently with the above constraints has it that gas ejected
by early star generations surrounds the parent galaxy in a halo
or slowly expanding shell until it collides with a similar cloud
or with the ambient ICM. If the gas is eventually brought to rest
relative to the cluster center, part of the kinetic energy of galaxies is effectively degraded into internal energy of the gas to
the result that $3\ v^2/2 \simeq 3\ kT/2\mu m$. Once collisions start, and
the ambient gas density rises to $n \simeq 10^{-4}$ cm^{-3} , the ablation
process becomes faster than a crossing time provided that the
mass loss rate within the galaxies is less than a few

$M_\odot$ yr^{-1}/$10^{11} M_\odot$ (Gisler 1979). Norman and Silk 1979, in particular, try to satisfy in full the Butcher-Oemler constraint stress-
ing that the gas can be stably retained as a massive halo by
many galaxies, to be released into the ICM only at a rather late
epoch z $\simeq$ 0.5 by mutual collisions and by ram-pressure stripping against the ICM accumulating in the cluster cores. Such comparatively abrupt and late injection modes cause an X-ray evolution different from that computed by Perrenod 1978 on the basis
of a continuously decreasing injection: the keV emission should
be cut-off at z $\gtrsim$ 0.5 and replaced by the softer and weaker
emission associated with the halos. More generally, the Butcher-
-Oemler effect in its straight interpretation is not easily compatible with the amount of ICM required by the Bremsstrahlung interpretation of the X-rays under any model if the blue galaxies
are to coexist with the emitting ICM; it might pose an interesting challenge to current ideas on galactic evolution.

The view that the heating is, after all, comparatively local-
ized in (recent) time and in space, or is controlled by local conditions anyway, may be examined in the context of static models
considering the limiting case where

$$kT(r) = \tau^{-1} \mu m \, v^2(r) \qquad\qquad (6.1)$$

holds everywhere (Cavaliere and Fusco-Femiano 1979).Eq.(4.5) then
yields

$$\frac{n}{n_0} = (\frac{\rho_G}{\rho_{Go}})^\tau \, (\frac{v^2}{v_0^2})^{\tau-1} \qquad\qquad (6.2)$$

After this "kinematic" model, as far as $v^2(r)$ decreases the
fall-off of $n(r)$ is steeper or slower than that of $\rho_G(r)$ according to whether $\tau > 1$ or < 1 holds, respectively. A single
shape parameter τ rules the gas, and hence the brightness, distribution both in and outside the core, while n_0 determines the
central emission. This 2-parameter model is currently under study,
taking for $v^2(r)$ the King's expression Eq.(5.6); it is seen
from the asymptotic expressions quoted there that $n \sim w^{\tau \, 7/2 \, - \, 1}$
so that $\tau > 2/7$ is required for convergence, while with increasing τ the ICM mass decreases down to $\simeq 10^{-2} M_V$ for $\tau \to 1$.
This is just one particular example of how the polytropic parame-
ter γ , with its indirect significance besides the limiting cases $\gamma = 1$ or $\gamma = 5/3$, may be eliminated in favor of a definite heating hypothesis. Other, more general models where T is
weighted with ρ_G , $T = T(v^2, \rho_G)$, are easily tractable in solving
Eq.(4.5) and their results may be tested against the data.

Heating associated to ejection entails $\tau \simeq 1$, which agrees
within the uncertainties with the average over the original data
listed by Mushotzky et al. 1978; no additional heating is stricly
required yet. Two other mechanisms are worth mentioning, however,

as they relate to tentative lines of evidence. Mushotzky et al.
1978 remark that lower temperatures corresponding to $\tau \simeq 1.5$
would obtain if inflow down a static potential minimum were all
that mattered, since then the potential energy shared between the
compressional work and the internal energy would produce
$kT_0 \simeq 3/5 \ \mu m v_0^2$; they note that their data would indicate $\tau \simeq 1.5$
if the average velocity dispersions were corrected to a central
value, presumably higher by $\simeq 20\%$. Lea and Holman 1978 contend
that additional thermal energy released by radio sources can prevent the establishment of cooling conditions (see next Sect.) and
further gas collapse in those cluster cores that tend to have high
ICM densities, thus maintaining a steadily high X-ray luminosity;
thus they account for some correlation of high L_x with the steep
spectrum, and sometimes clearly diffuse, radio emission found by
Owen 1974 and by Erickson et al. 1978, notwithstanding the accept
ed dominance of Bremsstrahlung over Inverse Compton X-ray emission that implies relatively high magnetic field $(B \gtrsim 10^{-7} \ G)$
and low density of relativistic electrons. They argue that, since
each electron loses a power $\dot{E} \simeq 10^{-18} n$ erg s^{-1} in collisions
with thermal gas, an assumed extension of the electron distribu-
tion function down to $E_{min} \simeq 10 \ mc^2 \ B_{\mu G}^{-1}$ would suffice to convert
enough power as to balance the Brems. losses; to make up the
total energy, it suffices that a few cluster galaxies undergo al-
ternate, recurrent activity at a level $\simeq 10^{45}$ erg s^{-1} so as to
cover cumulatively the cluster lifetime.

If most radio-halos are actually comprised of remnants of TRG's
(Jaffe and Rudnick 1979), heating by radio-clouds implies that
eventually X-ray hot spots associated with the radio spots should
be resolved; on the other hand, heating associated with mass
ejection from galaxies may well leave behind outlying cooler
spots in its early phase.

7. COOLING

The sufficient condition for cooling effects to be negligible,
$t_c > t_H$, that is

$$n_{-3} < 3 \ T_8^{\frac{1}{2}} \ h \qquad\qquad\qquad (7.1)$$

is satisfied at the cluster centers by a modest margin: a factor
$\simeq 2$ after model fits, cf. Table 1. Some evidence of cooling has
been found in the central region of A 2142 by the Columbia Astrophysical Laboratory Group; direct evidence of local cooling
has been obtained with resolved spectroscopy in Perseus Cluster
in the vicinity of NGC 1275 by the Goddard Space Flight Center
Group, reported at this meeting, and by the MIT FPCS Group.

Theoretically, cooling is expected to set in first in the
densest region at the cluster centers and near large, slow-moving
galaxies, of which NGC 1275 is just a canonical instance: away
of any local minimum of the gravitational potential dp/dr =

= -g ρ < 0 holds in a static case, while the condition for convective stability, dS/dr $\geq$ 0 may be written as

$$\frac{d}{dr} \left(\frac{T^{\frac{1}{2}}}{n} \, p^{\frac{3\gamma-1}{6\gamma}}\right) \geq 0 \quad , \tag{7.2}$$

whence $dt_c/dr \geq 0$ follows. In addition, material just ejected
from galaxies may be both cooler and richer in heavy elements
than the average ICM, thus serving as a cooling agent. In these
conditions, an accretion flow would be driven by the pressure of
outer gas. Cowie and Binney 1977 argue that the weak inequality
$t_c \gtrsim t_H$ apparently holding at the cluster centers on a scale r_c
is not casual, rather it follows naturally if cooling regulates
an inflow into the core of outer gaseous material, which accretes
onto the central galaxies; substantial release of gravitational
energy would take place in the central region. Their steady state
solution of Eqs.(2.1), taken with $j = j_B$ and limited to the
subsonic range, requires a rather large mass inflow $\dot{M} \gtrsim 250 \, M_\odot \, yr^{-1}$
in Coma Cluster; the temperature has a minimum at the center, and
rises by a factor $\simeq 3$ out to $\simeq 2 \, r_c$. Fabian and Nulsen 1977
apply a similar analysis to the region near NGC 1275 and interpret the surrounding low-velocity filaments as condensations in
the accretion flow; a complete analysis of this thermal instability has been carried out by Mathews and Bregman 1978. While in
the immediate surroundings of NGC 1275 there is direct evidence,
as mentioned, of a temperature minimum at the galaxy, the cooling-
-accretion dispositions do not seem to apply on the scale of the
core radius (Gorenstein et al. 1978); it remains to be seen how
sources like that in A 2142 can be modeled in detail.

In this data situation, it is worth stressing that the presence
of cooling is subject to a _direct_ observational check, independent
to a large extent of any detailed modeling. In fact, the cooling
time is defined as the ratio of thermal energy density to emissiv
ity (per unit volume). The latter, integrated over the line of
sight and over a beam of angular radius θ_b around the cluster
center (operations denoted by < >), is measured by the corresponding bolometric flux:

$$F_x = \frac{j_0 \, r_c^3}{4\pi \, D_L^2} \, \left\langle \left(\frac{n}{n_0}\right)^2 \left(\frac{T}{T_0}\right)^{\frac{1}{2}} \right\rangle \tag{7.3}$$

(D_L = luminosity distance). The former is measured by the Sunyaev-
-Zeldovich effect, averaged for simplicity over the same beam:

$$\frac{\overline{\Delta T}}{T} = - \, 2.2 \, 10^{-34} \, n_0 T_0 r_c \, \frac{\theta_c^2}{\pi\theta_b^2} \, \left\langle \frac{n}{n_0} \frac{T}{T_0} \right\rangle \quad . \tag{7.4}$$

Thus the averaged cooling time can be expressed as

$$\overline{t_c} = 1.4 \ 10^{10} \ \frac{\theta_b^2}{(1+z)^4} \ \frac{|\overline{\Delta T/T}|}{F_x} \ A \ \ yr$$

$$A = \frac{<(\frac{n}{n_0})^2 \ (\frac{T}{T_0})^{\frac{1}{2}}>}{<\frac{n}{n_0} \ \frac{T}{T_0}>} \ \ ,$$

(7.5)

in terms of direct observables and of the shape factor A. The
fact that the central value t_{co} is a minimum implies that A
has an upper bound $A \leq 1$: cooling is   implied when the
Eq.(7.5) with $A = 1$ gives $t_c \leq t_H$. On the other hand, A is
expected to be dominated by the contributions of the core region
where pressure and emissivity are a maximum, and to vary but little whenever $\theta_b < \theta_c$; in fact, using the model of Sect.4 it is
seen that when τ and γ vary between the values $0.4 \leq \tau \leq 1.8$
and $1 \leq \gamma \leq 5/3$, and $\cdot \ \theta_b = \theta_c/2$, A is bound by $0.7 \lesssim \hat{A} \lesssim 1.8$
considering also the effects of the a beam throw $(\geq 3\theta_b)$ neces-
sary in microwaves. Within these wide limits, then, the static
models would be consistent with no cooling if, computing $\overline{t_c}$
with the lower bound of $\hat{A}$, $\overline{t_c} \ (\hat{A}_{low}) > t_H$ should result. Cavaliere, Danese and De Zotti are discussing with P. Boynton the
feasibility of such a measurement, which may be a by-product of
(and should be preliminary to) the efforts to deduce the Hubble
constant from the cluster X-ray luminosity self-calibrated (cf.
Cavaliere et al. 1979) with the associated microwave temperature
diminution ΔT of the cosmic radiation.

REFERENCES

Bahcall, N.A., 1974, Astrophys. J. 187, 439.
Bahcall, N.A., 1977, Ann. Rev. Astron. Astrophys. 15, 505.
Bahcall, J.N. and Sarazin, C.L., 1977, Astrophys. J. 213, L99.
Birkinshaw, M., Gull, S. and Northover, K.,1978, Nature 275, 40.
Butcher, H. and Oemler, A., 1978, Astrophys. J. 219, 18.
Cavaliere, A. and Fusco-Femiano, R., 1976, Astron. & Astrophys.
49, 137.
Cavaliere, A. and Fusco-Femiano, R., 1978, Astron. & Astrophys.
70, 677.
Cavaliere, A. and Fusco-Femiano, R., 1979 in preparation.
Cavaliere, A., Danese, L. and De Zotti, G., 1979, Astron. &
Astrophys. 75, 322.
Chandrasekhar, S., 1967, "An Introduction to the Study of Stellar
Structure", Dover Pub., London.
Chincarini, G. and Rood, H.J., 1971, Astrophys. J. 168, 321.
Cowie, L.L. and Binney, J., 1977, Astrophys. J. 215, 723.
De Young, D.S., 1978, Astrophys. J. 223, 47.

Erickson, W.C., Matthews, T.A. and Viner, M.R., 1978, Astrophys. J. 222, 761.
Fabian, A.C. and Nulsen, P.E.J., 1977, M.N.R.A.S. 180, 479.
Field, G.B., 1978, Preprint CFA N. 1132, to be published in Proc. Astronomische Gesellschaft.
Field, G.B. and Perrenod, S.C., 1977, Astrophys. J. 215, 717.
Gisler, G.R., 1979, Astrophys. J. 228, 385.
Gorenstein, P., Fabricant, D., Topka, K. and Harnden, F.R., 1978, Astrophys. J. 224, 718.
Gorenstein, P., Fabricant, D., Topka, K. and Harnden, F.R., 1979, Astrophys. J. 230, 26.
Gull, S.F. and Northover, K.J.E., 1975, M.N.R.A.S. 173, 585.
Gursky, H. and Schwartz, D., 1977, Ann. Rev. Astron. Astrophys. 15, 541.
Harris, D.E., 1977, Highlights of Astronomy, 4, 321.
Jaffe, W.J. and Rudnick, L., 1979, Preprint to be published in Astrophys. J.
King, I.R., 1972, Astrophys. J. 174, L123.
Lea, S.M., 1975, Astrophys. Letters 16, 141.
Lea, S.M. and Holman, G.D., 1978, Astrophys. J. 222, 29.
Lucy, L.B., 1974, Astronom. J. 79, 745.
Lynden-Bell, D., 1967, M.N.R.A.S. 136, 101.
Malina, R.F., Lea, S.M., Lampton, M. and Bowyer, C.S., 1978, Astrophys. J. 219, 795.
Mathews, W.G. and Bregman, J.M., 1978, Astrophys. J. 224, 308.
Mushotzky, R.F., Serlemitsos, P.J., Smith, B.W., Boldt, E.A. and Holt, S.S., 1978, Astrophys. J. 225, 21.
Norman, C. and Silk, J., 1979, Preprint to be published in Astrophys. J. Letters.
Oemler, A., 1974, Astrophys. J. 194, 1.
Ostriker, J. 1978, in IAU Symposium "The Large Scale Structure of the Universe" Longair and Einasto eds., 357.
Owen, F.N., 1974, Astron. J. 79, 427.
Peebles, P.J.E., 1970, Astronom. J. 75, 13.
Perrenod, S.C., 1978, Astrophys. J. 226, 566. Also Ph. D. Thesis, Harvard University, 1977.
Quintana, H. and Havlen, R.J., ESO preprint to be published in Astron. Astrophys.
Rees, M.J., 1978, in "Observational Cosmology" Maeder and Tammann eds., Geneva Observatory.
Rood, H.J., Page, T.L., Kintner, E.C. and King, I.R., 1972, Astrophys. J. 175, 627.
Sarazin, C.L. and Bahcall, J.N., 1977, Astrophys. J. Suppl. 34, 451.
Schallwich, D. and Wielebinski, R., 1978, Astron. & Astrophys. 71, L15.
Silk, J. and White, S.D.M. 1978, Astrophys. J. 226, L103.
Smith, B.W., Mushotzky, R.F. and Serlemitsos, P.J. 1979, Astrophys. J. 227, 37.
Sunyaev, R.A. and Zeldovich, Ya.B., 1972, Comm. Astrophys. Sp. Phys. 4, 173.

Takahara, F., Ikeuchi, S., Shibazaki, N. and Hoshi, R., 1976, Prog. Theor. Phys. 56, 1093.
Tarter, J.C., 1978, Astrophys. J. 220, 749.
van der Bergh, S., 1977, Vistas in Astron., 21, 71.
White, S.D., 1976, M.N.R.A.S. 177, 717.
Yahil, A. and Ostriker, J., 1973, Astrophys. J. 185, 787.

APPENDIX: Time-scales (yr) used in the text

Hubble time $\qquad\qquad t_H = 1/H_0 = 20 \cdot 10^9 \, h^{-1}$

$\quad$ ($H_0 = 50$ km s^{-1} Mpc^{-1} throughout, $\; h = H_0/50$).

Age of the Universe $\quad t_0 H_0 = 2/3 \; (q_0 = \frac{1}{2}) - 1 \; (q_0 = 0)$

Cluster Age $\qquad\qquad\quad t_{cl} \lesssim t_0$

Crossing time $\qquad\qquad t_{cr} = R/<v^2>^{\frac{1}{2}} = 6 \cdot 10^8 \, R_{Mpc}/v_8$

$\quad$ ($<v_{//}^2>^{\frac{1}{2}} = v_8 \, 10^8$ cm s^{-1}).

Dynamical time $\qquad\qquad t_d = (3\pi/32 G \rho_{min})^{\frac{1}{2}} = \pi (3/10)^{3/2} \, G \, M^{5/2}/|E|^{3/2}$

$\quad$ (Spherical Symmetry, E = total energy of the cluster)

$$t_d = 1.5 \, R_V/<v^2>^{\frac{1}{2}}$$

$\quad$ (assuming E = const, and virial equilibrium with virial
$\quad$ radius R_V).

Relaxation time $\qquad\quad t_r = \dfrac{<v^2>^{3/2}}{4\pi \, G^2 \, M_G^2 \, N_G \, \ln\Lambda} = \dfrac{20 \cdot 10^9}{\ln\Lambda} \dfrac{v_8^3}{M_{12}^2 \, N_3}$

$\quad$ (galaxy mass $M_G = 10^{12} M_{12} \, M_\odot$, galaxy density $N_G = 10^3 \, N_3$
$\quad$ gal Mpc^{-3}).

Dynamical friction $\quad t_f = t_r$ with $N_G \, M_G = \rho_G$

$$* \qquad\qquad * \qquad\qquad *$$

Electron-electron $\qquad t_e \simeq 8 \cdot 10^6 \, T_8^{3/2}/\ln\Lambda \; n_{-3}$
collisions

$\quad$ ($\ln\Lambda \simeq 40$, $T = T_8 \, 10^8 \; °K$, $n = 10^{-3} \, n_{-3}$ cm^{-3}).

Ion-ion collision $\qquad t_i \simeq 43 \, t_e$

Electron-ion $t_{eq} \simeq 10^3 \ t_e$
equipartition

Gas dynamics $t_g = R/v$

Sound propagation $t_s = R/v_s = 6 \ 10^8 \ R_{Mpc}/T_8^{\frac{1}{2}}$

Thermal diffusion $t_D = 3 \ 10^9 \ n_{-3} \ R_{Mpc}^2/T_8^{5/2}$

(for B = 0. Even a minuscule $\underline{B}$ inhibits thermal conduction across the lines of force, while the path along the lines depends on the field geometry; tangling on a scale < 10 kpc effectively suppresses conduction).

Heating t_h

Cooling $t_c = 3nkT/j_B = 70 \ 10^9 \ T_8^{\frac{1}{2}}/n_{-3}$

X-RAY STRUCTURE OF THE COMA CLUSTER FROM OPTICAL DATA

James Binney

Magdalen College, and Department of Theoretical
Physics, Oxford, England.

1. INTRODUCTION

The work I am going to describe was carried out in Oxford and
Amherst in collaboration with O. Strimpel. A full account will be
found in Binney and Strimpel(1978) and in Strimpel and Binney(1979).

The projections of rich clusters of galaxies onto the sky are
rarely circularly-symmetric (Carter and Metcalfe, 1980). This
fact is well illustrated by the case of the best-studied rich
cluster, that in Coma, which has often been considered the prototype "spherical" cluster, even though several workers (Abell, 1977;
Schipper and King, 1978) have noted that it is actually elliptical
in shape with an axial ratio near 2:1.

The interaction of the aspherical shapes of rich clusters
with the emission of bremsstrahlung x-rays by the hot intracluster
gas provides an important probe of the structure and evolution of
rich clusters for two reasons: A) If the gas in a cluster such as
Coma has yet to cool appreciably, it is probable that the specific
entropy s of the hot gas is constant on surfaces of constant gravitational potential Φ; $s = s(\Phi)$ (Gull and Northover, 1975).
Therefore, studies of the radial and azimuthal variations of x-ray
emission from such clusters should enable us to map their potentials
and therefore their true mass distributions (which may well differ
from their distributions of luminosity). B) If studies of clusters
in which cooling has yet to become important can be used to establish the proper link between luminosity- and mass-distributions,
we could derive the potentials of clusters in which cooling is
important from optical data. These potentials might then be used
to model the hydrodynamics of the gas in even complex clusters
and thus to draw important conclusions regarding the interactions

R. Giacconi and G. Setti (eds.), X-Ray Astronomy, 239-243.

of galaxies with the IGM and to elucidate the role played by mass-loss in galactic evolution (Cowie and Binney, 1977).

2. MAPS OF THE COMA CLUSTER

The Coma cluster is the obvious candidate for study in connection with step (A) above because: i) it is the rich cluster for which the most complete optical data are available (e.g. Abell, 1977; Godwin and Peach, 1977; Gregory and Thompson, 1978); ii) the pre-HEAO-2 X-ray data and the absence of an active galaxy in the cluster core indicate that cooling might not yet be important in this cluster (Perrenod, 1979), and iii) the cluster is highly aspherical.

We assume that the cluster is approximately rotationally-symmetric with its approximate axis of symmetry in the plane of the sky. Whilst neither of these assumptions is likely to be exactly correct, one may argue that if the plane of the sky were to differ substantially from the plane formed by the cluster's shortest and longest principal axes, the cluster could have the observed 2:1 axial ratio in projection only if it were implausibly elongated in space. Furthermore, the prolate and oblate spheroidal models that one obtains by assuming that the symmetry axis is respectively the longer and the shorter of the cluster's principal axes on the sky, should bracket the predictions of the continuum of tri-axial models which are compatible with the projected data and our identification of the plane of the sky with the plane of the two extreme principal axes.

In the absence of any indication to the contrary, we assume that the galaxies are fair markers of the overall mass of the cluster; the idea here is that the cluster mass (which almost certainly does not reside in the visible galaxies (Faber and Gallagher, 1979)) is made up of a population of dark objects, some of which carry lights, and are called galaxies, but are otherwise kinematically indistinguishable from the other bodies. According to this hypothesis, the positions and velocities of all galaxies are to be assigned equal weight, irrespective of luminosity.

The first step in handling the data is to determine the centre and apparent principal axes of the cluster. We find that the data of Godwin and Peach(1977,GP) enable these to be determined to within ~ 1 and ~ 50 arcmin respectively. Next we match the GP data, which cover only the innermost 36 arcmin ($\sim 6r_c$), to Abell's data (Abell, 1977), which go out to 80 arcmin, adopting a completeness limit of $17^m_.5$ on the GP scale. A transform technique is then employed to generate from this data the Gregory and Thompson (1978) value of the r.m.s. galaxy velocity in the cluster and, an assumed $\sim 20\%$ contamination with randomly-distributed field galaxies, two models of the cluster potential: one appropriate to a prolate model of the cluster and one for the oblate model. Notice that our transform technique (which was inspired by a paper by Press, 1976) folds background-subtraction, deprojection and solution for the

gravitational potential into one simple step. We fill these
potentials with hot gas under two extreme assumptions regarding
the behaviour of $s(\Phi)$: i) an isothermal atmosphere at $T = T_o$ and
ii) an adiabatic $\gamma = 5/3$ atmosphere having central temperature
$T = T_c$. Finally the bremsstrahlung emission of these models is
calculated and its projection onto the plane of the sky displayed
as a contour map.

3. REALIZATION DEPENDENCE

If we were able to reobserve the Coma cluster $10^9 y$ from now, the
galaxies would have moved on to new positions and the potentials
we would derive by the procedure described above would differ from
those we find now. However, if the bulk of the mass of the cluster resides in a very numerous population of individually insig-
nificant particles (Jupiters, stellar-mass black holes, etc.), the
true potential would not have altered between our observations.
Therefore the fluctuations one observes in the derived potential
as the galaxies move from one realization of the true matter distribution to the next represent an irreducible uncertainty of the
true cluster morphology arising from the availability of only a
finite number of marker particles. It is important to discover
how great is this uncertainty for the available Coma data.
     Pseudo-clusters of 1000 particles (we use 1017 galaxy positions
in Coma) were constructed by assigning "galaxies" positions according to a suitable Coma-like spheroidal probability distribution,
and projected onto the "sky" to form sets of pseudo-data. The extent of realization-dependence was then explored by comparing the
potentials and X-ray distributions obtained from different sets of
pseudo-data. We conclude: The cluster mass one derives by our
technique is remarkable independent of the realization employed;
$\delta M/M < 2.5\%$. ii) The potential with respect to infinity fluctuates
by $\leq 10\%$ between realizations. However, for realistic central
gas-temperatures ($T \stackrel{\sim}{\scriptstyle\sim} 10^8 K$), even 10% fluctuations in Φ lead to
$\stackrel{\sim}{\scriptstyle\sim} 50\%$ fluctuations in the central X-ray brightness at fixed central
gas-density. This strong realization-dependence of the central
X-ray brightness (and with it the total luminosity of the cluster)
arises because gas at $T \stackrel{\sim}{\scriptstyle\sim} 10^8 K$ is very sensitive to fluctuations
in the potential difference between the centre and points near
the centre. These potential differences fluctuate by relatively
large amounts between realizations because the total number of
particles in the confining core is small ($\stackrel{\sim}{\scriptstyle\sim} 50$) and the statistical
determinacy of the core structure is correspondingly poor.
     These considerations indicate that it is important to formu-
late comparisons of theory and observation as far as possible in
terms of quantities which do not depend sensitively on the behaviour of the brightness at the very centre. For example, the X-ray
core radius (i.e. the radius at which the surface brightness in
X-rays falls to one half of the central value) is very ill-
determined theoretically because the central brightness is

uncertain. We have therefore defined an alternative measure of
the degree of central concentration of the X-ray emission; we
define $B_{31/55}$ to be the ratio of the X-ray surface brightnesses
at points 0.31 Mpc and 0.55 Mpc from the cluster centre along a
line inclined at 45° to the principal axes of the cluster. We
find that $B_{31/55}$ fluctuates between realizations by $\leq$ 7% for
realistic parameters and is a sensitive test of whether the cluster
is prolate or oblate in shape; the prolate models have much deeper and steeper potential wells than do the oblate models, with
the result that $B_{31/55}$ tends to be twice as great in the prolate
as in the oblate case (for fixed γ, T_o, etc.).

4. RESULTS OF THE MODELS OF COMA

1) The mass of Coma out to 30 arcmin from the centre is
$2.5 \times (\sigma/944)^2 \times 10^{15}M_\odot$ and $2.3 \times (\sigma/944)^2 \times 10^{15}M_\odot$ in the prolate
and oblate cases respectively. Thus the mass is neither geometry-
nor realization-dependent. It does, however, appear to rise nearly linearly with the radius r to which one includes galaxies;
$M = (r/35' + 0.26) \times (\sigma/944)^2 \times 10^{15}M_\odot$ in the prolate case.
2) The shape of the X-ray isophotes is also nearly independ-
ent of both geometry and realization; one finds that $\varepsilon \equiv (1-b/a)$
falls from $\varepsilon \sim 0.3$ near the centre to $\varepsilon \sim 0.2$ at r = 20 arcmin,
and is then nearly constant out to the furthest point we can de-
termine. There is a slight suggestion that the major axis of the
X-ray isophotes may twist very near the centre (r < 10 arcmin) or
that the isophotes become relatively round there. Both adiabatic
and isothermal models show the same ε-behaviour.
3) $B_{31/55}$ depends on the central temperature and on γ, but
is typically twice as great for a prolate model as for the corresponding oblate model. The X-ray core radius for isothermal
$T_o = 10^8$K models is ill-defined, but is smaller than the galaxy
core radius $(r_{c,X} \sim 4'; r_{c,G} \sim 6')$ for the prolate model and comparable to the galaxy core radius in the oblate case.
4) The central gas densities of the T = 10^8K isothermal
models required to generate the UHURU X-ray luminosity are $n_e =$
3×10^{-3} cm^{-3} in the oblate case and $n_e = 6 \times 10^{-3}$ cm^{-3} in the
prolate case. The latter value implies a central cooling time
which is appreciably shorter than the Hubble time. This suggests
that the cluster may be more nearly oblate than prolate.

REFERENCES

Abell, G.O., 1977, Astrophys. J., 213, 327.
Binney, J.J., and Strimpel, O., 1978. Mon.Not.Roy.Astr.Soc., 185,
473.
Carter, D., and Metcalfe, N., 1980. Mon.Not.Roy.Astr.Soc., submitted.
Cowie, L.L., and Binney, J.J., 1977, Astrophys. J., 215, 723.

Faber, S.M., and Gallagher, J.S., 1979, Ann.Rev.Astron.Astrophys., 17, in press.
Godwin, J.C., and Peach, J.V., 1977, Mon.Not.Roy.Astr.Soc., 181, 323.
Gregory, S.A., and Thompson, L.A., 1978, Astrophys. J., 222, 784.
Gull, S.F., and Northover, K.J.E., 1975, Mon.Not.Roy.Astr.Soc., 173, 585.
Perrenod, S.L., 1979, Astrophys. J., 226, 556.
Press, W.H., 1976, Astrophys. J., 203, 14.
Schipper, L., and King, I.R., 1978, Astrophys. J., 220, 798.
Strimpel, O., and Binney, J.J., 1979. Mon.Not.Roy.Astr.Soc., 188, 883.

X-RAY EMISSION FROM CLUSTERS AND THE FORMATION OF GALAXIES

James Binney

Magdalen College and Department
of Theoretical Physics
Oxford, England.

1. INTRODUCTION

In these remarks I hope to show that the discovery by X-ray astronomers in 1976 that the gas in rich clusters of galaxies has a nearly solar abundance of iron (Mitchell et al.,1976; Mushotzky et al.,1978) fits rather nicely with recent developments in other areas of extragalatic astronomy, and might indeed have been anticipated on the basis of papers published prior to 1976. Many of the ideas I shall be discussing will be found in Binney and Silk (1978).

2. THE G-DWARF PROBLEM

Van den Bergh (1962) and Schmidt (1963) long ago showed that the relative paucity of metal-poor F- and G-dwarfs in the solar neighbourhood must be considered rather surprising if one believes that the heavy elements in the Galaxy were synthesized by disk stars. Indeed, if M_g is the gas-mass, M_s the stellar-mass and H the mass of heavy elements in the galactic interstellar material, so that $z=H/M_g$ is the fractional metallicity of the ISM (which it is reasonable to assume is well-stirred and therefore homogeneous), then if a mass $\delta M_s=-\delta M_g$ of stars are made out of the ISM, one has for the small increase δH in the body of interstellar metals

$$\delta H = y \, \delta M_s - z \, \delta M_s = (z-y) \, \delta M_g \tag{1}$$

Here y, the heavy-element "yield", is the fraction of the mass of any generation of stars which is returned to the ISM by supernovae as heavy elements. Of course, the lifetimes of the extremely massiv

R. Giacconi and G. Setti (eds.), X-Ray Astronomy, 245-251.

(M>7 $M_\odot$) stars that are thought to produce heavy-elements are so
short that the heavy elements produced by the formation of a generation of stars are returned to the ISM effectively instantaneosly.
But as $\delta H = \delta(M_g z)$, one has on eliminating δH from equation (1)

$$\delta M_g/M_g = - \delta z/y \tag{2}$$

The yield y will depend sensitively on the relative rates of
formation of high-mass, metal-producing stars to lower-mass stars.
The simplest assumption (for which there is some direct evidence
(Searle et al.,1973; Wielen,1974) is that y is independent of z.
Then equation (2) yields the well known relation of Searle and
Sargent (1972)

$$z(t) = y \ln M_t/M_g(t) \tag{3}$$

Here $M_t = M_g + M_s$ is the total galactic mass. At present $M_t \sim 10\ M_g$,
so

$$\exp\left[z(t_o)/y\right] \simeq 10 . \tag{4}$$

Now consider how many unevolved stars there should be whose
metallicities are less than say 0.1 $z_\odot$ (taking z (Orion) $\sim z_\odot$,
Peimbert and Torres-Peimbert (1977)). Clearly the mass of all stars
formed with z < 0.1 $z_\odot$ is given by

$$M_s\ (0.1\ z_\odot) = M_t\left[1 - \exp(-0.1\ z_\odot/y)\right] \approx 0.21\ M_t \tag{5}$$

by equations (3) and (4). For any realistic initial mass function
most of this mass will have gone into stars with M < 0.7 $M_\odot$ which
have yet to evolve off the main sequence. Therefore as M_s (now)
$\lesssim M_t$, equation (5) says that if y does not depend on z, this type
of model predicts that more than 20% of all dwarf stars should
have z < 0.1 $z_\odot$. Actually Bond (1970), Clegg and Bell (1973) and
others find that <1% of F and G stars in the solar neighbourhood
have z < 0.1 $z_\odot$. Indeed one finds that stars with z < 1/3 $z_\odot$ are
already very rare near the Sun.
    Two ways out of this difficulty immediately suggest them-
selves: A) The yield y may have been higher in the past, due to
a bias at early times towards the formation of massive stars;
B) The galactic disk may not be a closed system such that $M_g + M_s = M_t$ is constant. Thuan, Hart and Ostriker (1975) have examined a
two-parameter family of models designed to implement proposal (A).
However, none of these models was able to reproduce satisfactorily
the counts of metal-poor dwarfs because these counts require such
a large bias towards the formation of massive stars at early times,
that in order to form the right number of dwarf stars by the present epoch, the rate of formation of dwarf stars in these models
actually has to increase as the Galaxy gets older. This increase
in the dwarf formation-rate (and thus the rate of conversion of

gas into stars) leads to the galactic gas supply running out very
soon after the present time. This last seems intrinsically improb-
able and cannot be reconciled with the very large number of spirals
which, like the Galaxy, have $M_g/M_t \simeq 0.1$. Thuan et al.(1975) thus
conclude that one-zone models in which the galactic disk forms a
closed system, can be ruled out. At any rate it is clear that a
one-zone model requires a very strong variation of y with z if it
is to escape the dual perils of too many metal-poor stars and a
rising rate of gas consumption, and that this suggests that one
examine instead proposal (B).

One may open up the galactic disk to two types of infalling
gas: i) primordial metal-poor gas (Binney and Silk,1978; Quirk
and Tinsley,1973; Larson,1974a) and ii) metal-rich ejecta from
pregalactic stars (Truran and Cameron,1972) or halo stars (Ostriker
and Thuan,1975). I think there is a strong case for believing with
Ostriker and Thuan (1975) that any infalling material consisted
of metal-rich ejecta from spheroidal-component stars: i) Observations of globular clusters, the galactic bulge and elliptical galaxies (e.g. Burstein,1979) indicate clearly that material with
$z \gtrsim z_\odot$ was produced by spheroidal-component stars long before the
disk existed. It is inconceivable that these stars did not pollute
the proto-disk material with heavy elements. ii) The clouds that
produce the heavy element absorption features in quasar spectra
clearly do not have primordial abundances and yet they lie at
great distances from the nearest galaxy (e.g. Sargent,1977). iii)
I shall argue next that in rich clusters the X-ray observations
are detecting the metal-rich proto-disk material.

3. THE INTRACLUSTER MEDIUM AS PROTO-DISK MATERIAL

Ostriker and Thuan explain the scarcity in the solar neighbourhood of G-dwarfs with $z < 1/3\ z_\odot$ by arguing that the metallicity
of the material from which the disk stars formed was raised to
$z \sim 1/3\ z_\odot$ by dying halo stars before the majority of disk stars
had formed. If this supposition is correct, and if the material
which should have been used to form disks about the spheroidal
components of rich clusters were somehow prevented from condensing
into disks, by being shock heated to form the $T \sim 10^8 K$ intracluster medium, then one would have a natural explanation of the striking fact that all the intracluster media for which z has been determined (Bahcall and Sarazin,1977) have the same metallicity
$z \simeq 1/3\ z_\odot$, at which the population of solar-neighbourhood G-dwarfs
cuts off. And what makes this line of reasoning rather compelling
is that it also explains a) why there is about as much hot gas in
clusters as there is and b) why this gas did not settle to form
disks, whereas the gas around the Galaxy and M31 mostly did. In
this section I shall deal with point (a) before turning to point
(b) in the next section.

The case of the Local Group, whose overall luminosity is

dominated by the contributions of two Sb - Sbc galaxies (M31 and
the Galaxy) is probably fairly typical of the situation in most
small groups of galaxies (which in turn account for most of the
luminosity of the Universe). De Vaucouleurs' photometry of M31
(de Vaucouleurs,1958) shows that in this Sb galaxy the disk has
about twice the luminosity of the spheroidal component. Let me
adopt a mass-to-light ratio of 3 $M_\odot/L_\odot$ as characteristic of the
disk of M31. Then, if the spheroidal component raised the metallicity of all this proto-disk material to $\sim 1/3\ z_\odot$, which corresponds
roughly to H = 5 x $10^{-3}\ M_{disk}$, we have that for each $L_\odot$ of the
spheroidal-component luminosity of M31, a mass H = 5 x 10^{-3} x 3
x 2 $M_\odot$ = 3 x $10^{-2}\ M_\odot$ of heavy elements must have been produced by
spheroidal-component stars early in the history of M31.

Now, apply this rate of production of heavy elements to the
spheroidal components of the Coma cluster. In a great cluster like
Coma essentially all the light derives from spheroidal components.
The total luminosity of Coma is probably $\sim 10^{13}\ L_\odot$ (Bahcall,1977),
so that one might expect 3 x $10^{11}\ M_\odot$ of heavy elements in the intracluster gas if this was also enriched by the spheroidal-component
stars. Bahcall and Sarazin (1977) conclude from the X-ray data that
there are 5 x $10^{11}\ M_\odot$ of heavy elements in the hot gas of Coma
(with an uncertainty of about a factor two). Thus the mass of
heavy elements in the Coma cluster agrees (to within the admittedly
large uncertainties) with the Ostriker and Thuan hypothesis that
galactic disks were pre-enriched by spheroidal-component stars.

4. ANGULAR MOMENTUM AND FREE-FALL TIMES

Recent clustering simulations can explain why the proto-disk
material belonging to galaxies near the cores of rich clusters
failed to form disks. It has long been argued that the angular
momentum of disk galaxies derives ultimately from tidal interactions with neighbouring proto-galaxies, but exactly how much angular momentum can be conveniently acquired in this way has until
recently been the subject of some debate. Now two developments
have made it possible to make a pretty good estimate of the amount
of spin actually imparted by tidal interactions: i) large numerical
simulations of the aggregation of particles in the expanding Universe to form rotating bound units (e.g. Efstathiou and Jones,
1979) and ii) observational measurements of the rotation velocities
of giant elliptical galaxies are in good agreement with one another
and with Peebles original estimate (Peebles,1969) of the tidally
induced value of the parameter $\lambda = \sqrt{2}\ J\ E^{\frac{1}{2}}/GM^{5/2}$, which was introduced by Lynden-Bell (1966) ($\lambda \sim 0.08$).

The quantity λ is a dimensionless measure of the angular momentum of a proto-galaxy. It is ~ 1 for a self-gravitating, centri-
fugally-supported disk and 0 for a spherical pressure-supported
body. Thus, if the galactic disk were principally constrained by
its own gravity, one would have $\lambda_{now} \approx 1$ and, as $\lambda \sim R^{-\frac{1}{2}}$ for a

system which contracts under conservation of angular momentum, one would obtain for the original radius R_1 of the cloud from which the disk formed, $R_1 \approx R_o/(0.08)^2 \gtrsim 1$ Mpc; that is, if there were no heavy halo, the solar material would have fallen in from a position well beyond M31! The impossibility of this scenario is evident when one calculates the free-fall time of a 10^{11} $M_\odot$ galaxy from such a radius; $t_{ff} \sim 5 \times 10^{10}$ y. A more acceptable picture is obtained if one supposes that the Galaxy does have a heavy halo whose mass increases about linearly with radius. Then the galactic circular velocity is constant at all radii and $\lambda_{initial}$ becomes approximately equal to the ratio of the initial cloud rotation speed to the circular velocity. If a small part of the cloud falls from $R=R_1$ under conservation of angular momentum ($v \sim R^{-1}$), its tangential speed will reach the circular velocity when it has fallen to $R_o \approx \lambda_{initial} \times R_1$. Therefore with a suitable heavy halo one obtains $R_{1\odot} \approx 120$ kpc. One should probably regard this second estimate of $R_{1\odot}$ as a lower limit, because it is uncertain that the halo has a flat rotation curve right out to 120 kpc.

At any rate these estimates of the initial λ-value of the Galaxy indicate that the solar-neighbourhood material must have fallen in from a very great radius, and that the substantial amount of disk material outside the solar circle must derive from even greater radii. The galaxies in rich clusters are typically less than a few hundred kpc from one another. Therefore it is likely that in these clusters the acquisition by spheroidal components of high angular-momentum disk material will have been disturbed by the overall collapse of the cluster. The gas will have been heated within one free-fall time of the cluster to $T \sim 10^8$ K, and has for the most part yet to cool effectively.

5. PRODUCTION OF HEAVY ELEMENTS

I have argued that there is a universal metals/halo-luminosity ratio that holds both for small groups of galaxies and for rich clusters. Furthermore, the well-known line-strength/absolute magnitude correlation and the occurrence of strong metallicity gradients in many population II systems (for example, the galactic globular-cluster system) strongly suggest that the principal seat of early heavy-element production was in the visible spheroidal components of galaxies. This argument has been strengthened recently by the discovery by Faber et al.(1980) that the line-strengths of elliptical galaxies are very tightly correlated with true velocity dispersion (that is with velocity dispersion corrected for shape effects); the existence of a tight correlation of metallicity with velocity dispersion, and therefore with depth of potential well, very strongly suggests that the galactic wind picture of Larson (1974b) and others is essentially correct. Furthermore, one may show that prompt enrichment of the proto-disk and proto-intergalactic material by the spheroidal components of groups and clusters is consistent with the conventional theory of stellar evolution in such systems.

Consider the formation $\sim 10^{10}$y ago of a population of stars with a power-law IMF $N(M) \sim M^{-n}$. Stars more massive than $M_T = 0.7\ M_\odot$ will have completed their evolution and have left only dark remnants. Some of these stars, those more massive than Mp say, will have returned a substantial fraction, α, of their mass to the ISM in the form of heavy elements. Essentially all the luminosity of the population is now emitted by stars (currently on the giant and horizontal branches) whose masses lie in a narrow range δM_T around M_T, whilst most of the mass of the population is contributed by stars of mass $<M_T - \frac{1}{2}\delta M_T$, which do not contribute appreciably to the overall luminosity of the system.

As Ostriker and Thuan have emphasized, the amount of heavy elements produced by such a population is related to the present luminosity via the slope n and the ratio of masses Mp/M_T; if one assumes n > 2 and that Mp is appreciably smaller than the mass of the largest stars formed, then one has approximately

$$H/L = \frac{K\alpha}{(n-2)}\ (M_T/Mp)^{n-2} \tag{6}$$

where K is given by the fraction β of their mass which stars now on the horizontal and giant branches will radiate before they die as white dwarfs through

$$K = \frac{2\ \tau_H}{c^2\beta} \tag{7}$$

Here τ_H is the Hubble time. Equations (6) and (7) enable one to derive theoretically the metals to halo-luminosity ratio which I have shown is $\sim 3 \times 10^{-2} M_\odot/L_\odot$ for the Local Group and for the Coma cluster. If one adopts the Salpeter IMF slope n = 2.35, takes $\alpha = 0.1$, $(M_T/Mp) = 0.1$ and $\beta = 0.3 \times 0.007$, then one finds H = 1.8×10^{-2} in excellent agreement with the observational value. If one calculates the total mass of gas ejected by dying stars, one finds that there is too little of this to account for the observed disk and IGM masses (Vigroux,1977); this confirms that the role of the spheroidal components was to act as polluters of fundamentally primordial material, and is in agreement with the angular momentum argument outlined above and with the presence in the ISM of deuterium (which would have been destroyed by astration).

Actually this model is too naive. Several generations of massive stars must have been formed near the centres of giant galaxies in order that the mean metallicities of the presently-evolving stars in those regions can be as high as they are. Yet it is encouraging that such diverse investigations as studies of gravitationally-driven clustering, stellar evolution and X-ray astronomy can be fit into a coherent picture of galaxy formation and evolution.

REFERENCES

Bahcall, J.N., and Sarazin, C.L., 1977, Astrophys. J. (Letters), 213, L99.

Bahcall, N.A., 1977, Ann. Rev. Astron. Astrophys., 15, 505.

Binney, J.J., and Silk, J., 1978, Comments Astrophys., 7, 139.

Bond, H.E., 1970, Astrophys. J. Suppl., 22, 117.

Burstein, D., 1979, Astrophys. J., 232, 74.

Clegg, R.E.S., and Bell, R.A.M., 1973, Mon. Not. Roy. Astr. Soc., 163, 13.

de Vaucouleurs, G., 1958, Astrophys. J., 128, 465.

Efstathious, G., and Jones, B.J.T., 1979, Mon. Not. Roy. Astr. Soc., 186, 133.

Faber, S.M., Terlevich, R., Davies, R.L., and Burstein, D., 1980, in preparation.

Ikeuchi, S., Sato, H., Sato, T., and Takeda, H., 1972, Prog. Theor. Phys., 48, 1885.

Larson, R.B., 1974a, Mon. Not. Roy. Astr. Soc., 166, 585.

Larson, R.B., 1974b, Mon. Not. Roy. Astr. Soc., 169, 229.

Lynden-Bell, D., 1966, Proc. IAU symp. No.31, Radio astronomy and the galactic systems, H. van Woerden (ed.), (London: Academic Press).

Mitchell, R.J., Culhane, J.L., Davison, P.J.W., and Ives, J.C., 1976, Mon. Not. Roy. Astr. Soc., 176, 29.

Mushotzky, R.F., Serlemitsos, P.J., Smith, B.W., Boldt, E.H., and Holt, S.S., 1978, Astrophys. J., 225, 21.

Ostriker, J.P., and Thuan, T.X., 1975, Astrophys. J., 202, 353.

Peebles, P.J.E., 1969, Astrophys. J., 155, 393.

Peimbert, M., and Torres-Peimbert, S., 1977, Mon. Not. Roy. Astr. Soc., 179, 217.

Quirk, W.J., and Tinsley, B.M., 1973, Astrophys. J., 179, 69.

Sargent, W.L.W., 1977, The evolution of galaxies and stellar populations, Tinsley, B.M. and Larson, R.B. (eds.), (New Haven: Yale University Observatory).

Schmidt, M., 1963, Astrophys. J., 137, 758.

Searle, L., and Sargent, W.L.W., and Bagnuolo, W.G., 1973, Astrophys. J., 179, 427.

Thuan, T.X., Hart, M.H., and Ostriker, J.P., 1975, Astrophys. J., 201, 756.

Truran, J.W., and Cameron, A.G.W., 1972, Astrophys. Space Sc., 14, 179.

van den Bergh, S., 1962, Astron. J., 67, 486.

Vigroux, L., 1977, Astron. Astrophys., 56, 473.

Wielen, R., 1974, Highlights. Astron., 3, 395.

RADIO PROPERTIES OF EXTRAGALACTIC X-RAY SOURCES

P. Katgert

Sterrewacht, Leiden, The Netherlands

1. INTRODUCTION

The classes of extragalactic objects identified with X-ray sources
are: clusters of galaxies, Seyfert galaxies, other types of
(active) galaxies, QSO's, BL Lac objects and, possibly, superclusters. For a review of the properties of extragalactic X-ray
sources one is referred to Gursky and Schwarz (1977). The relative
occurrence of the different classes of objects among the high-
latitude ($|b| > 20^{\circ}$) X-ray sources in the 4U (Forman et al. 1978)
and 2A (Cooke et al. 1978) catalogues is given in Table 1. The
results from the two catalogues are not at all independent,
because they have a large number of sources in common. The higher
identification rate in the 2A catalogue reflects the generally
smaller error boxes in this catalogue, compared with those in the
4U catalogue.
    Given the uncertainties in the X-ray positions, identifi-
cations can only seldom be made on positional coincidence alone.
Classes of objects that are not expected to be X-ray emitters may
therefore have been missed as identifications. Better positions
for the unidentified high-latitude sources are essential in this
respect. Below the limits of the 4U and 2A catalogues additional
classes of objects can be expected. Fabian and Rees (1978) suggest
e.g. that young galaxies and remote radio galaxies may be found
at the levels reached by the Einstein Observatory.

2. RELATIONSHIPS BETWEEN X-RAY AND RADIO EMISSION

The relation between radio and X-ray emission may be either direct

253

Table 1

Identification statistics of high-latitude X-ray sources

catalogue	4U		2A	
X-ray sources	152		87	
	probable	possible	probable	possible
clusters	21%	11%	24%	13%
superclusters		4		
Seyfert galaxies	5	1	5	6
other (active) galaxies	2	1	3	6
QSO's	1		1	
BL Lac objects	1		1	
other (Galactic, LMC)	7	3	8	4
unidentified	43%		29%	

or indirect. In the case of emission from clusters, the relation is probably indirect because the X-ray emission is due to thermal bremsstrahlung from a hot intracluster gas. The presence of this gas (and its properties) is also revealed by observations of radio sources in clusters. In quite a few cases there is evidence for interaction between the hot intracluster gas and the reservoirs of relativistic particles in the radio sources that produce the synchrotron radio emission. It has also been suggested that relativistic electrons escaping from the cluster radio galaxies may play an important role in heating the intracluster gas (Lea and Holman, 1978).

A direct relation between X-ray and radio emission may exist when the same relativistic electrons that produce the radio emission by synchrotron radiation in a magnetic field also produce X-rays by scattering against a photon field. The Lorentz factor γ of electrons radiating at frequency ν(MHz) in a magnetic field B (μG) is $\gamma \sim 500 \ (\nu/B)^{\frac{1}{2}}$. Inverse Compton scattering of these electrons against photons of energy ε_{ph} produces photons of energy $\sim \gamma^2 \ \varepsilon_{ph}$. For instance, photons near the maximum of the 3K blackbody spectrum of the microwave-background have energies $\varepsilon_{ph} \sim 10^{-3}$ eV. For these to be scattered into the soft X-ray domain ($\varepsilon_{ph} \sim 1$ KeV) requires relativistic electrons with $\gamma \sim 10^3$. Such electrons would also emit synchrotron radiation at frequencies around ~ 10 B MHz.

The ratio of the energies lost by the relativistic electrons through the synchrotron and inverse Compton processes respectively is equal to the ratio of the energy densities in the magnetic field and the photon field. For instance, electrons in both a magnetic field of ~ 3 μG and a 3K blackbody photon field lose equal amounts of energies by the two processes. A detailed dis-

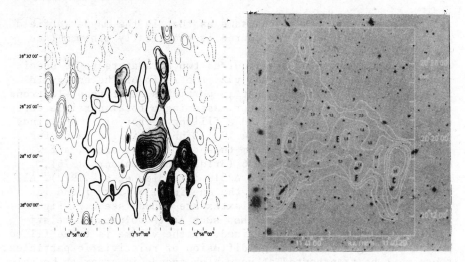

Fig. 1. Cluster halo sources in Coma (left) and A1367 (right).

cussion of the X-ray intensity to be expected from various classes
of radio sources as a result of inverse Compton scattering on the
3K background has recently been given by Harris and Grindlay (1979).

3. RADIO EMISSION FROM CLUSTERS OF GALAXIES

3.1 The extended steep-spectrum halo sources

Some clusters contain a diffuse, extended radio source for which
no single optical counterpart can be found. The first such source
to be found is Coma C, which was first mapped by Willson(1970) at
408 MHz. Figure 1a is a map at 610 MHz by Valentijn (1978a) with
a resolution of $\sim 1' \times 2'$. From the visibility curve a half-power
diameter of $33' \pm 5'$ and a total flux density of 1.2 ± 0.5 Jy at
610 MHz are deduced. A recent observation at 430 MHz with the
Arecibo dish (Hanish et al. 1979) (resolution $\sim 9'$) yields $45' \pm 5'$
and 4.5 ± 1.0 Jy at 408 MHz, for the extended source alone. The
linear size of the source is about 0.7 Mpc.
    Another example of a diffuse unidentifiable source is the one
found by Gavazzi (1978) in the cluster A1367 (Fig. 1b). This source
is considerably smaller than Coma C, with a linear size of at most
a few hundred kpc, and also weaker ($S_{610} \sim 0.4$ Jy). The spectral
index between 610 and 1415 MHz is about 1.5, i.e. comparable to
that of Coma C which is about 1.3.
    As is well known, the observed properties of a radio source
do not allow a unique determination of magnetic field strength and
relativistic particle density in the source. Assuming the synchrotron process to be responsible, one may however determine the so-
called equipartition magnetic field B_{eq}, for which the energy in

relativistic particles is the same as that in the magnetic field and the total energy in the source is practically minimum (see e.g. Moffet, 1975).

The calculation of B_{eq} requires two parameters to be specified which cannot be observed: viz. the ratio of the energy in relativistic electrons and that in particles other than electrons, and the so-called filling factor. Assuming a total particle energy twice that of the electrons and a filling factor of 1, one finds a value of B_{eq} in Coma C of $\sim 4 \ 10^{-7}$ G and a similar value for the source in A1367. Apart from these two halo sources, a few others may exist, e.g. in A2256 (Bridle and Fomalont, 1976), A2142 (Harris et al.,1977) and A2319 (Grindlay et al.,1977). All these halo sources (apart from the one in A1367) are found in clusters where tailed radio galaxies are found (see e.g. Figure 2). It is not entirely clear how such large volumes of relativistic particles can be maintained. If the volume is to be filled by the radio galaxies through diffusion of relativistic particles, these must be transported at very high speeds in order to be able to fill such large volumes within the timescales on which radiative losses occur. The existence of halo sources in clusters has therefore also been interpreted as evidence for reacceleration of relativistic particles in intracluster space.

3.2 Individual radio galaxies in clusters

A large number of Abell (and other) clusters have been observed at radio wavelengths with sufficiently high resolution that individual radio sources can be identified with individual galaxies and that structures of the extended radio galaxies can be obtained. Many of the extended sources in clusters appear not to have the classical symmetric double structure characteristic of strong extragalactic radio sources. Instead they are often asymmetric,

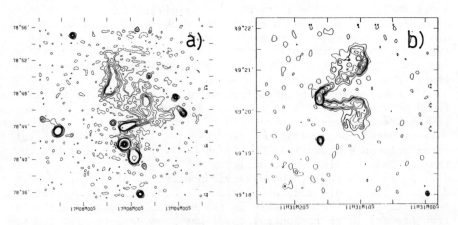

Fig. 2. Tailed radio sources in A2256 (left) and A1314 (right).

with so-called head-tail structures, the classical example of which is the radio galaxy NGC 1265 (see e.g. Miley et al.,1975 and Owen et al.,1978). This source has a clear double-stranded structure, which is found in all head-tail galaxies, if observed with sufficient resolution.

Figure 2a shows a map of the cluster A2256, made by Bridle et al. (1979) with the Westerbork Telescope at 21 cm, in which quite a few head-tail galaxies can be seen.

The head-tail sources in A2256 that are not clearly resolved, probably have a narrow angle between the two strands ("opening-angle"). In Figure 2b a 6 cm map of the head-tail radio galaxy IC 708 is shown (Vallee et al.,1979), which has a wider opening angle and, in addition, funny "wings" at the ends of the strands.

The usual interpretation of these structures is that they represent basically symmetric double sources that are distorted as a result of the movement of the parent galaxy through a hot intracluster medium. In the "independent-blob" model by Jaffe and Perola (1973), the shape of the "tails" depends on the relative velocity of the galaxy with respect to the medium, the ejection velocity, the mass of the ejected blob, and the density of the surrounding medium.

Owen and Rudnick (1976) and Simon (1978) have noted that the sources with large opening-angles are associated with the (optically) dominating galaxies in the clusters. Valentijn (1978b) has studied this correlation in more detail and finds that the large majority of the sources with a large ($\gtrsim 90^{\circ}$) opening angle have $Mp \lesssim -20.7$ and $P_{610} \gtrsim 5 \times 10^{24} WHz^{-1}$. On the other hand, sources with small ($\lesssim 40^{\circ}$) opening angles almost all have $Mp \gtrsim -20.7$ and $P_{610} \lesssim 10^{24} WHz^{-1}$.

Assuming the mass-to-light ratio to be same for the parent galaxies of both wide- and narrow-angle tail sources, this would imply that the wide-angle tail sources are associated with more massive galaxies than are the narrow-angle tail sources. Whether the wide-angle tails are then mainly the result of low velocities of massive parent galaxies, or whether the latter can produce higher ejection velocities for the blobs than the less massive parent galaxies, cannot yet be decided (see e.g. Guindon and Bridle, 1978).

Since the head-tail sources so clearly show evidence of interaction between the radio source and the surrounding medium, they are a useful probe of the intracluster medium. Consider again, as an example, the source IC 708 shown in Figure 2a. Using the equipartition argument, one can estimate the magnetic field and the energy density (and hence the internal pressure) in different parts of the tails. Close to the nucleus one finds $B \sim 12$ μG, while farther down the tails $B \sim 7$ μG.

Because the galaxy has an estimated velocity of about 1200 km/s with respect to the medium, the parts of the tail close to the nucleus experience a considerable ram-pressure, which must exceed the internal pressure in the tail. Close to the galaxy this argument provides a lower limit to the external electron density of $\sim 4 \times 10^{-4}$ cm^{-3}. The more relaxed regions down the tail are assumed to be confined by thermal pressure of the surrounding medium. Equating internal and external pressure, and assuming the lower limit to the external density as obtained earlier to be valid, one then gets an upper limit to the external temperature of $\sim 7 \times 10^7$K. Because the above line of argument contains a few assumptions, it is interesting that the derived numbers are quite similar to typical values of temperature and density implied by X-ray data of other clusters.

3.3 The radio source population in clusters

It has long been considered an important question whether the radio sources in clusters, taken as a class, have properties different from those outside clusters. Obvious questions are e.g. 1) whether galaxies in clusters have a higher chance of producing a radio source than those outside, 2) whether galaxies in clusters are associated with brighter radio sources than those outside, 3) whether radio sources have systematically different sizes in- and outside clusters, and 4) whether the radio spectra of sources in- and outside clusters are different. Information on any of these points might help to understand the influence of the cluster environment on individual cluster members.

By comparing the luminosity functions of galaxies in- and outside clusters questions 1) and 2) can be studied, albeit not separately. Auriemma et al. (1977) have made this comparison for elliptical and SO galaxies. Their cluster sample only contains galaxies from central regions of Abell clusters, with typical galaxy densities of ~ 100/Mpc^3. The "field" sample consists of sources in smaller aggregates with galaxy densities of at most ~ 10/Mpc^3. Their result is shown in Figure 3 in the form of the fraction of surveyed galaxies with $M_p \leq -19$ in each of the two groups that was detected as a radio source with radio luminosity $P_{1.4}$. As can be seen, there is no evidence for significant diffe-

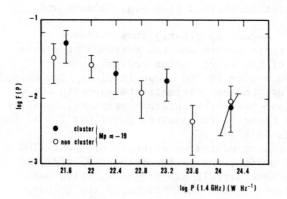

Fig. 3. Comparison of the fractional radio luminosity functions of elliptical and SO galaxies in and outside clusters.

rences between cluster and non-cluster populations. Recent determinations of the radio luminosity function (RLF) of E and S0
galaxies in the Hercules supercluster (Valentijn, 1978b) and the
cluster A1367 (Gavazzi, 1979) are in agreement with the determination by Auriemma et al.

For the spiral galaxies the situation is less clear. Jaffe
et al. (1976) studied the RLF of spiral and irregular galaxies in
the Coma cluster, and concluded that these are slighter "over-
luminous" as compared to the field spiral galaxies studied by
Cameron (1971). In the Hercules supercluster (Valentijn, 1978b)
the S and I galaxies are, if anything, slightly "underluminous"
with respect to the field, while a similar situation may be present
in A1367 (Gavazzi, 1979). There are indications that the presence
of optical emission lines may play a role.

With regard to the sizes of radio sources in- and outside
clusters the situation is even more uncertain. De Young (1972)
found evidence that the radio sources inside clusters may be
smaller than those outside. Hooley (1974) and Burns and Owen
(1977) did not find such a difference but the samples on which
these conclusions are based are small and systematic effects may
not be negligible. Lari and Perola (1978) compared the size distributions of 58 radio galaxies inside and 41 radio galaxies outside clusters. Because radio morphology and size have been found
to be correlated with luminosity (see e.g. Gavazzi and Perola,
1978) both samples only contain sources in the same, rather narrow,
luminosity interval. As can be seen from Figure 4 there is no
clear difference between the two distributions.

Recently, however, Stocke (1979) has suggested that the concept of cluster- and non-cluster samples is not very adequate for
the purpose of studying the influence of the intergalactic medium
on radio sources. Instead, he studied the relation between radio
source size and a measure of the mass-density in the vicinity of
the source, and found the results to be consistent with the
presence of a strong interaction between the radio galaxies and
their environment.

As far as the radio spectra are concerned, the situation is
relatively clear. Several authors (see e.g. Baldwin and Scott,

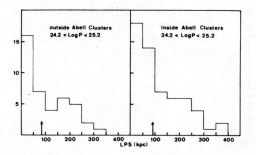

Fig. 4. The distributions
of largest physical size
(LPS) of radio sources
associated with elliptical
and S0 galaxies in and
outside Abell clusters
(with radio luminosities
in the range indicated).

1973 and Roland et al.,1976) have noted that steep-spectrum
sources are found more frequently in clusters than outside. Moreover, Lari and Perola (1978) have shown that the steep-spectrum
sources in clusters are more centrally condensed than the other
sources. Both effects can easily be understood in terms of confinement of the radio sources by intracluster gas, allowing radiative losses to show up as spectral steepening.

4. RADIO EMISSION FROM SEYFERT GALAXIES

Since Seyfert galaxies are generally rather weak radio sources,
a systematic study of their radio properties has become possible
only fairly recently. De Bruyn and Wilson (1978) discuss the
observational data for a more or less complete sample of 41
Seyfert galaxies, of which 29 have been classified as Type I and
12 as Type II. Given the large difference between the optical
spectra of the two types (see, however, Veron, 1979) it is
interesting to see whether the radio properties of the two classes
are also different.

Indeed, the detection rates of the two classes are different:
while 11 of the 12 Type II's are detected, only 10 of the 29 Type
I's were found. This result may be partly due to different distances for the two classes but, as can be seen from Figure 5,
there appears to be a real difference in the luminosities of the
two types. It is estimated that the average luminosity of Type I's
is at least a factor of 5 lower than that of Type II's.

None of the optically selected Seyfert galaxies that were
detected have the classical double radio structure. Quite a few
of the detected radio sources are unresolved, but this is generally consistent with the linear sizes of the resolved radio
sources in nearby Seyferts, which range from several hundred pc
to a few kpc. Radio spectral data are still rather limited, but
from the present data there do not appear to be significant spectral differences between the detected Type I's and Type II's.

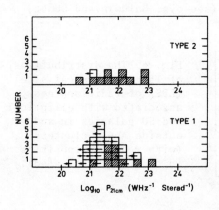

Fig. 5. The radio luminosity
distributions of fairly complete
samples of Seyfert galaxies of
Types I and II.

From the size of the emitting regions, and correlations between
radio continuum luminosities and forbidden-line luminosities, de
Bruyn and Wilson conclude that the radio emission originates in
the same general region as the forbidden, narrow lines, and that
there probably exists pressure equilibrium between the relativistic plasma and the thermal condensations in this region. As
mentioned earlier, the X-rays probably originate in a very much
smaller nuclear region (Elvis et al.,1978).

5. RADIO EMISSION FROM QUASI-STELLAR OBJECTS

As shown both by Tananbaum and Giacconi (this volume) QSO's have
evolved from a quantitatively minor constituent of the 4U and 2A
samples, to an important class of extragalactic X-ray sources in
the Einstein observations. Originally, QSO's were of course found
as the optical counterparts of strong radio sources but subsequent
work showed that the radio-emitting QSO's form only a (small)
fraction of the total QSO population.

Quite a few of the QSO's in the low-frequency radio surveys
have radio structures like the classical double sources associated with strong radio galaxies. On the other hand, QSO's found
in high-frequency surveys generally show a point-like radio source
coinciding with the optical QSO. Observations at high frequency
and with sufficient angular resolution of QSO's selected at low
frequencies do in almost all cases show the presence of a central
radio component (see e.g. Miley and Hartsuyker, 1978).

The relationship between the radio and optical emission from
QSO's has been studied in several ways. First, complete samples
of QSO's from the 3CR, 4C, Bologna, Parkes and Westerbork catalogues have been used (see e.g. Schmidt, 1970, Fanti et al., 1975,
Masson and Wall, 1977, Wills and Lynds, 1978 and De Ruiter,1978).
Second, radio observations have been made of samples of optically
selected QSO's (see e.g. Fanti et al.,1977) to determine the
fraction of QSO's that are radioemitting.

An important result was obtained by Schmidt (1970) when he
found that the redshift distributions of 18th magnitude radio-
loud and radio-quiet QSO's are not significantly different. This
led him to assume that at every redshift the fraction of QSO's
radioemitting with a given ratio of radio to optical luminosity
is the same; or rather, that there exists a universal "guillotine
factor" $G(R)$ which gives the fraction of QSO's with $P_{rad}/P_{opt} > R$.
Since P_{rad} is a monochromatic luminosity, $G(R)$ will be different
for different radio frequencies. However, to a first approximation,
the shape of $G(R)$ does not vary with radio frequency.

Figure 6, taken from the work by Wills and Lynds (1978),
shows $G(R)$ as determined for radio-loud quasars from the 3C and
4C catalogues. If one extrapolates $G(R)$ towards lower R one expects
essentially all QSO's to be radioemitting with $\log R_{178} \gtrsim 2.0$.
However, sensitive radio ·observations of samples of optically

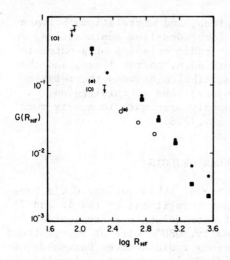

Fig. 6. The guillotine function G(R), i.e. the fraction of QSO's with a ratio between radio and optical luminosities exceeding R (with radio luminosity referring to 2700 MHz).

selected QSO's show quite a large fraction of these to be radio quiet with very low upper limits to R. Apparently, the concept of a universal G(R) function is at most applicable for relatively large optical and radio luminosities. This same conclusion was also reached by de Ruiter (1978) from a study of weak radio emitting QSO's found in deep Westerbork surveys.

6. RADIO OBSERVATIONS OF EINSTEIN OBSERVATORY DEEP SURVEY AREAS

As described by Giacconi (this volume), the Einstein Observatory is engaged in a deep X-ray survey, of several areas of sky of a few square degrees each, down to 10^{-2} to 10^{-3} Uhuru cts with both HRI and IPC. Of the northern ($\delta \gtrsim 30°$) survey areas, radio observations have been or will be made with the Westerbork Telescope at wavelengths of 21 and 49 cm. At 49 cm ∿ 250 radio sources are observed in an area of ∿ 3° diameter; at 21 cm ∿ 80 sources are detected in the central ∿ 1°.2 of the same area.

At the present moment a comparison between X-ray and radio data is possible for the Draco area at 17^h10^m, 71°10'. Of the 20 X-ray sources found, about 7 are probably identified with galactic stars and hence not expected (nor found) to be radio emitting. Of the remaining X-ray sources, 3 are coincident with a radio source, one may be and another one is coincident with an Sb galaxy (not a radio source) that is a member of a group of 3 spirals of similar apparent magnitudes, and of which the other two are radio emitting.

In Figure 7 we show contour maps of the 3 probable X-ray radio identifications and of the group of spiral galaxies. Crosses denote X-ray positions and their uncertainties, while optical objects have been indicated. The correct interpretation of the source 1E 17 12 10+71 14.4 is somewhat uncertain, since the radio

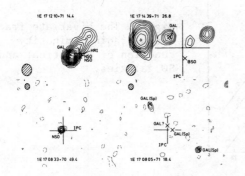

Fig. 7. Contour maps of three radio sources coincident with X-
ray sources from the HEAO-2 deep survey in Draco (X-ray positions
marked HRI or IPC), together with map of a region with three
comparably bright spiral galaxies.

source may be a head-tail source or an unequal triple. On the
first interpretation the optical identification is probably the
faint galaxy near the head, with the X-rays coming from the tail.
On the second interpretation, the optical identification may be
with a stellar object coinciding with central X-ray and radio
sources. The identification of 1E 17 14 39+71 26.8 is similarly
confused, and depends also on the interpretation of the four
radio peaks. Only the X-ray radio identification of 1E 17 08 33
70 49.4 is straightforward. This is a clear case where the radio
position allows a choice to be made between various optical
candidates.
     The results of the radio surveys carried out in the areas
of the deep X-ray survey will also be used for a study of the
evolution of radio sources and radio source populations. The deep
optical plates that are available for these areas will be used
for an optical identification program that will extend similar
work done for earlier Westerbork surveys. In the following
sections we discuss the background for these programs and the
results obtained so far.

7. THE TIME SCALE FOR RADIO SOURCE EVOLUTION

Already thirteen years ago Schmidt (1966) estimated the lifetime
of the strong extended radio sources associated with bright
elliptical galaxies. Using the fraction of bright elliptical
galaxies that contain a strong radio source, he calculated the
harmonic mean lifetime of the radio emitting phase in elliptical
galaxies as the product of this fraction and the age of the ellip-

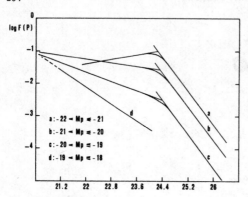

Fig. 8. The bivariate fractional radio luminosity
function of elliptical and
SO galaxies.

tical galaxies. It is implausible, however, that the strong and
weak radio sources have the same lifetimes, if only because the
strong sources would use up a sizeable fraction of the energy-
equivalent of the galaxy mass if they were to radiate at the
observed level over the harmonic mean life-time of, say, 10^9 years.
Lifetimes for different radio luminosities were estimated by
Schmidt and by Van der Laan and Perola (1969) assuming various
dependences of lifetime on luminosity.

Very detailed, and direct, information on this question has
been obtained by Auriemma et al. (1977) who determined the local
($Z \lesssim 0.1$) bivariate (radio-optical) luminosity function of elliptical (and SO-) galaxies. More recently, Meier et al. (1979) have
extended and confirmed this work. Figure 8 summarizes the results
in the form of the fraction of ellipticals of various absolute
optical magnitude classes that is radio emitting with radio
luminosity P_r. This figure is based on data for about a hundred
elliptical radio galaxies both in- and outside clusters. Using
Schmidt's argument one derives lifetimes of between $\sim 10^6$ and
$\sim 10^9$ years for strong and weak elliptical radio galaxies respectively.

An interesting feature of the fractional bivariate luminosity
function in Figure 8 is that for $\log P_{1.4}$ (WHz^{-1}) $\gtrsim 24.4$ the
slopes of the fractional radio luminosity functions for different
M_p are practically identical. This means that the relative proportions of, say, $\log P = 24.5$ and $\log P = 26$ radio galaxies is
the same for $M_p = -21.5$ and $M_p = -19.5$ elliptical galaxies. It
also means that the average absolute magnitude of $\log P = 24.5$
and $\log P = 26$ radio galaxies is the same, viz. $<M_p> = -20.3$
($H_o = 100$ km/s Mpc). The dispersion around this average value
is $\sim 0\overset{m}{.}3$ for radio luminosities $\log P \gtrsim 24.4$.

Below the "break" in the radio luminosity functions at $\log
P \sim 24.4$, this constancy of $<M_p>$ does no longer hold. Instead,
there $<M_p>$ depends on $\log P$ approximately as follows: $<M_p> \sim
-20.3 - 0.4*$ ($\log P-24.4$), and the dispersion around this mean is
larger than $\sim 0\overset{m}{.}3$. From the fractional radio luminosity functions
above the break, one finds that the fraction of radio galaxies at

a given radio luminosity depends on optical luminosity as $P_{opt}^{3/2}$.
This means that the probability for an elliptical galaxy to
produce a radio source of a given radio luminosity depends rather
strongly on the optical luminosity, and hence mass, of a galaxy.
Given an optical luminosity, a galaxy may produce a radio source
within a large range of radio luminosities; the relative probability of the latter is given by the fractional radio luminosity
function, which depends on P_{rad} as $P_{rad}^{-1.2 \pm 0.1}$ (with logarithmic
P_{rad}-differential).

8. THE REDSHIFT DEPENDENCE OF THE RADIO LUMINOSITY FUNCTION

8.1 Models based on radio source count analysis

The bivariate luminosity function for elliptical galaxies as
determined by Auriemma et al. (1977) and Meier et al. (1979) describes the properties of the elliptical radio galaxies at the
present epoch. For an understanding of the evolution of radio
sources one should have similarly detailed data at other epochs,
and also for other types of radio source. At present it is very
hard to determine the bivariate luminosity at higher redshifts,
i.e. earlier epochs. However, if one restricts oneself to the
monovariate radio luminosity function (i.e. summing all optical
luminosity classes) the situation is better. Already from the
radio source counts alone one can put useful limits on the
possible redshifts dependences of the radio luminosity function
(RLF). Such limits can become rather tight by the addition of
optical data.
    One small complication in the use of radio source counts –
which represent the number of radio sources per unit solid angle
as a function of flux density at a given frequency – is their
frequency dependence (see e.g. the compilation made by Wall, 1978).
This reflects the fact that the spectral composition of radio
source samples depends on selection frequency and, to a lesser
extent, flux density. In other words: the mix of extended, steep-
spectrum sources and compact, flat-spectrum sources depends on
frequency and flux density (see e.g. Fanaroff and Longair, 1973
and Kulkarni, 1978).
    Another, related problem is that the radio source counts
involve not only elliptical radio galaxies but also quasars and
spiral galaxies. From the local radio luminosity function of
spiral galaxies it appears that these contribute only at very low
flux density levels, unless they were very much more luminous in
the past. As to the relative contributions of quasars and radio
galaxies at different flux densities, these may be largely determined by observational definitions (see e.g. de Ruiter, 1978)
especially when optical spectra are not available.
    Figure 9 shows a log P-Z diagram, meant to illustrate the
relation between the source count and the radio luminosity function

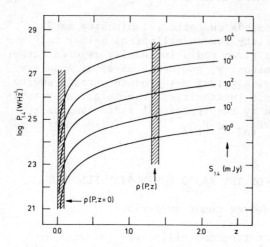

Fig. 9. The (log P,Z)-diagram, illustrating the relation between radio source count and redshift-dependent radio luminosity function.

at various redshifts. It is immediately clear that there exist numerous ways of populating this diagram with sources if the only constraint consists of integrals along lines of constant flux density. And even if a unique solution for the population of this diagram could be derived from observations, the interpretation in terms of changing source densities and relevant time-scales would involve a geometric cosmological model.

Generally, radio source count analyses have involved some assumptions about the functional form of the redshift dependence of the RLF $\rho(P,Z)$, the numerical details of which could then be solved for. It has become customary to describe a source count model by the enhancement function $E(P,Z) = \rho(P,Z)/\rho(P,Z = 0)$ rather than $\rho(P,Z)$. In Table 2 some of the main functional forms that have been used for $E(P,Z)$ are summarized. In most models there is not only a critical luminosity P_c below which the RLF is redshift independent, but also a cut-off redshift Z_c beyond which the RLF is either set equal to that at Z_c or equal to zero. Of course, all models require a local RLF $\rho(P,Z = 0)$ to be specified. A recent determination of the total local RLF (i.e. including quasars and spiral galaxies) was presented by Fanti and Perola (1977).

As is well-known, satisfactory modelling of source counts at both low and high frequencies requires $E(P,Z)$ to acquire values of $\sim 10^3$ at $Z \sim 2-3$, at least for the luminous radio sources. The detailed behaviour of $E(P,Z)$ remains largely undetermined from such modelling, although the numerical details of a given model can be determined quite uniquely, once the <u>form</u> of $E(P,Z)$ has been chosen.

Every source count model of course predicts redshift- (and hence: luminosity-) distributions at each flux density level. As an example, Figure 10 shows the predicted strong-source luminosity distribution and the predicted redshift distributions for a model (with $E(P,Z)$ of Type 2) that successfully describes the 1.4 GHz

Table 2

Summary of most frequently used enhancement functions E(P,Z)

1) $E(P,Z) = E(Z) = (1+Z)^a$		$P > P_c$	Longair (1966)
	1	$P \leq P_c$	
2) $E(P,Z) = (1+Z)^{a(P)}$		$P > P_c$	Schmidt (1972)
or $\exp(a(P)\tau)$			Wall et al. (1977)
	1	$P \leq P_c$	Katgert (1977)
3) $E(P,Z) = E(Z)$		$P > P_c$	
	1	$P \leq P_c$	Robertson (1978, 1979)

source count as summarized by Willis et al. (1977) and Oosterbaan (1978). Also shown is the observed strong-source luminosity distribution given by Wall et al. (1977).

A different approach to source count modelling is due to Robertson (1978, 1979). This so-called "free-form" analysis leaves the form of E(Z), applying to all sources above a critical luminosity P_c, to be determined. This method is based on iterative adjustment of E(Z) and also takes into account the strong-source luminosity distribution. This "best-fit" E(Z) is not very well represented by the usual functions $(1+Z)^a$ or $\exp(a \tau)$ where τ is the lookback time). For $Z \gtrsim 0.6$, or equivalently $E(Z) \gtrsim 10^2$, E(Z)

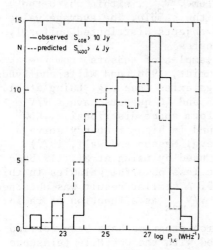
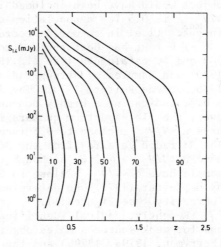

Fig. 10. Observed and predicted luminosity distributions of strong source sample (left), and predicted redshift distributions as a function of flux density, as the percentage of sources with redshift less than Z (right).

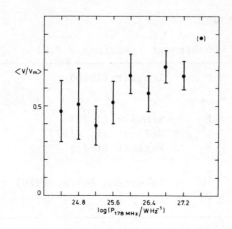

Fig. 11. The parameter V/V_{max} as a function of radio lumino-sity, for a complete sample of 65 3CR radio galaxies.

is reasonably approximated by the function $\exp(8.6\ \tau)$ but towards lower redshifts $E(Z)$ decreases faster than this exponential. Below $Z \sim 0.2$ $E(Z)$ is not significantly different from unity.

8.2 Direct determinations of the evolution of radio source populations

For samples of objects complete to well-defined optical and radio limits, and with redshifts, the V/V_{max}-test (Schmidt, 1968) can be used to measure deviations from a uniform space distribution. In this test, the volume of space out to the distance of an object, V, is compared with the maximum volume, V_{max}, within which the object would have been included in the sample. The average value of V/V_{max} is 0.5 for a population of objects distributed uniformly in space (i.e. in co-moving coordinates).

The test has been applied to samples of quasars from the 3CR and 4C catalogues (see e.g. Schmidt, 1968, and Wills and Lynds, 1978) and samples of strong radio galaxies (see e.g. Laing et al. 1978, and de Ruiter, 1978). The 3CR and 4C quasars have $< V/V_{max} >$ ~ 0.7, indicating a highly non-uniform space distribution. How-ever, the flat-spectrum quasars found in high-frequency surveys have a $< V/V_{max} >$ close to 0.5 (see e.g. Masson and Wall, 1977). The 33 radio galaxies from the 3CR used by Laing et al. (1978) have a $< V/V_{max} > \sim 0.6$, although the less powerful galaxies in this sample have a $< V/V_{max} >$ close to 0.5. A similar result was obtained by de Ruiter (1978) who determined V/V_{max} as a function of radio luminosity (see Figure 11).

Recently, optical identification programs have been carried out for weak source samples observed with the One-Mile telescope (Perryman, 1979a, 1979b) and the Westerbork telescope (de Ruiter et al.,1977, Willis and de Ruiter, 1977, Katgert et al.,1979a). The results of the latter have been used to estimate the RLF of radio galaxies out to redshifts of 0.5-0.7 (Katgert et al.,1979b). From the magnitudes of the radio galaxies, redshifts were derived

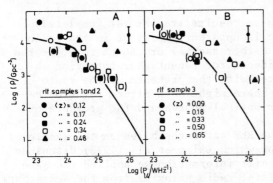

Fig. 12. Estimates of the radio luminosity function of elliptical galaxies based on iden-tification programs of weak radio sources from Westerbork sur-veys, carried out on IIIa-J (left) and 127-04 (right) plates.

assuming constant absolute magnitudes M_b = -20.5 and M_r = -22.4 (H_o = 100) and K-corrections due to Pence (1976). With the red-shifts radio luminosities can be derived from the flux densities, and finally $\rho(P,Z)$ of the radio galaxies can be calculated.

The result is shown in Figure 12 for identifications on deep IIIa-J plates (left) and deep 127-04 red-sensitive plates (diffe-rent sample of objects). The main conclusions from these data are: 1) Not only the density of <u>powerful</u> radio sources increases strongly with Z, but also that of the intermediately strong ones (by factors of 10-30 at Z $\sim$ 0.5); 2) Out to redshifts of 0.2-0.3 the RLF is practically constant; 3) There are quantitative diffe-rences between RLF estimates based on blue and red magnitudes, which are considered evidence for colour evolution of the ellip-tical radio galaxies. An analysis of the apparent colours of the faint radio galaxies indicates that at $m_r \gtrsim 19$ (Z $\gtrsim 0.4$) the colours are 1-2 magnitudes bluer than predicted from e.g. the spectrum of the giant elliptical M87.

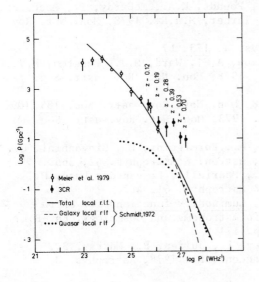

Fig. 13. Estimates of the radio luminosity function of elliptical galaxies, based on a completely identified sample of 33 3CR radio galaxies.

The implications of these results are not yet completely
clear. The blue excess that is apparently present in the radio
galaxies at z ∿ 0.5 can be either thermal or non-thermal. In the
former case it could be evidence for recent star formation which
might be indirectly connected with the presence of the radio
source. In the latter case there might be a direct relation between a non-thermal nuclear optical source and the presence of
the radio source.

Finally Figure 13 shows RLF estimates based on the sample of
33 strong radio galaxies for which Laing et al. (1978) discussed
$<V/V_{max}>$. Ignoring the redshift information we applied the same
procedure as used for the faint radio galaxies from the Westerbork
survey. Again, there are indications that the RLF of these powerful radio galaxies is constant out to z ∿ 0.2-0.3, beyond which
redshifts there is a strong increase.

REFERENCES

Auriemma, C., Perola, G.C., Ekers, R., Fanti, R., Lari, C., Jaffe,
W.J., Ulrich, M.H., 1977, Astron. & Astrophys. 57, 41.
Baldwin, J.E., Scott, P.F., 1973, Mon. Not. Roy. astr. Soc. 165,
259.
Bridle, A.H., Fomalont, E.B., 1976, Astron. & Astrophys. 52, 107.
Bridle, A.H., Fomalont, E.B., Miley, G.K., Valentijn, E., 1979,
Astron. & Astrophys., in press.
de Bruyn, A.G., Wilson, A.S., 1978, Astron. & Astrophys. 64, 433.
Burns, J.O., Owen, F.N., 1977, Astrophys. J. 217, 34.
Cameron, M.J., 1971, Mon. Not. Roy. astr. Soc. 152, 429.
Cooke, B.A., Ricketts, M.J., Maccacaro, T., Pye, J.P., Elvis, M.,
Watson, M.G., Griffiths, R.E., Pounds, K.A., McHardy, I., Maccagni,
D., Seward, F.D., Page, C.G., Turner, M.J.L., 1978, Mon. Not. Roy.
astr. Soc. 182, 489.
De Young, D.S., 1972, Astrophys. J. 173, L7.
Elvis, M., Maccacaro, T., Wilson, A.S., Ward, M.J., Penston, M.V.,
Fosbury, R.A.E., Perola, G.C., 1978, Mon. Not. Roy. astr. Soc.
183, 129.
Fabian, A.C., Rees, M.J., 1978, Mon. Not. Roy. astr. Soc. 185, 109.
Fanaroff, B.L., Longair, M.S., 1973, Mon. Not. Roy. astr. Soc.
161, 393.
Fanti, C., Fanti, R., Ficarra, A., Formiggini, L., Giovannini, G.,
Lari, C., Padrielli, L., 1975, Astron. & Astrophys. 42, 365.
Fanti, C., Fanti, R., Lari, C., Padrielli, L. van der Laan, H.,
de Ruiter, H., 1977, Astron. & Astrophys. 61, 487.
Fanti, R., Perola, G.C., 1977, Luminosity functions for extra
galactic radio sources in Radio Astronomy and Cosmology, Ed. D.L.
Jauncey (Reidel, Dordrecht), p. 171.
Forman, W., Jones, C., Cominsky, L., Julien, P., Murray, S.,
Peters, G., Tananbaum, H., Giacconi, R., 1978, Astrophys. J.
Suppl. 38, 357.

Gavazzi, G., 1978, Astron. & Astrophys. 69, 355.
Gavazzi, G., 1979, Astron. & Astrophys. 72, 1.
Gavazzi, G., Perola, G.C., 1978, Astron. & Astrophys. 66, 407.
Grindlay, J.E., Parsignault, D.R., Gursky, H., Brinkman, A.C.,
Heise, J., Harris, D.E., 1977, Astrophys. J. 214, L57.
Guindon, B., Bridle, A.H., 1978, Mon. Not. Roy. astr. Soc. 184,
221.
Gursky, H., Schwarz, D.A., 1977, Ann. Rev. Astron. Astrophys. 15,
541.
Hanish, R.J., Matthews, J.A., Davis, M.M., 1979, Astron. J. 84,
946.
Harris, D.E., Bahcall, N.A., Strom, R.G., 1977, Astron. & Astro-
phys. 60, 27.
Harris, D.E., Grindlay, J.E., 1979, Mon. Not. Roy. astr. Soc. 188,
25.
Hooley, T., 1974, Mon. Not. Roy. astr. Soc. 166, 259.
Jaffe, W.J., Perola, G.C., 1973, Astron. & Astrophys. 26, 423.
Jaffe, W.J., Perola, G.C., Valentijn, E.A., 1976, Astron. &
Astrophys. 49, 179.
Katgert, P., 1979, Populations of weak radio sources, Ph.D. Thesis
Leiden University.
Katgert, P., de Bruyn, A.G., Willis, A.G., 1979, Astron. & Astro-
phys. Suppl. 36, 213.
Katgert, P., de Ruiter, H.R., van der Laan, H., 1979, Nature 280,
20.
Kulkarni, V.K., 1978, Mon. Not. Roy. astr. Soc. 185, 123.
van der Laan, H., Perola, G.C., 1969, Astron. & Astrophys. 3, 468.
Laing, R.A., Longair, M.S., Riley, J.M., Kibblewhite, E.J., Gunn,
J.E., 1978, Mon. Not. Roy. astr. Soc. 184, 149.
Lari, C., Perola, G.C., 1978, Radio properties of Abell clusters
in The large-scale structure of the universe, Eds. M.S. Longair
and J. Einasto (Reidel, Dordrecht), p. 137.
Lea, S.M., Holman, G.D., 1978, Astrophys. J. 222, 29.
Longair, M.S., 1966, Mon. Not. Roy. astr. Soc. 133, 421.
Masson, C.R., Wall, J.V., 1977, Mon. Not. Roy. astr. Soc. 180, 193.
Meier, D.L., Ulrich, M.H., Fanti, R., Gioia, I., Lari, C., 1979,
Astrophys. J. 229, 25.
Miley, G.K., Wellington, K.J., van der Laan, H., 1975, Astron. &
Astrophys. 38, 381.
Miley, G.K., Hartsuyker, A.P., 1978, Astron. & Astrophys. Suppl.
34, 129.
Moffet, A.T., 1975,Strong nonthermal radio emission from galaxies
in Galaxies and the Universe, Stars and Stellar Systems Vol. IX
(Univ. Chicago Press), p. 211.
Oosterbaan, C.E., 1978, Astron. & Astrophys. 69, 235.
Owen, F.N., Rudnick, L., 1976, Astrophys. J. 205, L1.
Owen, F.N., Burns, J.O., Rudnick, L., 1978, Astrophys. J. 226,
L119.
Pence, W., 1976, Astrophys. J. 203, 39.
Perryman, M.A.C., 1979, Mon. Not. Roy. astr. Soc. 187, 223.

Perryman, M.A.C., 1979, Mon. Not. Roy. astr. Soc. 187, 683.
Robertson, J.G., 1978, Mon. Not. Roy. astr. Soc. 182, 617.
Robertson, J.G., 1979, Astron. & Astrophys., in press.
Roland, J., Véron, P., Pauliny-Toth, I.I.K., Preuss, E., Witzel, A., 1976, Astron. & Astrophys. 50, 165.
de Ruiter, H.R., 1978, Faint extragalactic radio sources and their optical identifications, Ph.D. Thesis Leiden University.
de Ruiter, H.R., Willis, A.G., Arp., H.C., 1977, Astron. & Astrophys. Suppl. 28, 211.
Schmidt, M., 1966, Astrophys. J. 146, 7.
Schmidt, M., 1968, Astrophys. J. 151, 393.
Schmidt, M., 1970, Astrophys. J. 162, 371.
Schmidt, M., 1972, Astrophys. J. 176, 303.
Simon, A.J.B., 1978, Mon. Not. Roy. astr. Soc. 184, 537.
Stocke, J., 1979, Astrophys. J., 230, 40.
Valentijn, E.A., 1978a, Astron. & Astrophys. 68, 449.
Valentijn, E.A., 1978b, Ph.D. Thesis Leiden University.
Vallée, J.P., Wilson, A.S., van der Laan, H., 1979, Astron. & Astrophys., in press.
Veron, P., 1979, Seminar given at the School (Astron.Astrophys., to be published).
Wall, J.V., 1978, Mon. Not. Roy. astr. Soc. 182, 381.
Wall, J.V., Pearson, T.J., Longair, M.S., 1977, Interpretation of source counts and redshift data in evolutionary universes in Radio Astronomy and Cosmology, Ed. D.L. Jauncey (Reidel, Dordrecht), p.269.
Willis, A.G., Oosterbaan, C.E., Le Poole, R.S. de Ruiter, H.R., Strom, R.G., Valentijn, E.A., Katgert, P., Katgert-Merkelijn, J.K., 1977, Westerbork surveys of radio sources at 610 and 1415 MHz in Radio Astronomy and Cosmology, Ed. D.L. Jauncey (Reidel, Dordrecht), p. 39.
Willis, A.G., de Ruiter, H.R., 1977, Astron. & Astrophys. Suppl. 29, 103.
Wills, D., Lynds, R., 1978, Astrophys. J. Suppl. 36, 317.
Willson, M.A.G., 1970, Mon. Not. Roy. astr. Soc. 151, 1.

X-RAY EMISSION FROM ACTIVE GALAXIES

K.A. Pounds

X-Ray Astronomy Group
University of Leicester, U.K.

1. INTRODUCTION

The period since ∿1976 has seen a rapid development in the
study of X-ray emission from active galaxies. At that time, based
mainly on sources in the third Uhuru (3U) catalogue (Giacconi et
al.,1974), only 4 active galaxies had been reliably identified as
X-ray emitters (viz. 3C 273, NGC 4151, NGC 5128 and M 87), with
the possible association of a 3U source with 4 others (NGC 1275,
Cyg A, 3C 390.3 and M 82). By the end of 1978, this number had increased to more than 50, forming a second major class of extragalactic X-ray emitter, comparable in number (and range of luminosity) to the galaxy clusters. Many of these new identifications
are from the Ariel V Sky Survey and are based on sources listed
in the "2A" catalogue (Cooke et al.,1978a). Other recent additions
have come from X-ray observations with the SAS-3, OSO-8 and HEAO-1
satellites, together with the revised and considerably extended
Uhuru (4U) catalogue (Forman et al.,1978).

It is now clear that powerful X-ray emission is a common feature of active galaxies of several different types, with type I
Seyferts, high excitation/narrow emission line galaxies, BL Lacertids and QSO's prominent in the objects located to date. The present paper will review the identification and classification of
these X-ray active galaxies, with the discussion roughly following
the historical development of the topic. Little will be said about
X-ray spectra since this matter is dealt with by Dr Holt, nor
about emission mechanisms and models of galactic nuclei, which
are the subject of a lecture by Professor Rees.

R. Giacconi and G. Setti (eds.), X-Ray Astronomy, 273-290.

2. SEYFERT GALAXIES

2.1 Identification

It is instructive to recall the first (post '3U') Seyfert
galaxy identified from the Ariel V survey, namely the bright
southern galaxy NGC 3783. This barred spiral, classified as a
Type I Seyfert by Khachikian and Weedman (1974), was found to be
the dominant object in the "error box" of the Ariel V source 2A
1135-373 (Fig.1). Optical spectra obtained with the 3.8 m Anglo-
Australian Telescope (AAT) revealed the presence of λ 6374 of
[FeX] and other features indicating an unusually high degree of
ionization and supporting the X-ray identification (Cooke et al.,
1976). Nevertheless, with an initial X-ray error box of 0.12 deg^2
the Seyfert identification could be considered probable rather
than certain. Subsequently a systematic search of many other 2A
error boxes produced a remarkably high proportion of catalogued
Seyfert galaxies, significantly in excess of the number expected
by chance coincidence. A statistical assessment by Elvis et al.
(1978) confirmed the great majority of Seyfert I identifications
were indeed valid, thereby establishing this new class of powerful
X-ray emitter.

The present total (31) of identified X-ray Seyferts, all of
type I, is listed in Table 1. Of these it is interesting to note
that 7 have been found to be Type I Seyfert galaxies as a result
of an optical search of the source error boxes. The first new
Seyfert to be discovered in this way, MCG 8-11-11, previously
listed as a barred spiral galaxy, was found by Ward et al. (1977)
at 14^m to be by far the brightest extragalactic object in the
0.093 deg^2 error box of 2A 0551+466, and a Type I Seyfert. A continuing search of the error boxes of unidentified 2A sources by
the same group, using both objective prism plates and slit spectra
of the brightest candidates, quickly led to the discovery (Ward
et al.,1978) of 3 further Type I Seyferts, viz. ESO 113-IG45,
MCG-2-58-22 and ESO 141-G55. Notably, ESO 113-IG45 is optically
the most luminous (and the brightest) Seyfert in the Southern
Hemisphere, comparable with a low luminosity QSO, while MCG-2-58-
22 has Balmer lines broader than any other known Seyfert except
3C 382. The identification of ESO 113-IG45 with 2A 0120-591 was
also proposed, independently, by Ricker (1978) on the basis of
positional coincidence and the Seyfert-like properties of this
galaxy reported by Fairall (1977).

Included in Table 1 are several additional X-ray Seyferts
identified over the past year, 4 with refined source positions
in the 4U catalogue, 7 from HEAO-1 observations and 1 further 2A
source identification (2A 1326-310 = MCG-6-30-15) on the basis of
a substantially improved error box obtained by SAS-3. An improved
SAS-3 X-ray position (Delvaille et al.,1978a) has also confirmed
the 2A identification of IC 4329A, distinguishing this Seyfert
from the nearby cluster of galaxies. A rapidly increasing number

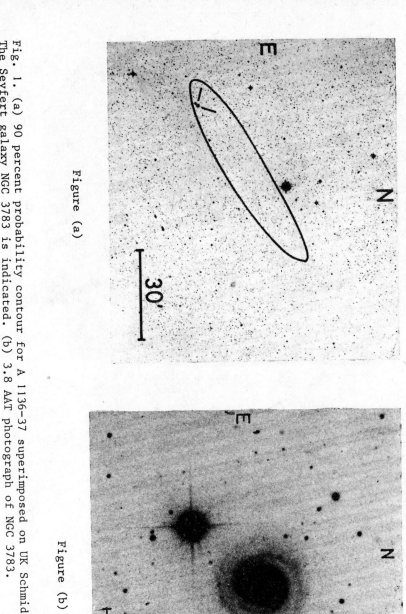

Figure (a)

Figure (b)

Fig. 1. (a) 90 percent probability contour for A 1136-37 superimposed on UK Schmidt Telescope plate. The Seyfert galaxy NGC 3783 is indicated. (b) 3.8 AAT photograph of NGC 3783.

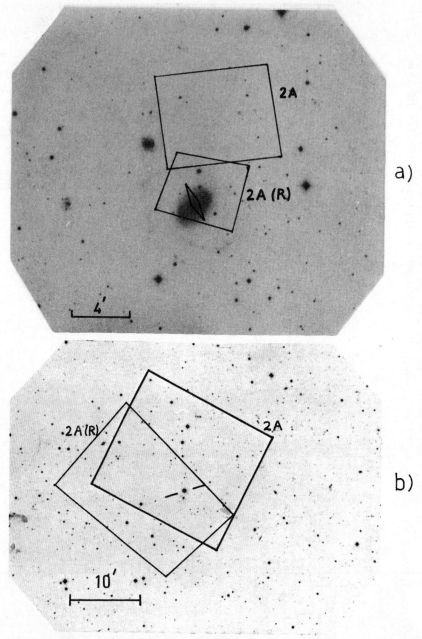

Fig. 2. Two further X-ray Seyferts. (a) NGC 4151 lies outside the 2A error box but inside the revised 2A (R) box while the HEAO-1 (A3) position overlays the galactic nucleus. (b) Ariel V error boxes for 2A 2302-088 superimposed on a Palomar blue plate. The new Seyfert MCG-2-58-22 is indicated.

Table 1   Seyfert Type I Identifications (31)

3U catalogue identifications	(3)	NGC 4151, 1275[a], 3C 390.3[b]
2A catalogue identifications	(10)	NGC 3783, 5548, 6814, 7469
		MK 79, 279, 376, 509
		IC 4329A, MCG 8-11-11*
2A identifications - post catalogue	(4)	MCG -2-58-22*, ESO 141-G55*
		ESO 113-IG45*, MCG -6-30-15[c]
New SAS-3 and HEAO A-3 identifications	(3)	III Zw 2[d], NGC 3227[e], 3C 120[f]
New 4U identifications	(4)	MK 335, 506, 541, 1040*
HEAO-2 A-2 sources[g]	(7)	NGC 4593, 7213*, 3C 111, 382
		MK 464, ESO 103-G35*, ESO 140-G43

Notes

(a) Resolved from Perseus cluster in Fabian et al (1974)

(b) Copernicus position confirmed identification (Charles et al, 1975)

(c) Improved SAS-3 position (Pineda et al, 1978)

(d) Schnopper et al, (1978a)

(e) Griffiths et al, (1979a)

(f) Schnopper et al, (1977)

(g) Marshall et al, (1978)

*   Indicates Seyfert properties established after X-ray source detected.

of precise ($\sim$arc min) positions, obtained with the SAS-3 or HEAO-1 modulation collimator experiments (see Griffiths et al.,1979a) has confirmed the great majority of these X-ray Seyferts; indeed it is remarkable that one of the 13 original 2A X-ray Seyferts have been disallowed by the more precise positions.

The X-ray luminosities of the Seyferts in Table 1 extends from 2×10^{42} ergs s^{-1} (NGC 3227) to 3×10^{45} ergs s^{-1} (III Zw 2), a range closely comparable with that of clusters. Fig. 3 shows the distribution of L_x for detected Seyferts against red shift z, and it is interesting to note that the range of luminosities covered by the Seyferts extends more than four-fifths of the way from NGC 3227 to the quasar 3C 273. A similar demonstration of H_β and Ly-α intensities has been given by Weedman (1976), who argues that this supports both the contention that QSO's are (simply) distant Seyferts, at $z \gtrsim 0.1$, and a cosmological interpretation of the red shifts. Optically the first distinguishing property of a Seyfert galaxy is its extremely bright nucleus, followed by the breadth of its emission lines. It now seems clear that a third feature of Seyfert galaxies is their powerful X-ray emission and it would be

surprising if it were found that this was not associated with the
Seyfert nucleus. The observed variability for several Seyferts
supports this view (Section 2.2).

Elvis et al.(1978) have made a detailed comparison of 15
X-ray Seyferts with their optical, infra-red and radio emission.
They find the X-ray luminosity to be correlated with the infrared
power (at 3.5 μm and 10 μm), the optical continuum power, the FWZI
of the broad emission lines (Fig.4 shows a more recent plot of this)

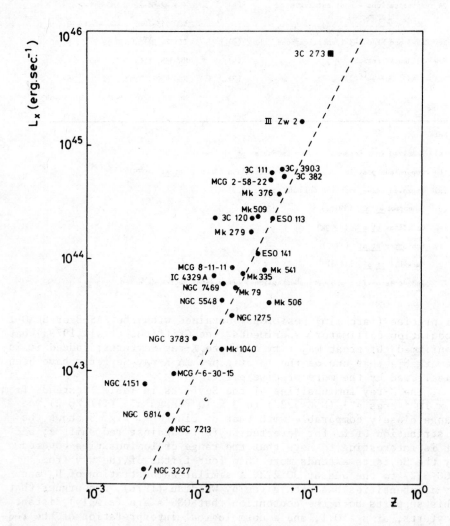

Fig. 3. X-ray luminosity versus red shift distribution for the
Seyfert galaxies listed in Table 1, together with the QSO 3C 273.
The dashed line corresponds approximately to a sensitivity of
0.5 Ariel cts sec^{-1}.

and with the luminosity in the H_α emission line, but not significantly with the radio power or the luminosity in the forbidden
optical emission lines. It is suggested that these correlations
strongly support the view that the X-ray emission originates in
the dense inner nuclear region, of extent $\leq$ 0.1 pc, which is peculiar to Type I Seyferts. In passing, it seems clear that the
high success rate of identifying X-ray Seyferts in the 2A catalogue is due, in part, to the X-ray to optical continuum correlation, in that the detected X-ray Seyferts are also among the
brightest known.

2.2 Variability in Seyfert X-radiation

Establishing variability in the intensity of faint X-ray
sources is difficult, particularly when such evidence is based on
a comparison of data from different instruments. Nevertheless,
there is now no doubt that some X-ray Seyferts do vary on quite
short timescales and it can be anticipated that over the next few
years observations of greater precision will show this to be a
feature common to the class. Charles et al. (1975) first showed
an apparent factor-of-four variation for 3C 390.3 between Uhuru
(1971) and Copernicus (1973) measurements. In the same way Ives
et al.(1976) found evidence for a factor 2.5 increase in the
2 - 6 keV emission of NGC 4151 from 1971 to 1974. More recently,
Elvis (1976) reported in Ariel V observations of NGC 4151 a flare
with a factor 1.7 $\pm$ 0.2 increase in less than 3 days during
August 1975. Also from the sky survey data, Ward et al.(1977)
have found MCG 8-11-11 to vary by at least a factor 3 on a timescale of 30 days or less, while comparison of Uhuru and Ariel V
SSI data on Mk 279 (Tananbaum et al.,1978) indicates variability
on a timescale $\leq$ 3 years.

The best studied X-ray Seyfert remains NGC 4151, the first
to be discovered, and also the brightest. Tananbaum et al.(1978),
from Uhuru, have reported flares in the X-radiation from NGC 4151
at an implied rate of 1 or 2 per day, with one event showing an
order-of-magnitude change in $\sim$ 730 secs. Extended observations
with OSO-8 and with the Ariel V Sky Survey have not confirmed
such large-scale variability (Mushotzky et al.,1978; Lawrence,
1979) but do show flares on a 1 - 2 day timescale to be a feature
of the NGC 4151 emission. Fig. 5 shows a sample of the Ariel V
observations and Lawrence (1979) has recently discussed these and
similar data in terms of a shot noise emission model for NGC 4151.
He concludes that the power emitted in these short term flares
indicate an efficiency for the source in excess of 1%. It may be
anticipated that future, more sensitive observations of this type
should establish the characteristic timescales of Seyfert nuclei
and provide strong clues as to their nature.

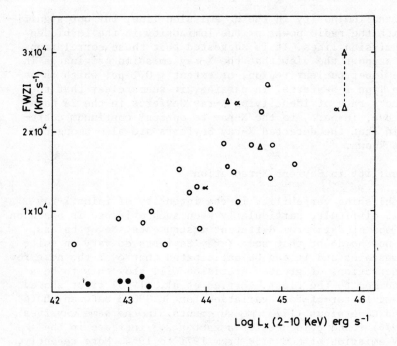

Fig. 4. Permitted line widths (FWZI) in H_β or H_α versus X-ray luminosity for type I Seyfert's (o), HEXELG's (●) and QSO's (Δ).

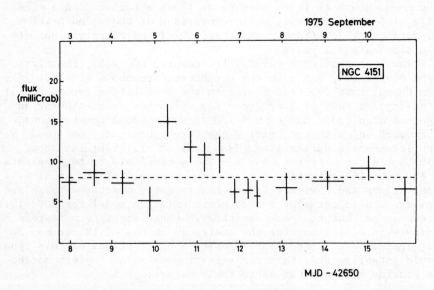

Fig. 5. A typical X-ray flare from NGC 4151 recorded by the Ariel V survey instrument.

3. HIGH EXCITATION, NARROW EMISSION LINE GALAXIES (HEXELG'S)

3.1 Identification and classification

The past eighteen months has seen an increasing number of X-ray identifications with active galaxies having high excitation forbidden line spectra, but without the broad permitted lines characteristic of Type I Seyferts. This number, now 10 if M82 is included (Table 2a), suggests strongly that powerful X-ray emission is also a feature of these objects. Perhaps the prototype of this kind of galaxy is NGC 5506, (see Wilson et al.,1976). This galaxy has long been a candidate identification for the X-ray source 3U 1410-03 (=2A 1410-029) (Bahcall et al.,1975), and this has now been confirmed by SAS-3 RMC observations (Ricker et al., 1979) which produced a 1.1 arc min radius error circle (90% confidence) centred only $\sim$12 arc sec north of the galaxy nucleus.

The optical search of 2A error boxes by Ward et al.(1978) produced two other HEXELG's, NGC 7582 and NGC 2992 and gave the first indication of a new class of X-ray active galaxy. A resulting PST search of the Ariel V sky survey data in the vicinity of 25 other bright emission line galaxies revealed a further positive detection, NGC 1365. The nearby galaxy M 82 is morphologically very similar to NGC 5506 (Wilson et al.,1976) and shows emission lines in places, although its nucleus is obscured by dust. An arc min error box from HEAO-1 (A3) confirms the identi-

Table 2   Summary of other X-ray/Active Galaxy Identifications (22)

(a) High excitation/narrow emission line galaxies (10)

Galaxy	X-ray source	$\text{Log}_{10}L_x$ (2 -10 keV)	λ 5007 [O III] /H$_\beta$
NGC 526A	2A 0120-353	43.5	18
NGC 1365	A 0331-36	42.3	1.9
NGC 1685	H 0447-637		
NGC 2110	2S 0549-074	43.15	$\leq$ 6.5
NGC 2992	2A 0943-140	43.3	14
NGC 5506	4U 1410-03	43.3	10
NGC 6221	H 1649-595		
NGC 7582	2A 2315-428	42.8	3
MCG-5-23-16	2A 0946-310	43.4	$\geq$20
M 82	4U 0954+70	40.8	0.45

(b) BL Lacertids (6)

			$\text{Log}_{10}L_{opt}$ (0.3 - 0.7μ)
MK 421	2A 1102+384	44.3	44.8
MK 501	4U 1651+39	44.3	44.5
1219+305	2A(R) 1219+305	-	-
3C 371	H 1801+698	43.9	44.5
PKS 0548-32	H 0548-32	44.8	44.5
PKS 2155-304	H 2155-304	-	-

(c) QSO's (3)

3C 273	4U 1226+02	45.8	46.4
MR 2251	2A 2251-178	44.8-45.2	45.0
QSO 0241+622	2S 0241+622	44.5	45.7

(d) Radio galaxies, etc (3)

NGC 5128	4U 1322-42	41.6-42.3	
M 87	4U 1228+12	43.3	
3C 445	H 2216-027		

fication of M 82[+] and shows the X-ray source to lie in the central
part of the galaxy (Griffiths et al.,1979b).

Four most recent additions have resulted from SAS-3 or HEAO-1
observations. MCG-5-23-16 and NGC 2110 were found by optical spectroscopy of bright galaxies in the arc min SAS-3 error boxes of 2A
0946-310 (Schnopper et al.,1978b) and S 0549-746 (Bradt et al.,
1978) respectively, whilst a further precise location, with the
HEAO-1 modulation collimator experiment (A3) allowed the emission
line galaxy NGC 526A to be identified with the Ariel source 2A
0120-353 (Griffiths et al.,1979b).

Galaxies of the HEXELG type have been characterised by Wilson
(1979) as having spiral or irregular morphologies, with forbidden
emission lines from ionisation levels up to [Fe VII] and with line
widths in the range 300-600 Km/sec for both forbidden and permitted
lines. Recently the detection of broader wings in several cases,
e.g. NGC 5506, 2992 and 526A have led to alternative classifications as Seyfert type II's (see references in Griffiths et al.,
1979b). Very little is so far known about the X-ray emission of
classified type II Seyferts except that it is much weaker than for
type I's. From the X-ray studies reported here it appears that the
strength of the high excitation lines is the best optical indication of X-ray power. Fig.6 shows an impressive correlation of
X-ray luminosity with the ratio of λ 5007 [O III] to H_β for the
narrow emission line galaxies, supporting the view that the optical emission lines are excited by a central ionising source.

A further point of interest in relation to the HEXELG's is
the high proportion that have nearby or even visibly interacting
companion galaxies. This is so for NGC 526A (B), NGC 2992 (2993),
NGC 5506 (5507) and M 82 (81), while NGC 7582 is a member of the
Grus Quartet. Several authors have suggested the possibility of
tidal interaction or of accretion triggering the activity in the
nucleus of a galaxy in such cases and the prevalence of powerful
X-ray emission amplifies this point (Griffiths et al.,1979b).

+ Although its X-ray identification is now certain, the inclusion
of M 82 in the category of X-ray active galaxies is open to doubt.
Whilst the galaxy is morphologically similar to others in the
emission line galaxy group, its nucleus is not seen optically
(due to dust) and the X-ray luminosity is about two orders of magnitude below the next weakest, NGC 1365. It may be that the X-ray
emission of M 82 is mainly stellar and it is noted that the mass
of the central region in M 82 is some 50 times that of our Galaxy
(implying a similar L_x-to-mass ratio). On the other hand, it is
intriguing to consider X-ray emission from galactic nuclei extending all the way from QSO's down to 'normal' galaxies, a sequence on which M 82 would lie between the 'normal' galaxies and
the HEXELG's. In this regard it is also of interest that the Einstein Observatory finds $L_x \sim 10^{38}$ erg sec^{-1} for the nucleus of M 31,
some two orders of magnitude greater than the source associated
with the nucleus of our own Galaxy.

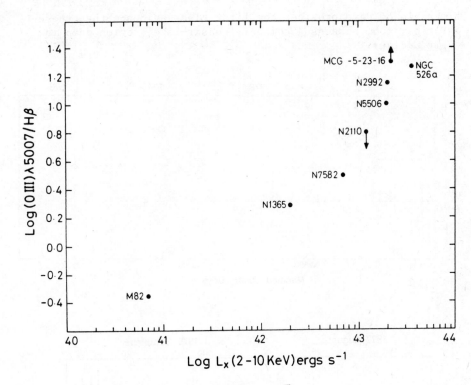

Fig. 6. The 'excitation measure' $\left[O\ III \right]$ /H_β is compared with the X-ray luminosity of several HEXELG's.

3.2 Variability

Confirmation that the powerful X-ray emission of the HEXELG's arises, as in Seyfert I's, from the galactic nucleus has come from the detection of variability in several cases. The best example to date is for NGC 5506 which shows no long term trend in the Ariel V survey data from 1975-8, being consistent (Fig.7a) with a steady source of 1.55 ± 0.1 c/s, or 1.3 x 10^{43} ergs sec (2-10 keV) at 38 Mpc. This is also consistent with the Uhuru flux of 3.0 ± 0.3 c/s. However, closer examination of the Ariel data for NGC 5506 during the two most extended observations, and those with the highest statistical precision, revealed a remarkable short-lived flare (Fig.7b). This flare, lasting ∿4 days contained ∿8 x 10^{48} ergs, intermediate in energy between similar X-ray flares observed from NGC 5128 (Cen A) and MK 421 (Marshall and Warwick, 1979).

In summary, it is now clear that strong X-ray emission is a characteristic of high excitation, narrow emission line galaxies with luminosities possibly from ∿7 x 10^{40} (M 82) to ∿5 x 10^{43} erg s^{-1} (NGC 526A), lying to the lower end of the Seyfert Type I range. Further, the fact that several are variable X-ray emitters on a timescale of days provides the first clear evidence that these galaxies, too, possess a compact 'quasar-like' region deep in their nucleus.

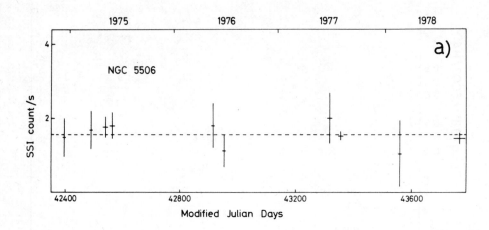

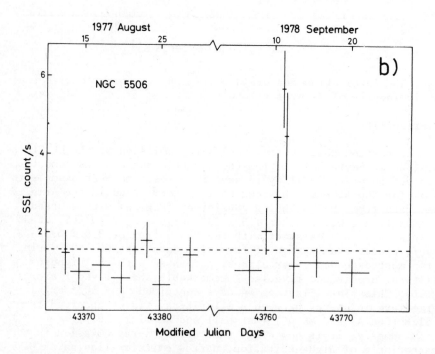

Fig. 7. Ariel V observations of NGC 5506 show (a) no long-term variability, but (b) a pronounced flare during an extended observation in September 1978.

4. BL LACERTIDS

The first positive identification of X-radiation from an object of this class came with the identification of the bright BL
Lac-type object, MK 421, with the X-ray source 2A 1101+38. The
identification was proposed by Ricketts et al.(1976) on the basis
of positional coincidence, albeit with an X-ray error box 0.15
deg^2, and highly variable X-ray emission during May 1975. Subsequent Ariel V observations have shown the X-ray source to persist
at a generally faint level ($\sim$0.8 Ariel c/s) with a further flaring
in late December 1977. No simultaneous optical observations have
yet been obtained, but the rapidly improving location has subsequently confirmed the identification with MK421. The revised
Uhuru (4U) catalogue contains a source, 4U 1651+39, identified
with a second BL Lac-type object, MK 501. It is interesting to
note that MK 421 and 501 are in the brightest (in V) 3 or 4 in
the 32 BL Lacertae objects listed by Stein et al.(1976).

Once again, the examination of a new X-ray source 'error box'
quickly led to the discovery of a previously unknown active galaxy.
Wilson et al.(1979) used a revised location of 2A 1219+305 (area
0.028 deg^2) and noted a weak radio source coincident with a 16 mag
blue stellar object as a likely counterpart. They obtained optical
spectra on the 2.5 m Isaac Newton Telescope, finding an intense
blue continuum with no emission or absorption lines, characteristic
of a BL Lacertid. Preliminary observations at 5 GHz with the VLA
showed the object to have a flat radio spectrum and unresolved
extent strongly supporting its proposed BL Lac nature.

Within the last year or so HEAO-1 (A2) observations have been
reported proposing 3 additional BL Lac X-ray identifications
(Marshall et al.,1978). The present number, now 6, are listed in
Table 2b. Again, 5 of the sources in the table have now been precisely located with the HEAO-1 (A3) scanning modulation collimator,
the $\sim$arc min positions confirming the corresponding identifications.

The ill-understood nature of BL Lac-type objects, with their
compact appearance, featureless blue continuum and high optical
variability, possibly lying somewhere between Type I Seyfert galaxies and QSO's, makes the establishment of powerful X-ray emission
especially interesting.

5. X-RADIATION FROM QSO'S

Until last year 3C 273 was distinguished as the nearest and
brightest QSO and also the only certain QSO identified as an X-ray
source. A PST search of 65 radio-bright QSO's using the Ariel V
sky survey data produced only one marginal additional detection
(White and Ricketts,1977). Upper limits to the X-ray luminosity
ranged from $\sim$0.15 to $\sim$50 times that of 3C 273.

Recently, a refined SAS-3 position of the source 2A 2251-179
has led to its unambiguous identification with a bright (B $\sim$16)

quasi-stellar object (Ricker et al.,1978). Given the preliminary
designation MR 2251-178, this object - at a red shift of 0.068 -
has an X-ray luminosity varying (at 2-20 keV) from 5 x 10^{44} to
1.6 x 10^{45} ergs sec^{-1} and which exceeds the optical luminosity by
a factor of ~10. Now, a third X-ray QSO has been reported (Apparao
et al.,1978), again from the optical search of a small SAS-3 RMC
error box. As yet unnamed, the quasi-stellar object identified
with the X-ray source 2S 0241+622 (=4U 0241+61) is, at a red shift
of 0.0438, the "nearest known QSO" and represents a further example
of the increasing number of active galaxies whose outstanding properties have been recognised only after the discovery of an adjacent X-ray source.

It is noteworthy that the 3 X-ray QSO's in Table 2c are the
3 brightest, after correction for extinction, in the list of 633
QSO's given by Burbidge et al. (1977). This emphasises the point
made in connection with the Seyfert type I identifications,
namely that there exists a good correlation between the optical
continuum and X-ray luminosities, a fact that has been recognised
to forecast the discovery of many new X-ray QSO's by the Einstein
Observatory.

Although the number of identified X-ray QSO's is still insufficient to allow a useful study of the relation of the X-ray
emission to other properties - or to provide a useful limit to
the likely evolutionary effects - the increased sensitivity of
the HEAO-1 sky survey may do so, and certainly the Einstein Observatory will.

6. RADIO GALAXIES, ETC.

Although several active galaxies combine powerful X-ray and
radio emission (e.g., NGC 1275, 3C 390.3, 3C 273) the majority of
extragalactic X-ray sources are, at most, weak radio emitters. In
fact, only for one object, Cen A, is a physical relation between
X-radiation and radio emission established to date. Although Cen
A is tentatively listed in Table 2a under the subhead 'radio galaxies', the major X-ray emission arises from the nucleus of the
central, giant elliptical galaxy NGC 5128 (Delvaille et al.,1978b).
X-ray observations spanning almost a decade - a more complete
'light curve' than at any other wavelenght - have shown the 2-10
keV emission to vary by an order-of-magnitude, with factor-of-two
changes within a few days. A compilation of data up to 1977 is
given by Lawrence et al.(1978) and additional OSO-8 observations
are reported by Mushotzky et al.(1978). The latter authors also
report radio flare changes in phase with the X-ray variability
and argue for a dominant X-ray production by Compton scattering
of the radio photons by a single relativistic electron population.
X-radiation has recently been detected also from the large radio
lobes of Cen A (Cooke et al.,1978b). Again, the emission is likely
to be by the so-called inverse Compton process, though in these

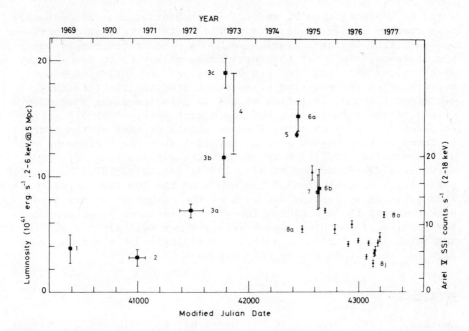

Fig. 9. Composite X-ray light curve of NGC 5128. Data point reference list in Lawrence et al.(1977).

extended regions the scattered photons are more likely to be of the microwave background. The derived magnetic field, of 0.7 μG, is close to the equipartition value and indicates Cen A is merely the first crude example of an X-ray map of extended radio sources that will be obtained over the next few years.

NGC 5128 is the only clear example so far of a long term variation in the X-ray emission of an extragalactic object. Fig.9 shows a compilation of (2-10 keV) fluxes over almost a decade, with a long term increase evident over ∿5 years. The energy content (2-10 keV) in this outburst is ∿2 x 10^{50} ergs and both this quantity and the timescale involved invite the intriguing suggestion that the cause of the outburst may have been the capture and accretion of an object of stellar mass by a massive black hole residing deep in the nucleus of NGC 5128.

7. FINAL COMMENTS

Bradt (1979) has chosen to include several type I Seyferts in his list of X-ray QSO's, while Wilson and Penston (1979) have recently argued for IC 4239A (also in the X-ray Seyfert list of Table 1) being named as the 'nearest QSO'. This description appears largely semantic (as on the border of the HEXELG's and Seyfert type II's) and would seem merely to emphasise the point first

noted by Weedman, from H_β and L_α line observations, and strongly
supported by the X-ray data (e.g. Fig.3) that there is a continuous
sequence from Seyfert galaxies to QSO's, the latter simply being
too distant (faint) for the surrounding galaxy to be seen. X-ray
observations with the Einstein Observatory and future deep sky
surveys offer the exciting prospect of studying the (possibly)
different rates of evolution of active galaxies along such a sequence. Without evolution, the luminosity function of <u>all</u> active
galaxies derived from Ariel V measurements of most of the sources
included in this review account for 10-15% of the background radiation (Warwick, 1978). However, several authors (e.g. Pye and
Warwick, 1979; Schnopper et al., 1978b; Schwartz, 1979) have noted th
possibility of evolution of the QSO's of the low luminosity emission
line galaxies accounting for most of the enigmatic cosmic X-ray
background in the keV range. The greatly enhanced point source
sensitivity of the Einstein Observatory may well yield a sufficient
sample of X-ray active galaxies, particularly QSO's out to quite
high red shifts, to verify this proposal.

REFERENCES

Apparao, K.M.V., Bignami, G.F., Maraschi, L., Helmken, H., Margon,
B., Hjellming, R., Bradt, H.V. and Dower, R.G., 1978, Nature, <u>273</u>,
450.
Bahcall, J.N., Bahcall, N.A., Murray, S.S. and Schmidt, M., 1975,
Astrophys. J. (Letters), <u>199</u>, L9.
Bradt, H.V., Burke, B.F., Canizares, C.R., Greenfield, P.E., Kelley,
R.L., McClintock, J.E., van Paradys, J., Koski, A.T., 1978,
Astrophys. J. (Letters), <u>226</u>, L111.
Bradt, H.V., 1979, Ann. New York Acad. Sci., to be published.
Burbidge, G.R., Crowne, A.H., and Smith, H.E., 1977, Astrophys. J.
Supp., <u>33</u>, 113.
Charles, P.A., Longair, M.S., and Sanford, P.W., 1975, Mon. Not.
Roy. Astr. Soc., <u>170</u>, 17P.
Cooke, B.A., and Pye, J.P., 1976, Mon. Not. Roy. Astr. Soc., <u>177</u>,
21P.
Cooke, B.A., Ricketts, M.J., Maccacaro, T., Pye, J.P., Elvis, M.,
Watson, M.G., Griffiths, R.E., Pounds, K.A., McHardy, I., Maccagni,
D., Seward, F.D., Page, C.G., and Turner, M.J.L., 1978a, Mon. Not.
Roy. Astr. Soc., <u>182</u>, 489.
Cooke, B.A., Lawrence, A., Perola, G.C., 1978b, Mon. Not. Roy.
Astr. Soc., <u>182</u>, 661.
Delvaille, J.P., Geller, M.J., and Schnopper, H.W., 1978a,
Astrophys. J. (Letters), <u>226</u>, L69.
Delvaille, J.P., Epstein, A., and Schnopper, H.W., 1978b, Astrophys.
J. (Letters), <u>219</u>, L81.
Elvis, M.S., 1976, Mon. Not. Roy. Astr. Soc., <u>177</u>, 7P.
Elvis, M.S., Maccacaro, T., Wilson, A.S., Ward, M.J., Penston, M.V.,
Fosbury, R.A.E., Perola, G.C., Mon. Not. Roy. Astr. Soc., <u>183</u>, 129.

Fabian, A.C., Zarnecki, J.C., Culhane, J.L., Hawkins, F.J.,
Peacock, A., Pounds, K.A., Parkinson, J.H., 1974, Astrophys. J.
(Letters), 189, L59.
Fairall, A.P., 1977, Mon. Not. Roy. Astr. Soc., 180, 391.
Forman, W., Jones, C., Cominsky, L., Julien, P., Murray, S.,
Peters, G., Tananbaum, H., and Giacconi, R., 1978, Astrophys. J.
Supp., 27, 357.
Giacconi, R., Murray, S., Gursky, H., Kellogg, E., Schreier, E.,
Matilsky, T., Koch, D., and Tananbaum, H., 1974, Astrophys. J.
Supp., 27, 37.
Griffiths, R.E., Briel, U., Schwartz, D.A., Schwarz, J., Doxsey,
R.E., and Johnston, M.D., 1979a, Mon. Not. Roy. Astr. Soc., 188,
813.
Griffiths, R.E., Doxsey, R.E., Johnston, M.D., Schwartz, D.A.,
Schwarz, J., Blades, J.C., 1979b, Astrophys. J. (Letters), 230, L21.
Ives, J.C., Sanford, P.W., and Penston, M.V., 1976, Astrophys. J.
(Letters), 207, L159.
Khachikian, E.Ye., and Weedman, D.W., 1974, Astrophys. J., 192,581.
Lawrence, A., Pye, J.P., and Elvis, M., 1977, Mon. Not. Roy. Astr.
Soc., 181, 93P.
Lawrence, A., 1979, Mon. Not. Roy. Astr. Soc., to be published.
Marshall, F.E., Boldt, E.A., Holt, S.S., Mushotzky, R.F., Pravdo,
S.H., Rothschild, R.E., Serlemitsos, P.J., 1978, Astrophys. J.
Supp., 40, 657.
Marshall, N., and Warwick, R.S., 1979, Mon. Not. Roy. Astr. Soc.,
189, 37P.
Mushotzky, R.F., Serlemitsos, P.J., Becker, R.H., Boldt, E.A., and
Holt, S.S., 1978, Astrophys. J., 220, 790.
Pineda, F., Delvaille, J., Huchra, J., and Davis, M., 1978, IAU
Circ. No. 3202.
Pye, J.P., and Warwick, R.S., 1979, Mon. Not. Roy. Astr. Soc.,
187, 905.
Ricker, G.R., Clarke, G.W., Doxsey, R.E., Dower, R.G., Jernigan,
J.G., Delvaille, J.P., MacAlpine, G.M., and Hjellming, R.M.,
1978, Nature, 271, 35.
Ricker, G.R., 1978, Nature, 271, 334.
Ricker, G.R. et al., 1979 Astrophys. J., to be published.
Ricketts, M.J., Cooke, B.A., and Pounds, K.A., 1976, Nature, 259,
546.
Schnopper, H.W., Epstein, A., Delvaille, J.P., Tucker, W., Doxsey,
R., and Jernigan, G., 1977, Astrophys. J. (Letters), 215, L7.
Schnopper, H.W., Delvaille, J.P., Epstein, A., Cash, W., Charles,
P., Bowyer, S., Hjellming, R.M., Owen, F.N., and Cotton, W.D.,
1978a, Astrophys. J., 222, L91.
Schnopper, H.W., Davis, M., Delvaille, J.P., Geller, M.J., and
Huchra, J.P., 1978b, Nature, 275, 719.
Schwartz, D.A., 1979, (Cospar) X-ray Astronomy, Baity and Peterson
(eds.), (Pergamon Press, Oxford).
Stein, W.A., O'Dell, S.L., and Strittmatter, P.A., 1976, Ann. Rev.
Astron. Astrophys., 14, 173.

Tananbaum, H., Peters, G., Forman, W., Giacconi, R., Jones, C.,
and Avni, Y., 1978, Astrophys. J., 223, 74.
Ward, J., Wilson, A.S., Disney, M.J., Elvis, M., and Maccacaro,
T., 1977, Astron. Astrophys., 59, L19.
Ward, M.J., Wilson, A.S., Penston, M.V., Maccacaro, T., Elvis, M.,
and Tritton, K.P., 1978, Astrophys. J., 223, 788.
Warwick, R.S., 1979, Proc. R. Soc. Lond. A, 366, 391.
Weedman, D.W., 1976, Astrophys. J., 208, 30.
White, G.J., and Ricketts, M.J., 1977, Mon. Not. Roy. Astr. Soc.,
181, 435.
Wilson, A.S., Penston, M.V., Fosbury, R.A.E., Boksenberg, A.,
1976, Mon. Not. Roy. Astr. Soc., 177, 673.
Wilson, A.S., 1979, Proc. R. Soc. Lond. A, 366, 461.
Wilson, A.S., Ward, M.J., Axon, D.J., Elvis, M., and Meurs, E.J.A.,
1979, Mon. Not. Roy. Astr. Soc., 187, 109.
Wilson, A.S., and Penston, M.V., 1979, Astrophys. J., to be
published.

EINSTEIN OBSERVATIONS OF ACTIVE GALAXIES

H. Tananbaum

Harvard-Smithsonian Center for Astrophysics
Cambridge, Massachusetts

In this paper we present a detailed discussion of the Einstein observations of Centaurus A (NGC5128). We then briefly describe the Einstein imaging results on Seyfert galaxies. We conclude with a discussion of our observations of time variations in Seyfert galaxies and quasars.

The results on Centaurus A are based on the recent work of Schreier et al. (1979). Figure 1 is a 4-meter CTIO photograph of Centaurus A (NGC5128), which lies at a distance of ∼ 5 Mpc. This is an elliptical galaxy dominated by a very prominent dust lane. NGC5128 is a remarkable galaxy in that it has a bright infrared nucleus, a nuclear radio source, a pair of inner radio lobes, and a pair of aligned outer radio lobes at about 5° distance. NGC5128 has also been known to be an X-ray source for about 10 years. The X-ray source has been identified with the galaxy nucleus on the basis of two features: (1) the spectrum shows a very substantial amount of low energy absorption (Tucker et al. 1973), and (2) the X-ray source is variable. Winkler and White (1975) and Lawrence, Pye, and Elvis (1977) have shown, for example, that the intensity can vary by about a factor of 1 1/2 or 2 on a timescale of 1 to 2 days.

Figure 2 is the Einstein Imaging Proportional Counter (IPC) picture for this source. The data are dominated by a point source of ∼ 2 cts/sec. This detector has about 1 arc minute resolution, and the figure shows evidence for an elongation in the northeast and even more prominently in the southwest direction. The data suggest the possibility that this excess emission is associated with the inner radio lobes. In fact, in Figure 3 we see the same X-ray data, with an inset for

R. Giacconi and G. Setti (eds.), X-Ray Astronomy, 291-310.
Copyright © 1980 by D. Reidel Publishing Company.

Fig. 1.   CTIO 4-meter photograph of Centaurus A (NGC5128).

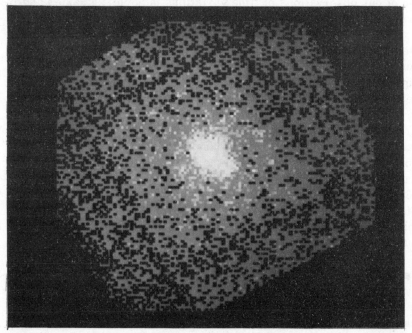

Fig. 2.   Einstein IPC image of central region of Cen A, showing central source plus excess counts southwest of nucleus.

the 1410 MHz radio data, showing the alignment of the X-ray elongation with the inner radio lobes. It actually turns out that the center of the 1410 MHz emission is displaced from the excess X-ray emission by about 1', which corresponds to $\sim$ 1 1/2 Kpc at the distance of NGC5128. Radio data at 400 MHz actually show the centroid in much better agreement with the excess X-ray emission, which corresponds to about 1% of the counts seen from the point source in the center. Figure 4 is an azimuthal display of the X-ray data seen in Figures 2 and 3. For an annulus located from 3' to 5' away from the nucleus, we have summed the X-ray counts and displayed the result in six angular intervals. At the position of the southwest radio lobe we see a 4.4σ excess X-ray emission. At the position of the northeast lobe there is a 2.7σ excess emission. Because of the higher statistical significance we concentrate the following discussion on the southwest lobe.

For the X-ray emission from the radio lobes the most acceptable model involves inverse Compton radiation powered by the electrons that are producing the radio emission. Since the source is quite extended in size, the synchrotron or the self-synchrotron models will not work. For the inverse Compton model, we require a soft photon source at temperature T to provide the photons to scatter the electrons producing the synchrotron radio emission. One obvious radiation field, which we know is present, is the microwave background, the 2.7^{o}

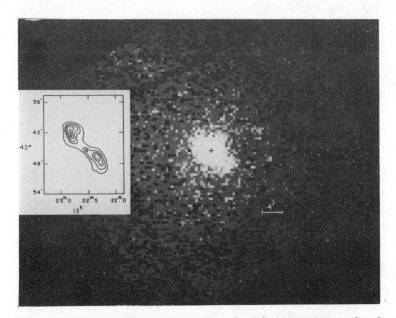

Fig. 3.   Same IPC picture as Fig. 2 with 1410 MHz radio data for central region shown on same scale in inset.

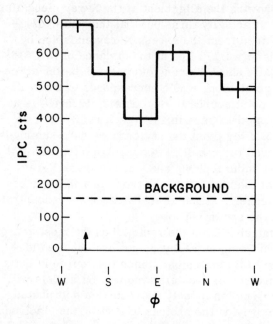

Fig. 4. The azimuthal distribution of IPC counts in an annulus 3 to 5
arc minutes from the nucleus. The arrows show the positions
of the inner radio lobes.

blackbody radiation. Equation (1) is the standard formalism (e.g. see
Tucker 1975) for the inverse Compton process:

$$B^{\alpha + 1} = (2.5 \times 10^{-19}) (5.0 \times 10^3)^{\alpha} T^{3 + \alpha} \frac{F(\alpha)}{a(\alpha)} \times$$

$$(\frac{E_S}{E_X})^{\alpha} \frac{dF_S/d\nu}{dF_X/d\nu} \tag{1}$$

Here the magnetic field strength is B, the spectral index observed in
the radio is α, the soft photon temperature is T, $F(\alpha)$ and $a(\alpha)$ are
numerical functions of α, the ratio of the synchrotron to the X-ray energy is given by E_S/E_X, and the synchrotron flux density divided by
the X-ray flux density is given by $\frac{dF_S/d\nu}{dF_X/d\nu}$. From our observations of

the southwest lobe we now determine the value of the X-ray luminosity
in the 0.3 to 3.0 keV band as 2×10^{39} ergs/sec. We take the radio
data published by Christiansen et al. (1967) to get the radio flux and α.
We use the temperature of 2.7°K, and we find $B = 4 \times 10^{-6}$ gauss. In
fact, this is a minimum value of the field because some of the X-ray

emission could be coming from a process other than this inverse Compton scattering. For purposes of comparison, the equipartition field is 50 to 100 microgauss for the inner lobes. The equipartition field is a convenient formalism for the minimum system energy in which the energy is divided equally between particles and fields. What is significant is that we actually have in this case a direct estimate of the field as 4 microgauss. Also available in the literature already are the Ariel V data published by Cooke, Lawrence, and Perola (1977) on the outer 5° lobes, where the detected X-ray emission is from a slightly higher energy band. They find 3.7×10^{41} ergs/sec for the X-ray luminosity and using equation (1) with the 2.7°K radiation field they determine a field strength of 0.7×10^{-6} gauss. The equipartition field in that particular region was estimated at 0.8×10^{-6} gauss.

        We also have data on NGC5128 obtained with the High Resolution Imager (HRI). Figure 5 shows the HRI data plotted in counts per square arc second as a function of distance from the center of the source. The solid curve is the observed counting rate, the dotted curve has the background, determined elsewhere in the detector, subtracted from the data. The x's represent the expected response to a point

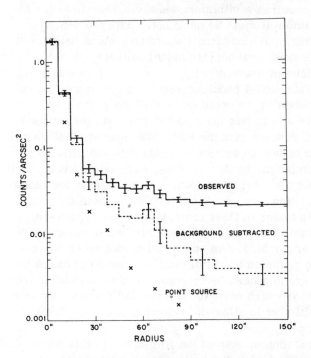

Fig. 5.   The radial distribution of surface brightness for Cen A (HRI cts arcsec^{-2}), with and without subtraction of background. X's represent response to point source.

source. The intrinsic luminosity of the central (point) source in the 0.3 to 3.0 keV band is $\sim 3 \times 10^{42}$ ergs/sec, but due to the low energy absorption intrinsic to the source only a fraction of this flux reaches us. Clearly, there is an excess emission which extends at least 2' out from the center. When we subtract the contribution of this point source we find that that excess emission has a total counting rate of about 0.04 ± 0.01 cts/sec which is essentially equal to what we measure in the central point source. Since we are now outside of the nuclear region, we assume that this source has absorption corresponding only to that due to gas along the line of sight from us to NGC5128, or $\sim 10^{21}$ N_H/cm^2. We can compute the luminosity for the extended source and we find $\sim 2 \times 10^{40}$ ergs/sec, or about 1% of the intrinsic emission from the central source.

There are several possible mechanisms we consider for the extended emission. One possibility is a large cloud which is illuminated by the central source, electron scatters the X-rays, and then is observed as an extended X-ray source. A more detailed calculation of this model immediately leads to a problem. The cloud has a large size; the radius of 2' corresponds to a radius of 3 Kpc. For the electron scattering, the cloud requires a minimum density. Then in order for the cloud itself to be stable it must be quite hot. With the density required for the scattering, the cloud itself would be a much brighter X-ray source than we observe, making this model unlikely. A second possibility is a hot gas cloud of lower density, not powered by scattering of the central source. This model requires several x 10^8 $M_\odot$ of hot material maintained somehow in the potential well of the galaxy. A third possibility is that there are in fact many individual discrete sources which we are unable to resolve with the HRI. We estimate that we require $\gtrsim 300$ sources in order not to resolve them with the HRI. If we divide the observed luminosity by 300 sources, each would have a luminosity slightly less than 10^{38} ergs/sec, which is acceptable for individual galactic sources. However, we know very little about the output of elliptical galaxies in terms of their individual stellar type X-ray sources. Also, this emission would be substantially more than we observe for our Galaxy or for M31. An additional model which has been suggested involves the diffusion of cosmic ray electrons out of the nuclear region with X-rays produced by inverse Compton scattering from starlight of other photons which exist in the extended region. Present observations are insufficient to allow the selection of one of these models over the others.

Figure 6 is a contour map of the HRI data. In this figure the extended 2' source is sufficiently diffuse that it cannot be seen. The map is dominated by the central X-ray source, and a jet-like feature which also can be seen as a bump at $\sim 1'$ distance in Figure 5. The new

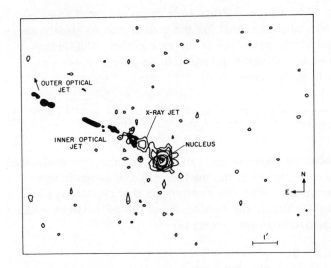

Fig. 6.   Iso-intensity contour map of the HRI image around the nucleus.
The X-ray jet to the northeast is clearly seen.   The dark
shapes to the northeast show the positions of diffuse features
in the optical jet discovered by Dufour and van den Bergh (1978).

feature is about 1' away from the nucleus; it has a diameter of about
15"; and it may have a length of the order of 1'.   When these X-ray contours are superimposed on the optical data recently published by Dufour
and van den Bergh (1978), we see that the X-ray data line up very well
with the optical jet discovered by Dufour and van den Bergh.   The jets
point in the general direction of the northeast inner radio lobe, although
there is actually a slight misalignment.   The X-ray emission from this
region is $\sim 3 \times 10^{39}$ ergs/sec which corresponds to 10% of the overall
extended emission.

In a recent paper, Blandford and Rees (1978) suggested that
the radio lobes might require an energy source to remain powered.
This raises the possibility that the X-ray and optical jet provide the
required power.   Schreier et al. (1979) have carried out an order of
magnitude calculation to test this picture of powering the inner lobe via
a matter stream.   They assume that the power in such a jet must be at
least equal to that required to replenish the inner lobe, $P \gtrsim 5 \times 10^{40}$
ergs/sec.   Equating the power with the kinetic energy flux of a matter
stream, they calculate a mass transfer rate:

$$\dot{M} = 10^{26} v_8^{-2} P_{41.7} \epsilon_{.1}^{-1} \text{ g s}^{-1}$$

(2)

where $v_8 = v/10^8$ cm s^{-1}, $P_{41.7} = P/5 \times 10^{41}$ ergs/sec, and an efficiency $\epsilon_{.1} = \epsilon/0.1$ has been assumed for the conversion of kinetic energy in the jet into relativistic particles at the radio lobe. Continuity arguments can be used to calculate the particle density in a conical jet at a distance r (kpc) from the source:

$$n = \frac{\dot{M}}{m_H v \pi (r\,\theta/2)^2} = 1.3\, P_{41.7}\, v_8^{-3}\, r^{-2}\, \theta_{0.25}^{-2}\, \epsilon_{.1}^{-1}\ \mathrm{cm}^{-3} \quad (3)$$

where $\theta_{0.25}$ is the opening angle in units of 0.25 radians, the estimated angle subtended by the X-ray jet at the nucleus (15 arc seconds at a distance of 60 arc seconds). With the assumption that the X-rays are due to thermal bremsstrahlung, the luminosity of the jet between radii of 1 and 2 kpc (the approximate jet extent) is:

$$L_x \simeq 10^{-27} \int_{r=1}^{r=2} n^2\, \pi\, r^2\, \theta^2\, T^{1/2}\, dr$$

$$= 4 \times 10^{39}\, T_7^{1/2}\, P_{41.7}^2\, v_8^{-6}\, \theta_{0.25}^{-2}\, \epsilon_{.1}^{-2}\ \mathrm{ergs/sec.} \quad (4)$$

With a temperature of 10^7°K, and galactic absorption, the observed flux of 0.0029 cts s^{-1} (0.3 to 3.0 keV) converts to a luminosity of $\sim 3 \times 10^{39}$ ergs/sec (0.3 to 3.0 keV), in good agreement with the above expression. Line emission may be important if $T_7 \sim 1$.

The strongest dependence of the luminosity expression is on the flow velocity. Velocities of about 10^8 cm s^{-1} have been inferred from the structure and dynamics of the jet in 3C31 (Blandford and Icke 1978) and from depolarization measurements of 3C449 (Perley, Willis, and Scott 1979). One can also estimate the flow velocity from the opening angle:

$$v = c\, \theta^{-1}$$

$$v_8 = 1.2\, T_7^{1/2}\, \theta_{0.25}^{-1} \quad (5)$$

where c is the sound speed in the gas. These direct and indirect estimates suggest that $v_8 \sim 1$, lending support to this origin for the X-ray emission in the jet.

The cooling time for matter in the jet, due to bremsstrahlung radiation, is:

$$t_c = \frac{2 \times 10^{11} \, T^{1/2} \sec}{n} = 4 \times 10^7 \, T_7^{1/2} \, P_{41.7}^{-1} \, \text{x}$$

$$v_8^3 \, r^2 \, \theta_{0.25}^2 \, \epsilon_{.1} \text{ years} .$$

(6)

Additional cooling due to line emission and/or inhomogeneities in the stream may decrease the cooling time by an order of magnitude, leading to cooling time estimates as low as 10^6 years. This should be compared with the flow time of matter from the nucleus to the jet $t_f = r/v = 10^6 \, r_{kpc} \, v_8^{-1}$ years. Thus, the possibility of thermal instability exists, suggesting that the inner optical jet is due to emission from matter in the stream which has cooled and is moving ballistically outward ($\sim 10^6$ $M_\odot/10^6$ year). This interpretation is consistent with optical observations (Dufour and ven den Bergh 1978; Graham 1975; Osmer 1978). The stream would have to avoid cooling too soon (radii less than 1 kpc), which might imply greater velocities in the stream closer to the nucleus. It should be noted that most of the energy in the above model for the jet resides in a dense subrelativistic plasma. A mechanism for reacceleration of electrons is required to provide for synchrotron emission in the lobe. The above cooling timescales indicate that little X-ray emission is produced within the lobes by bremsstrahlung. A higher and possibly variable stream velocity may lead to shocks and X-ray emission as suggested for the M87 jet (Rees 1978; Blandford and Konigl 1979), but the Einstein observations are consistent with much of the energy being fed into the northeast inner radio lobe via a dense subrelativistic plasma. The lack of an observed jet to the southwest may be due to a factor ~ 2 higher velocity, or to an intrinsic variability in the jet production mechanism.

We next turn to our X-ray imaging observations of type I Seyfert galaxies. Figure 7 shows the Einstein HRI observations of 4 type I Seyfert galaxies: Mkn 509, which is a rather luminous X-ray source emitting in excess of 10^{44} ergs/sec; Mkn 79 emitting several x 10^{43} ergs/sec; and NGC6814 and NGC4151, both emitting several x 10^{42} ergs/sec. The NGC4151 data show an additional source in the field, just above the "NGC4151" label, which is associated with a quasar discovered by Arp (1979). The quasar has a redshift of 1.8 and is about 20.3 magnitude. Its X-ray luminosity is in excess of 10^{47} ergs/ sec, so it is among the most luminous X-ray objects that we have observed.

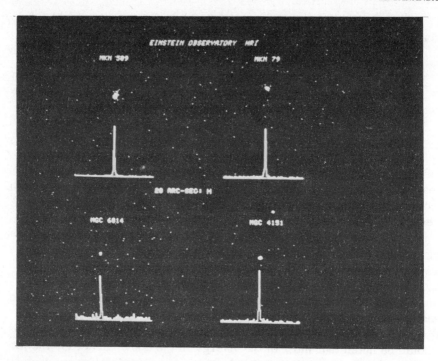

Fig. 7. HRI images of four type I Seyfert galaxies.

The HRI observations had two basic purposes. The first was to determine the locations of the X-ray sources -- in all four cases the locations are within 5" of the visible light nucleus of the Seyfert galaxy. The second objective was to determine if there was any evidence for extended X-ray emission -- the data for all four sources are in fact perfectly well fit by a point source. One-dimensional projections are shown below the point-like images of each of the sources in Figure 7. Figure 8 demonstrates this situation in more detail for NGC4151. As in Figure 5 for NGC5128, we have plotted the counts per square arc second versus the radius. The point response is shown by the dashed line; the data are shown with the error bars. In this relatively short exposure it is not necessary to subtract background. The data are fit well by a point source. With this exposure we can set a limit of a few x 10^{41} ergs/sec for any extended emission from NGC4151.

Our primary scientific objective in studying the type I Seyfert galaxies is to search for variability in the X-ray emission. Figure 9 shows the IPC picture for the type I Seyfert galaxy NGC6814, which is quite similar in its X-ray luminosity and spectral characteristics to NGC4151. The field contains a second, as yet unidentified, weaker X-ray source. From the imaging data we see that we can take

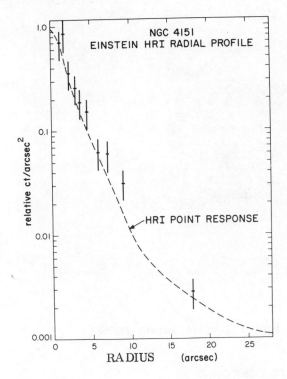

Fig. 8. The radial distribution of surface brightness for NGC4151 (HRI cts arcsec^{-2}). Dashed line indicates point response; data are shown with error bars.

the counts in a circular region around NGC6814, determine the background elsewhere in the detector, and then carry out timing studies. The imaging capability and the simultaneous background determination allow us to carry out such studies with much greater precision than was previously possible in X-rays. Also the IPC is the instrument to use for these timing studies. Firstly, it has a higher quantum efficiency so it will collect more photons than the HRI. Secondly, the IPC has some spectral resolution which allows us to check the spectrum of the source during any observed variations.

Figure 10 is a plot of the NGC6814 counting rate in counts per 100 seconds versus time in K-sec. The data cover almost 2 days. In this figure the data are actually summed in 10000 sec bins. The background, as measured elsewhere in the counter, is shown at the bottom of the figure at less than 1 count per 100 seconds for the same area. The background is quite stable and is measured each time we measure the source intensity. The upper part of the figure shows the source data, the counts occurring in a three arc minute radius circle

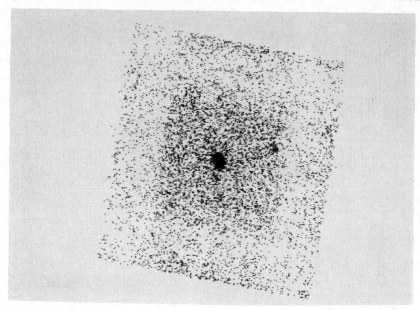

Fig. 9.   IPC picture of type I Seyfert galaxy NGC6814.

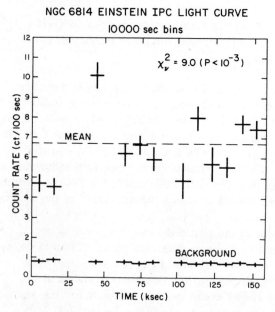

Fig. 10.   IPC counting rate (cts/100 sec) for NGC 6814 as a function of time in K-sec. Data are averaged over 10000 sec intervals. Background counting rate for equal detector area is indicated at bottom.

centered on NGC6814. Most of the bins are not full exposures for
the entire 10000 sec since the source is occulted by the earth for
part of each orbit. The plotted rates are corrected for the actual
exposure, and the error bars indicate the uncertainty in the rate
based on the actual number of counts. The mean of all the data is
shown by the dashed line. A chi squared test to see if this
source is constant gives χ^2_ν = 9 per degree of freedom; the probability that the source is constant is negligible.

In Figure 11 we have divided the same data with finer timing
resolution. The data are now summed into 2000 second intervals.
The event described above is broken into 3 data points which show
an increase and decrease. There is a suggestion of a second event
about one day later. The characteristic shape seems quite similar,
although these data are not as statistically significant. A chi
squared test for this region is consistent with a constant intensity only at the 3% confidence level. These two events seem to
have a characteristic timescale of the order of 20000 seconds.

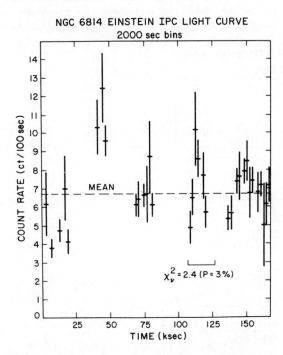

Fig. 11. Same as Fig. 10, except data are averaged over 2000
second intervals.

It does not appear that the source decreases to zero intensity at
the end of a flare, and the increase appears to be about 50%
above the mean. Ideally, we need continuous data over several
days for at least a few of the sources such as NGC6814 in order
to determine the characteristics of these kinds of events. Figure
12 uses the spectral information to search for spectral changes
associated with changes in count rate. We have computed a hardness ratio, which is the 2.3 to 3.5 keV counting rate data divided by the 1.2 to 2.3 keV data. The ratio is plotted as a function of the count rate. The error bars are as indicated in the
figure. At the high rates there is no significant change in the
hardness ratio which leads us to believe that these variations
represent a change in the emission measure of the source, rather
than a change in the low energy absorption at the source.

   Figure 13 shows the IPC data for 3C120, a luminous Seyfert I
or N-type galaxy. The counting rate data are shown in the figure
and the mean is given by the dashed line. The data span is about
1 1/2 days with a large interruption in the middle. The chi
squared is 1.2 per degree of freedom; the probability that the
source is constant during this particular observation is 30%.
This shows that at least some of the time our data indicate a
constant source behavior. On the other hand, Holt (1979) has
reported Einstein Solid State Spectrometer observations indicating
changes in the spectral shape of 3C120 on a timescale of weeks.
This suggests the importance of carrying out observations on many

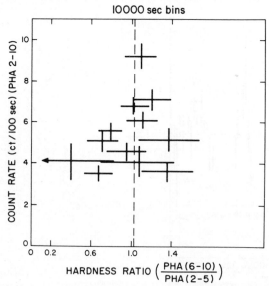

Fig. 12. Hardness ratio for NGC6814 (2.3 to 3.5 keV counts/1.2
to 2.3 keV counts) plotted versus count rate.

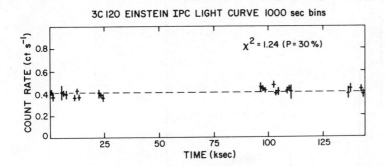

Fig. 13. IPC count rate versus time for 3C120.

different occasions and for relatively long periods of time in
order to be able to get the overall signature for objects such as
3C120.

        Our variability studies are not limited to Seyfert galaxies.
Figure 14 shows HRI observations of the quasar OX169, reported by
Tananbaum et al.(1979), with counts per thousand seconds plotted
with error bars versus time. The dotted lines at the bottom indicate the background rate. A chi squared test indicates that
the probability that this source is constant is less than 10^{-4}.
The largest single intensity change corresponds to about a factor
of three in $\sim$ 6000 seconds. Alternatively, one could attempt to
fit these data by a linear rate of change for the source intensity.
Individual points fit fairly well, and the measured slope corresponds to approximately a factor of 2 in 15000 seconds. This may
be a better way of characterizing the variability of this source.
The sophistication of our analysis is limited at present. Improvements in technique are required, since the short timescales observed for the variability in these Seyferts and quasars are very
important for testing different models.

        Our HRI counting rate data for 3C273 are shown in Figure 15.
We observed the source for about two days with a 12-hour interruption to perform boresight alignment checks. The average intensity for the first day of observation is 0.691 $\pm$ 0.006 counts
s^{-1}, while the average for the second part of the observation is
0.754 $\pm$ 0.005 counts s^{-1}. Thus, the data show an increase of
about 10%, which corresponds to an 8σ change in the average level.
These data are based on Einstein quick look data. The production
data have now been analyzed by Henry et al.(1979) who find an additional low intensity data point at the very beginning of the
second part of the observation. This indicates that the 10%
change actually occurred in approximately 1 orbit, or 6000 seconds.
The luminosity of 3C273 is significantly larger than the sources
we have discussed previously. Therefore, the 10% increase corresponds to a very large luminosity change in a very short time.
        We attempt to place these observations of time variations in

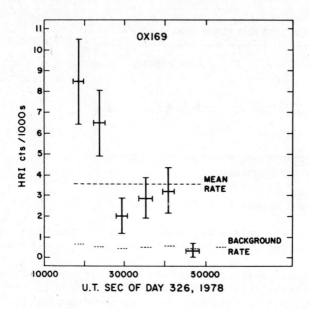

Fig. 14. HRI count rate versus time for quasar OX169.

a preliminary theoretical context. Cavallo and Rees(1978) provide
a formalism which allows one to relate the luminosity in an
outburst, ΔL, that occurs in the time, Δt, to the efficiency of
matter conversion, η, and the matter available to power the outburst, m:

$$\Delta L \Delta t \simeq \eta \, mc^2 \tag{7}$$

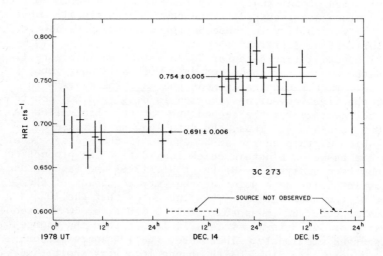

Fig. 15. HRI count rate versus time for quasar 3C273.

For a given mass and for a given outburst luminosity, Cavallo and
Rees show that the minimum timescale will occur for an electron
scattering depth of 1. On that basis they find:

$$(\eta/0.1) \geq \frac{\Delta L}{\Delta t} \frac{1}{2 \times 10^{41}} \tag{8}$$

Except for assumptions of spherical symmetry and uniform density,
this equation may otherwise be model independent. A further formalism has been described by Lightman, Giacconi, and Tananbaum
(1978) which is based on an unsaturated inverse Compton model.
In this thermal model, the size of the emitting region (R_{13} in
units of 10^{13} centimeters), is related to the timescale for the
outburst ($\Delta t/700$ in units of 700 seconds), and the electron temperature (T_9 in units of 10^9 degrees):

$$R_{13} \approx 0.6 \left(\frac{\Delta t}{700}\right) T_9^{-1/2} \tag{9}$$

In order to carry the calculation further a more specific model
is required, such as the unsaturated inverse Compton model coupled
with an accreting black hole. Insisting that the observed radiation not exceed the Eddington limit provides a relationship between the luminosity (L_{44} in units of 10^{44} ergs/sec) and the mass
of the central black hole (M_7 in units of 10^7 $M_\odot$). Then if we
assume that there are no relativistic motions and that the size
of the emitting region R_{13} indicated by the variability timescales
is at least as large as the Schwartzschild radius of the black
hole, we obtain:

$$.07 \; L_{44} \leq M_7 \leq 2T_9^{-1/2}\left(\frac{\Delta t}{700}\right) \tag{10}$$

The validity of the right hand side of this last expression may
be somewhat uncertain, if the flare region does not represent the
whole emission region around the black hole. On the other hand,
if the flare is comparable to or greater than the overall luminosity output of the source, then in fact the flare region is probably coincident with the overall emitting region and equation (10)
can be used to set an upper limit on the black hole mass.

Table 1 is a preliminary summary of the observations and
parameters for the sources we have discussed. For NGC6814, the
average X-ray luminosity is $\sim 4 \times 10^{42}$ ergs/sec. We observe 50%
changes in $\sim 2 \times 10^4$ sec. Equation (8) for the efficiency indicates a relatively low efficiency $\eta \sim 5 \times 10^{-4}$. Equation (9)
gives the size of the emitting region as 2×10^{14} cm. The minimum mass based on the X-ray Eddington limit is $3 \times 10^4 M_\odot$ and the
maximum mass is as high as 6×10^8. For 3C120 all we have is the
X-ray luminosity of order 3×10^{44} ergs/sec, which gives us a
minimum mass of 2×10^6 solar masses for this source in a black
hole model. For OX169 we have an average X-ray luminosity of

1.1×10^{44} ergs/sec. We actually observed changes of 1.5×10^{44} ergs/sec in ~ 6000 seconds, so this is a case where we may be observing changes comparable to the entire energy output of the source. The emission was decreasing in the change that we saw, and it is not clear whether one can apply equation (8) for the efficiency. If equation (8) does not apply, then the efficiency is $\sim 1.2\%$. The size of the emitting region is $\sim 5 \times 10^{13}$ cm, and the luminosity required from the Eddington limit would give us a minimum mass of order 8×10^5 $M_\odot$. The upper limit, based on the size, would be of the order 2×10^8 $M_\odot$. For 3C273 the observed low energy X-ray output is $\sim 1.7 \times 10^{46}$ ergs/sec. We see a change of 1.5×10^{45} ergs/sec, in ~ 6000, which gives us a very high efficiency of about 12%. Since only 10% of the luminosity is involved, we cannot really determine a size for the overall emitting region, although the rise time for the flare would indicate that it occurs in a region of $\sim 5 \times 10^{13}$ cm. The minimum mass is $\sim 10^8$ $M_\odot$.

Finally, the table includes data reported previously by Tananbaum et al. (1978) for NGC4151. The X-ray luminosity was a few x 10^{42} ergs/sec. Changes were observed that could be as large as a factor of 10 in as little as 700 seconds. The statistical significance of the observation was 2.8σ suggesting some caution be used in interpreting the data. The amplitude change based on the error bars also would have been consistent with factors of 2 or 3 increase in intensity, as well as the nominal factor of 10. Applying equations (8-10), we find an efficiency of 1.5%, a size of $\sim 6 \times 10^{12}$ cm, and a mass range between 2×10^5 and 4×10^6 $M_\odot$.

With the extension of the spectrum of NGC4151 up to 3 MeV as discussed by Lawrence (1979), a somewhat different conclusion was derived for the size of the emitting region and the mass of a central black hole. In part, this different result derives from the assumption that the low energy gamma rays come from the same region as the X-rays. A very important new observation would be a search for simultaneous X-ray and gamma ray variability to see how strongly these radiations are coupled. From the Ariel V keV X-ray data, Lawrence showed variations of ~ 2 on times of ~ 1 day. If the gamma rays come from the same region, one would expect their intensity to vary in the same way. Coupling of the X-ray and gamma ray variations would not be consistent with the 6×10^{12} cm size derived by Tananbaum et al. (1978), but would require the several x 10^{14} cm size derived by Lawrence. On the other hand, if the gamma rays arise in a larger region, outside of the X-ray emitting region, then the contradiction can be avoided. This has been proposed, for example, in a paper by Fabian and Rees (1979) in which they discuss 3C273. Fabian and Rees point out that in order to avoid the pair production catatrophe in 3C273, the 100 MeV gamma rays must be produced in a region $\gtrsim 10^{18}$ cm. On the basis of the observed variability, the X-rays are presumed to be produced in a much smaller region.

Table 1

Parameters of Variable Extragalactic X-Ray Sources

Source	L_x ergs/sec	ΔL_x ergs/sec	Δt sec	η	R cm	M_{min} $M_\odot$	M_{max}^* $M_\odot$
NGC6814	$(4)10^{42}$	$(2)10^{42}$	$(2)10^4$	$(5)10^{-4}$	$(2)10^{14}$	$(3)10^4$	$(6)10^8$
3C120	$(3)10^{44}$	-	-	-	-	$(2)10^6$	-
OX169	$(1.1)10^{44}$	$(1.5)10^{44}$	$(6)10^3$	$(1.2)10^{-2}$	$(5)10^{13}$	$(8)10^5$	$(2)10^8$
3C273	$(1.7)10^{46}$	$(1.5)10^{45}$	$(6)10^3$	$(1.2)10^{-1}$	-	$(1.1)10^8$	-
NGC4151	$(3-7)10^{42}$	$(1-3)10^{43}$	$(7)10^2$	$(1.5)10^{-2}$	$(6)10^{12}$	$(2)10^5$	$(4)10^6$

*If flare involves entire emission region around a black hole (more likely at Eddington limit).

For NGC4151 we still have a somewhat inconclusive situation. Additional detailed observations of the quasars and the Seyferts are required to obtain results beyond those presented in Table 1.

Acknowledgements: We are grateful for the assistance of our many colleagues who have contributed to the results described in this paper. We especially thank Ethan Schreier who has headed the analysis of the Einstein Centaurus A data, and Martin Elvis who has headed the analysis of the Einstein Seyfert galaxy data. This work was supported by NASA contract NAS8-30751.

REFERENCES

Arp, H.,1979., Private communication.
Blandford, R.D., and Icke, V., 1978, Mon.Not.Roy.Astr.Soc., 185, 527.
Blandford, R.D., and Konigl, A., 1979, Astron.Lett., 20, 15.
Blandford, R.D., and Rees, M.J., 1978, Phys. Scripta,17, 265.
Cavallo, G., and Rees, M.J., 1978, Mon.Not.Roy.Astr.Soc., 183, 359.
Christiansen, W.N., Frater, R.H., Watkinson, A., O'Sullivan, J.D., Lockhart, I.A., and Goss, W., 1976, Mon.Not.Roy.Astr.Soc., 181, 183.
Cooke, B.A., Lawrence, A., and Perola, G.C., 1978, Mon.Not.Roy. Astr.Soc.,182, 661.
Dufour, R.J., and van den Bergh, S., 1978, Astrophys.J.(Letters), 226, L73.
Fabian, A.C., and Rees, M.J., 1979, COSPAR/IAU Symposium on Non-Solar X-ray Astronomy, eds. L.E. Peterson and W.A. Baity (Oxford: Pergamon Press).

Graham, J.A., 1979, Preprint.

Henry, J.P., et al., 1979, in preparation.

Holt, S.S., 1979, this volume.

Lawrence, A., 1979, seminar presented at the School, Erice.

Lawrence, A., Pye, J.P., and Elvis, M., 1977, Mon.Not.Roy.Astr.Soc., 181, 93p.

Lightman, A.P., Giacconi, R., and Tananbaum, H., 1978, Astrophys.J., 224, 375.

Osmer, P.S., 1978, Astrophys.J.(Letters), 226, L79.

Perley, R.A., Willis, A.G., and Scott, J.S., 1979, Preprint.

Rees, M.J., 1978, Mon.Not.Roy.Astr.Soc., 184, 61p.

Schreier, E.J., Feigelson, E., Delvaille, J., Giacconi, R., Grindlay, J., Schwartz, D.A., and Fabian, A.C., 1979, Astrophys.J. (Letters), 234, L39.

Tananbaum, H., Peters, G., Forman, W., Giacconi, R., Jones, C., and Avni, Y., 1978, Astrophys.J., 223, 74.

Tananbaum, H., Avni, Y., Branduardi, G., Elvis, M., Fabbiano, G., Feigelson, E., Giacconi, R., Henry, J.P., Pye, J.P., Soltan, A., and Zamorani, G., 1979, Astrophys.J.(Letters), 234, L9.

Tucker, W.H., 1975, Radiation Processes in Astrophysics, Cambridge: MIT Press.

Tucker, W., Kellogg, E., Gursky, H., Giacconi, R., and Tananbaum, H., 1973, Astrophys.J., 180, 715.

Winkler, P.R., Jr., and White, A.E., 1975, Astrophys.J.(Letters), 199, L139.

EINSTEIN OBSERVATIONS OF QUASARS

H. Tananbaum

Harvard-Smithsonian Center for Astrophysics
Cambridge, Massachusetts

Before the launch of the Einstein X-ray Observatory, there
were three known X-ray emitting quasars. They were all relatively
nearby. Estimates on the X-ray emission from quasars as a group
were very uncertain because, based on three objects, our knowledge
was very limited. Figure 1 is a picture, obtained during the first two
months' operation of Einstein, of the quasar 3C279 at a redshift of
0.54. In this 2500 second exposure with the Imaging Proportional
Counter (IPC) we have no difficulty at all in detecting the quasar, the
bright X-ray source right in the middle of the field. The counting rate
is ~ 1/4 count per second, which indicates that this source is only a
small factor ($\lesssim 4$) below the thresholds of earlier surveys. Figure 2
is a similar IPC picture with a 2500 second exposure showing the quasar 3C47 at a redshift of 0.43. The quasar in this case is again the
central source in the image. The counting rate is ~ 1/7 count per second, about a factor of two lower than for 3C279. Also seen in the
picture as an obvious X-ray source is a foreground star that turns up
in the field.

The Einstein Observatory carries two imaging instruments
on board: the Imaging Proportional Counter (IPC), with which most of
the quasar observations are performed, and the High Resolution Imager
(HRI). The IPC has a higher quantum efficiency; therefore, it is the
more appropriate of our two detectors for observing strong sources.
In addition, we obtain some spectral information with the IPC. The
spatial resolution of the detector is ~ 1 arc minute, and its current
positional accuracy is also of that order. The HRI has a somewhat
lower quantum efficiency and essentially no spectral resolution. On

R. Giacconi and G. Setti (eds.), X-Ray Astronomy, 311-325.
Copyright © 1980 by D. Reidel Publishing Company.

Fig. 1. IPC image of quasar 3C279.

Fig. 2. IPC image of quasar 3C47.

the other hand, it has a spatial resolution and positional accuracy of a few arc seconds, as well as a lower background for point source observations.

Figure 3 is an example of an observation which we carried out with the HRI. In this case we were interested in studying a possible relationship between the quasar Mkn 205 and the galaxy NGC4319. This pair of objects has been discussed by Arp (1971). The quasar has a redshift of 0.07, and the galaxy NGC 4319, to which Arp observes a luminous bridge, has a redshift of 0.006. The two are separated by ~ 40 seconds of arc. The X-ray observations show a point source (367 counts in the HRI) coincident with the quasar Mkn 205. The $\llcorner$ symbol in the figure indicates the position of the galaxy NGC 4319 for which we have an upper limit of less than ten counts. We can convert the counts to luminosities if we assume that the redshifts are in fact valid distance indicators. For Mkn 205 the luminosity is 3×10^{44} ergs/sec in the 0.5 to 4.5 keV energy band, and for NGC 4319 the upper limit corresponds to 5×10^{40} ergs/sec. Also, there is no indication of a luminous bridge in the X-rays, which indicates that if there is such a bridge then it is not composed of very hot or energetic material capable of emitting X-rays.

Fig. 3. HRI image of Mkn 205. Location of galaxy NGC 4319 is indicated by $\llcorner$.

Fig. 4.   Einstein calibration data showing mirror plus HRI response
          to 1.49 keV point source.

Figure 4 shows pre-flight Einstein calibration data, taken
looking at a point source in a special facility built for the Einstein program at Marshall Space Flight Center. The figure shows the observation of a strong point source, and the picture is very much overexposed
to show the small amount of scatter that we have in the system. The
projection through the center of the source shows the full width at half
maximum which is 4 seconds of arc, our basic resolution. The figure
also shows 16 alternating bands radiating as spokes from the center.
These bands are shadows in the scattered flux, introduced by the mirror support structure. There are 16 structural ribs that support the
cylindrical mirror system. The figure shows the response expected
for the observation of a point source with very high counting statistics.
Figure 5 shows the HRI image obtained with ~80000 seconds
effective exposure on 3C273. The primary scientific objective was to
search for X-ray emission coming from the jet in 3C273. The X-ray
data are consistent with a point source located at the central object
which is known as 3C273B in the radio. The jet is located to the southwest, about 20 arc seconds from the central source. Any sensitive
search for X-ray emission from the jet requires very careful handling

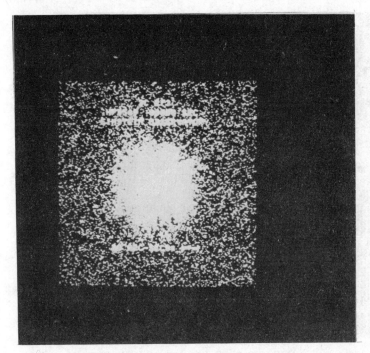

Fig. 5. HRI image of 3C273.

to remove the scattering seen in the calibration data. A preliminary
analysis demonstrates that any X-ray emission from the jet is certainly less than 1% of the X-ray emission from the central source.
[A more thorough analysis carried out after this lecture was presented
indicates that the jet may be a weak X-ray source with about the same
jet to central source ratio in the X-ray as in the optical. Further details are presented by Henry et al. (1979)].

We can also combine the 3C273 HRI data with 2 to 20 keV
data obtained simultaneously with the Einstein Monitor Proportional
Counter to examine the X-ray spectrum. The MPC data are well fit by
a power law with energy index of 0.5 in agreement with the HEAO A-2
spectrum published by Worrall et al. (1979). The HRI is sensitive
down to 1/4 keV and by combining the data from the two Einstein instruments, we find that for a slope of 0.5 the low energy absorption is
approximately 3×10^{20} hydrogens/cm^2. This is consistent with the
amount of material along the line of sight to 3C273 looking through our
own galaxy. In turn, this allows very little absorbing gas at 3C273.

Another objective of our quasar observations was the study
of the properties of quasars at larger redshifts. Figure 6 is an IPC
picture of the quasar B2 1225+31. The 0.5 to 4.5 keV luminosity is
$\sim 5 \times 10^{46}$ ergs/sec; at the time we made this observation this was the

Fig. 6. IPC image of quasar B2 1225+31.

most luminous quasar that we had seen. Based on our success in detecting B2 1225+31, we decided that it would be worthwhile to observe
even more distant quasars. About the same time a guest observer pro-
,posal was submitted by Peter Strittmatter, Malcolm Smith, Ray Wey-
mann, Jim Liebert, and Jim Condon to study distant quasars and certain other groups of quasars with interesting optical or radio properties.
We therefore formed a collaboration to carry out these studies. Figure 7 is one of the first results that we obtained on a still more distant
quasar. This is a picture of the quasar 0420-38, and again the quasar
is seen very nicely in the middle of the IPC field. This quasar has a
redshift of 3.1, and the 0.5 to 4.5 keV luminosity is ~ 8 x 10^{46} ergs/
sec. Of course when we detect emission from this quasar at an energy
as high as 3 keV, then the corresponding emission energy is 12 keV at
the source, and the overall X-ray luminosity is still higher. Figure 7
also shows at least three additional X-ray sources in this field that we
have not yet identified.

        Figure 8 shows a 20000 second exposure for the quasar 0537-
28, also at a redshift of 3.1. In this case the quasar was not centered
in the field, and the X-ray image is at the eastern or left side of the
picture. The exposure was taken for 20000 seconds because the quasar
is ~ 20th magnitude in the visible. Based on the discussion below con-

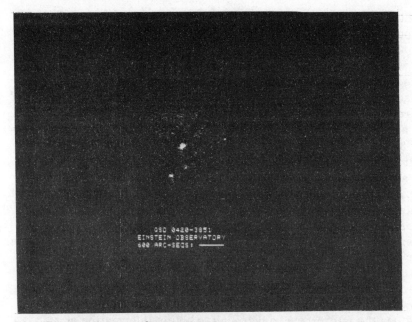

Fig. 7.   IPC image of quasar 0420-38.

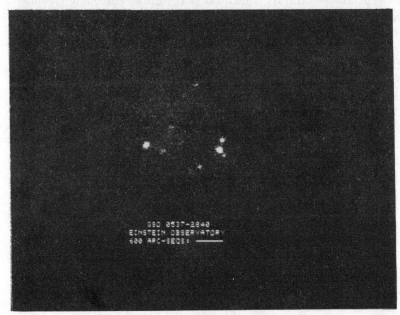

Fig. 8.   IPC image of quasar 0537-28 located to left (east) side of pic-
          ture. Other sources are unidentified except for 2 bright F-
          stars associated with the 2 strong X-ray sources at the right
          (west) side of the picture.

cerning the ratio of the X-ray to optical emission, we estimated the long exposure would be required to detect this optically faint quasar. However, it turned out that this quasar was extremely luminous in the X-ray, with an 0.5 to 4.5 keV luminosity of $\sim 1.3 \times 10^{47}$ ergs/sec, which is a factor of 8 times the X-ray luminosity observed for 3C273. For this exposure we note that we could still have detected the quasar if it were ten times fainter than it actually is. This means that if the quasar had its observed X-ray luminosity, but instead of being at a red-shift of 3.1 it were at a redshift of 5, we still would have been able to detect it with ease. This is a very important point, since from current optical surveys we only know that the quasars extend to redshifts of about 3.5. Observers are working very hard to see whether or not still more distant quasars exist. A possible alternative to the optical searches is the identification of objects which are found in our X-ray surveys. Since the number of X-ray objects per square degree is much smaller than the number of optically faint stellar objects, we may be able to improve the prospects for finding more distant quasars by selecting them based on their X-ray emitting properties. At present, we do not know much about what the Universe looks like at redshifts beyond 3 until we use the microwave background to obtain information at $z \gtrsim$ 1000, corresponding to the first million years in the Big Bang Cosmology. Thus the ability to explore from redshifts of 3 to 10, for example, by searching for distant quasars is a very important capability.

Figure 9 provides a summary of our quasar observations as of June 1979. We have plotted the quasar luminosity in the 0.5 to 4.5 keV band, ranging from 10^{43} to 10^{47} ergs/sec, versus the redshift. The three quasars which were previously known are shown by the two x's plus 3C273 and all are at low redshifts. The figure shows that we have extended the X-ray observations out to redshifts beyond 3. The absence of sources in the lower right portion of the figure is basically a statement of our sensitivity limit. For a given exposure we can only detect sources to the upper left of the imaginary line running from lower left to upper right in the figure. Different symbols are used to indicate some of the different features of the sources. Those sources marked with the triangles were detected by their X-ray emission and then later determined to be quasars with follow-up optical studies. These identifications are still in a preliminary stage, and of course it will be very interesting to complete and study our X-ray selected samples. Quasars which are strong radio sources are shown by squares and those that are radio quiet by filled circles in the figure. The data show a relatively well mixed set of X-ray sources independent of the radio emission properties.

The preponderance of squares in the figure results from a program to study the X-ray emission from a complete sample of 3CR

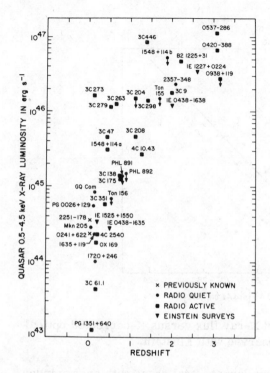

Fig. 9. Plot of X-ray luminosity (0.5 to 4.5 keV) in ergs/sec versus
quasar redshift for objects observed with Einstein.

radio quasars. Since optically selected quasars comprise more than
90% of all quasars, it is important for us to know if those quasars
which are optically selected have vastly different X-ray properties
from those selected by their radio properties. In Figure 9 the circles
are fairly well mixed with the squares, which suggests no obvious
dichotomy. However, complete optically selected samples will be required before we can reach a firm conclusion in this area.

Figure 10 is a plot of the K-corrected X-ray emission versus the K-corrected optical flux. We are interested in seeing whether
the fluxes that are observed in the X-ray and the optical are correlated.
However, since the objects that have been observed are at different
redshifts, we use the K-corrections to determine the flux that we would
have observed if there were no redshift in the system. For the X-ray
flux, we take the 0.5 to 4.5 keV band and integrate the X-ray emission, fit a spectrum, and then calculate the observed spectral density
at 2 keV/(1 + z), which corresponds to the source emission at 2 keV.
Since our X-ray data are somewhat limited we carry out the above

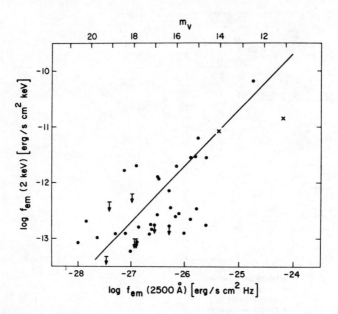

Fig. 10. Plot of K-corrected X-ray flux versus K-corrected optical
          flux for quasars observed with Einstein.

procedure by assuming a power law spectral shape with energy index
0.5. In the optical, a similar procedure is followed to determine the
source emission at 2500 Å. The data in Figure 10 show that even with
a considerable amount of scatter there is a correlation whereby the
optically brighter quasars are also brighter in the X-ray. The range
that one observes corresponds to about a factor of 250 in the ratio of
X-ray to optical flux. If we take the bandwidth of 0.5 to 4.5 keV in the
X-ray and the bandwidth of 3000 to 6000 Å in the visible, then the ratio
of X-ray to optical flux ranges from 0.02 to 5. The particular line
which is drawn in Figure 10 corresponds to an X-ray to optical ratio of
0.56.

    We can define the quantity:

$$\alpha_{ox} = -\log \left[\ell(\nu_x)/\ell(\nu_{opt}) \right] \quad \log \left[\nu_x/\nu_{opt} \right] \tag{1}$$

which is basically the energy slope of a power law connecting the optical emission at frequency ν_{opt} with the X-ray emission at frequency
ν_x. In equation (1), $\ell(\nu)$ is the luminosity density in ergs/sec - Hz at
frequency ν. In Figure 11 we have plotted α_{ox}, the optical to X-ray
slope, versus the redshift for the different quasars. Our observations

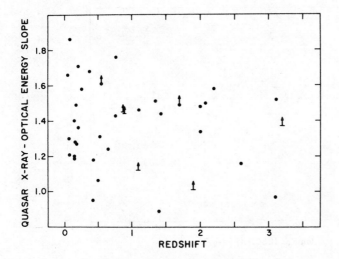

Fig. 11.   Plot of the optical-X-ray slope (α_{ox}) versus redshift for
quasars observed with Einstein.

show that α_{ox} ranges from 0.94 to at least 1.86. Again there is considerable spread, but there is no obvious trend for this index to change
with redshift, which is an important point in what follows. In Figure 12
we plot α_{ox}, the optical to X-ray slope, versus the intrinsic optical
luminosity at 2500 $\overset{o}{A}$ (in ergs/sec - Hz). Again our basic conclusion is
that there is no obvious trend for the slope to change with optical luminosity.

       With the assumptions that the optical to X-ray slope is inde-
pendent of redshift and independent of intrinsic optical luminosity,
Avni et al. (1979) have shown how to compute the effective average
optical to X-ray slope. The expression resembles equation (1) except
that ℓ_x / ℓ_{opt} is replaced by $< \ell_x / \ell_{opt} >$. Avni et al. show how to
compute this quantity from the X-ray observations including the upper
limits for quasars not detected. The result of this calculation is that
the value of the effective average optical to X-ray slope is 1.26 $\pm$ .09
(2σ). In this calculation we use all of the observations except those
quasars which have been selected by their X-ray properties. The X-
ray selected quasars have to be excluded from the average because in
selecting those objects which are brightest in the X-ray, we would not
be obtaining a fair or random sample of the average optical to X-ray
slope. Also we assume that the radio quasars have the same α_{ox}
distribution as the optically selected quasars.

       We are now in a position to combine the result for the aver-
age optical to X-ray slope with the optical source counts in order to

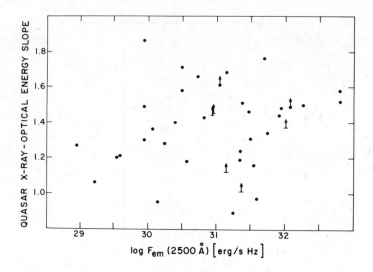

Fig. 12.   Plot of the optical-X-ray slope (α_{ox}) versus intrinsic optical
luminosity for quasars observed with Einstein.

estimate the contribution of the quasars to the X-ray background (see
Tananbaum et al. 1979 for additional details).   Figure 13 is taken from
a recent paper by Braccesi et al. (1979) and shows the optical source
counts, the number of sources per square degree brighter than a given
blue magnitude as a function of magnitude.   The key data points are the
Green and Schmidt (1978) sample at 15th and 16th magnitude and the
data at faint magnitudes which determine a power law fit for the source
counts.   The density of 19th magnitude quasars is about 5 per square
degree.   Braccesi et al. (1979) conclude that the counts can be described, from $m_B = 15.5$ to $m_B = 21.4$, by

$$\log N \ (< m_B) \ = \ 2.16 \ (\frac{m_B - 18.33}{2.5}) \ , \tag{2}$$

where $N \ (< m_B)$ is the number of objects per square degree with magnitude less than m_B.

Using this formula and the effective average α_{ox} we can calculate the expected contribution to the observed extragalactic 2 keV
X-ray background (Schwartz 1979) from quasars down to a given limiting magnitude m_{Lim}.   The resultant computed quasar X-ray background
is proportional to:

$$(10^{-2.605 \, \alpha_{ox}}) \ (1 + z_{eff})^{\alpha_o - \alpha_x} \ \frac{a}{(a-1)} \ 10^{(\frac{a-1}{2.5}) \ (m_{Lim} - B_o \frac{a}{a-1})} \tag{3}$$

where a and B_0 are taken from equation (2) as 2.16 and 18.33 respectively.

In equation (3), z_{eff} is the redshift for the typical quasar contributing to the X-ray background. For an assumed optical energy slope $\alpha_0 = 0.7$ and X-ray energy slope $\alpha_x = 0.5$, the result is relatively insensitive to our actual choice of $z_{eff} = 1.5$. The effective α_{ox} (1.26) and the slope of the optical counts (2.16) lead to a computed background that exceeds that actually observed if the optical counts extend beyond 20.3^m. Equation (3) shows how the result depends on α_{ox}, a, and m_{Lim}. We note that a larger effective α_{ox} could result if the radio quiet quasars, which represent $\gtrsim 90\%$ of the total number of quasars, but are not adequately represented in our sample, are fainter X-ray emitters than the radio observed. Also if more detailed studies show that α_{ox} depends on redshift or optical magnitude our conclusions could be slightly modified.

The above calculation shows that the diffuse X-ray background plus the observed emission from quasars will ultimately be useful for setting firm limits on the maximum number of faint quasars.

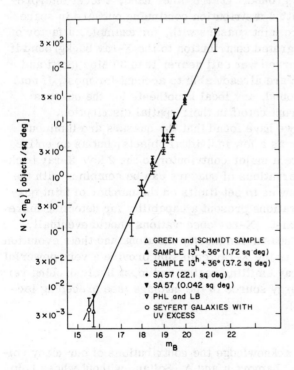

Fig. 13. Quasar optical source counts showing number of quasars brighter than blue magnitude versus blue magnitude.

Our data already strongly suggest that the slope of the optical counts does not exceed 2.16 and also that somewhat above 20^m the slope must flatten. [We recently received a paper from Setti and Woltjer (1979) which predicts a similar effect based upon the three pre-Einstein quasars.]

The suggestion from the X-ray data that the optical source counts flatten above 20^m also points out the importance of carrying out spectroscopic studies on the 20^m and 21.4^m objects indicated in Figure 13. The selection of these objects as quasars is based upon photo-electric measurements of their colors and has not been confirmed by spectroscopic observations. Since optical observers find many more stellar objects still fainter than 21.4^m, spectroscopic studies are absolutely required to determine the nature of these objects.

In a related context, Setti and Woltjer (1970) predicted that the observations of X-ray emitting quasars could be combined with optical source counts to restrict or eliminate non-cosmological or local explanations of the quasar redshifts. As discussed above, the source counts of Braccesi et al. (1979) to 20.3^m already can account for the extragalactic X-ray background. On the other hand, a local interpretation of the redshifts with a distribution continuing outward in space would require many more faint quasars with, for example, a factor of ~ 80 increase in the integrated contribution to the X-ray background if the quasars already observed were all nearer than 30 Mpc (Setti and Woltjer 1970). Since we are already able to account for most, if not all, of the X-ray background, any local hypothesis for the quasars would require a very abrupt cutoff in their spatial distribution.

In summary, we have found that the quasars are luminous X-ray emitters, not only as a few individual objects, but as an entire class. These objects are a major contributor to the 2 keV X-ray background. The X-ray observations of quasars can be combined with the X-ray background to allow us to set limits on the number of faint quasars. The X-ray observations present a capability for detecting more distant quasars if they exist. X-ray observations should eventually prove useful in studying quasar luminosity functions and their evolution. Finally, time variations in the X-ray emission provide a very powerful tool for studying the X-ray emitting region and most likely probing very close to the ultimate energy source in the quasars (see preceding lecture by Tananbaum 1979).

Acknowledgements

We gratefully acknowledge the contributions of our many colleagues, in particular G. Zamorani and A. Soltan, without whose help this work would not have been possible. This research was supported by NASA contract NAS8-30751.

References

Arp, H. 1971. Astrophys. Lett., 9, 1.

Avni, Y., Soltan, A., Tananbaum, H., and Zamorani, G. 1979.
   Submitted to Ap. J.

Braccesi, A., Zitelli, V., Bonoli, F., and Formiggini, L. 1979.
   Astron. and Astrophys., in press.

Green, R.F., and Schmidt, M. 1978. Ap. J. (Letters), 220, L1.

Henry, J. P., et al. 1979. In preparation.

Schwartz, D.A. 1979. COSPAR/IAU Symposium on Non-Solar X-ray
   Astronomy, L. E. Peterson and W. A. Baity, eds. (Oxford:
   Pergamon Press), pp. 453-465.

Setti, G., and Woltjer, L. 1970. Ap. and Sp. Sci., 9, 185.

Setti, G., and Woltjer, L. 1979. Astron. and Astrophys., 76, 4.

Tananbaum, H., Avni, Y., Branduardi, G., Elvis, M., Fabbiano, G.,
   Feigelson, E., Giacconi, R., Henry, J.P., Pye, J.P.,
   Soltan, A., and Zamorani, G. 1979. Ap. J. (Letters), 234.

Worrall, D.M., Mushotzky, R.F., Boldt, E.A., Holt, S.S., and
   Serlemitsos, P.J. 1979. Ap. J., 232, 683.

X-RAY SPECTRA OF ACTIVE GALACTIC NUCLEI

S.S. Holt

Laboratory for High Energy Astrophysics
NASA/Goddard Space Flight Center
Greenbelt, Maryland 20771 U.S.A.

1.0 ACTIVE GALACTIC NUCLEI

There are a variety of galaxies which are classified as nuclear-
active, ranging from those which have nuclei which are merely
distinguishable from a stellar component all the way to QSO's.
The subject of this lecture will be only those extreme cases
(Seyfert I, BL Lac and quasar) which constitute the most intense
compact X-ray sources which are presently known. The luminosities
of these sources in the 2-10 keV band range upward of 10^{42} ergs/s
(to more than 10^{46}ergs/s), while those sources classified Seyfert
II or NELG are typically less luminous.

The main distinguishing characteristic of these objects is
not their absolute luminosity, but the extent to which the nucleus
is a compact luminous entity. The prime defining characteristic
of Sy I nuclei is the very broad ($\gtrsim 10^4$ km/s FWZI) emission lines
which emanate from the central portion of the nucleus, upon which
the much narrower (< 10^3 km/s) lines from the surrounding region
are superposed: Sy II spectra contain only the "narrow" lines,
while Sy I contain both. The central broad-line-region (BLR) is
generally taken to have a characteristic dimension < .05 pc, based
upon the timescale for variability observed, and the surrounding
NLR may have a characteristic dimension of $\sim$ 1 kpc. Forbidden
lines are only observed from the NLR, so that the electron density
in the BLR optical filaments must be $\gtrsim 10^8$ cm^{-3}. The X-ray emission appears to correlate with continuum and line emission from
the BLR, so that we expect that this central region of these
nuclei is the site of the X-ray emission; indeed, the shortest
timescale for variability from at least one of them, NGC4151, is
less than one day. A recent review of the properties of X-ray
emitting Seyfert galaxies and their correlations with emission
in other bands may be found in Wilson (1979).

The current prevailing opinions place the energy source for
the luminosity of these objects in a central black hole (or

327

equivalent) of mass $\sim 10^8$ $M_\odot$. Quasars are assumed to be the most
extreme cases of Sy I, and BL Lac nuclei are also assumed to be
closely related (the latter objects differ from Sy I in four
respects: they are generally in elliptical rather than spiral
galaxies, they are strong variable radio emitters, the optical
emission lines are very weak or undetectable, and their optical
emission is highly polarized).

The X-ray emission from these nuclei, as in the case of
binary X-ray sources, cannot be assumed to emerge without substantial Compton scattering. Fabian (1979) has recently reviewed
the possible mechanisms which might be most directly responsible
for the X-rays observed. If there exists very hot plasma (T $\gtrsim$
10^{8o}K), as might be expected from shock heating or accretion onto
the central object, there will be a direct bremsstrahlung component
which will emanate from the nucleus. The presence of the non-
thermal continuum in the BLR will favor X-ray production via
Compton scattering from the high temperature electrons, however,
so that the spectrum may not be easily interpretable even if these
were the only possibilities. The presence of ultrarelativistic
electrons ($\gamma \gtrsim 100$) complicates the story further, as X-rays may
be produced from Compton interactions of these electrons with the
far infra-red synchrotron radiation which they produce in the
ambient magnetic field. The primarily produced X-radiation may be
further Compton-scattered before emerging from the source region,
as well.

All of the X-ray measurements are obtained over limited dynamic
ranges, so that the bulk of this lecture will be devoted to a
review of the search for consistent characteristics of the spectra
of active galactic nuclei which we might hope to ultimately recon-
cile with detailed models.

2.0 SEYFERT I SPECTRA

A sample of seven Seyfert I spectra obtained with HEAO A-2 in the
energy range 2-50 keV have recently been analyzed by Mushotzky et
al. (1979b). The spectra are generally better-fit by power-laws
than by bremsstrahlung continua, so that a power-law representation
will be utilized for all of them. Five of the seven "inverted"
spectra are displayed in Figure 1.

If we consider the spectra > 5 keV, all can be fit to single
power-law forms. Figure 2 illustrates the best-fit and range
of acceptable indices for each of them, as a function of source
luminosity. It appears from this sample that there is no obvious
correlation of spectral index (the label α is used in the Figure
for the photon spectral index, which is higher by unity than the
index of the energy spectrum) with luminosity, and that the entire
sample is consistent with an energy index of 0.6 - 0.7. The
simple average of all the indices measured from this sample is
0.65, but the weighted average is closer to 0.6 since a high
exposure to NGC4151 makes the formal statistical quality of that

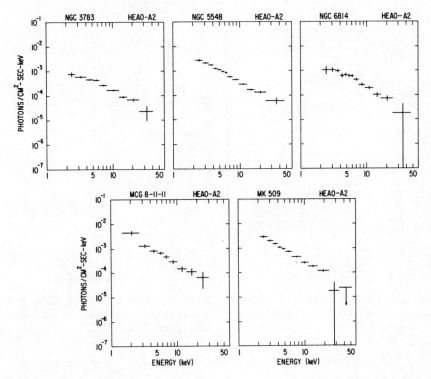

Fig. 1.   Incident X-ray spectra of five Seyfert I galaxies.

index better than the others. There is no evidence for steepening
at the highest energies we measure in any of the spectra (although
they must clearly steepen eventually).

There is clear evidence for variability in about half the
sample (the HEAO-1 satellite measures each at six month intervals,
so that all may be variable to some extent). The highest degree
of variability has been measured from the least luminous sources
(NGC3783 was measured to vary by more than a factor of two in less
than one week, and NGC4151 by a like amount in less than one day).
We cannot be sure that this apparent inverse correlation with
luminosity is real because of the small sample, but there does not
appear to be any change in spectral index for any of these sources
with intensity.

Below 5 keV there are two effects which deserve consideration.
The first is that substantial absorption attributable to nonionized
material has been measured only from NGC4151; this absorption
changes on short timescales and has been, therefore, attributed
to the optical filaments within the BLR (Ives, Sanford and Penston
1976). The only other objects for which absorption corresponding
to a column density in excess of $N_H \sim 10^{22}$ cm^{-2} has been inferred
are NGC3783 and NGC6814, so that we have an indication that the

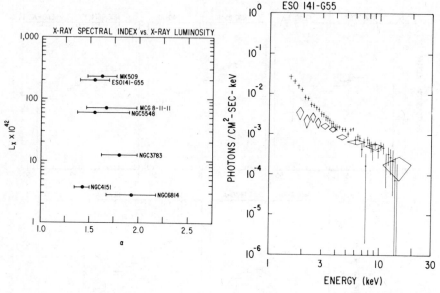

Fig. 2. Photon power-law index   Fig. 3. Two separate HEAO A-2
versus 2-10 keV X-ray luminosity   measurements of the spectrum of
for seven Seyfert I galaxies.   ESO 141-G55.

lower luminosity Sy I sources are more likely to exhibit such
an effect.

The other low energy characteristic of these objects is a
variable excess below 5 keV that is exhibited by at least some of
them. Figure 3 displays the spectra from two exposures of
ESO141-G55, which illustrates the steepening below 5 keV in one
(the crosses) while consistency with the other (the diamonds) above
5 keV is achieved. A single 0.8 - 4.5 keV measurement of ESO141-
G55 with the Einstein SSS measured this steep component, as well.

The overall characteristics of this sample, therefore, indicate
that NGC4151 is hardly prototypical. In fact, NGC4151 has an
anomalously hard spectrum, and anomolously large low energy
absorption (in both cases, representing the extreme for the sample).
The other Seyferts may, in fact, exhibit excesses rather than
deficiencies at low energies. All may be variable (no others have
been detected to vary on timescales as short as NGC4151, but this
may reflect a selection effect because NGC4151 has the highest
apparent magnitude), and all appear to preserve their characteristic
power law index above 5 keV if they do vary. The average index
for the total sample is consistent with the statistical errors for
each of the individual members of the class. Sy I objects analyzed
after completion of the study from which Figures 1-3 are taken,
NGC7469 and MCG-2-58-22, have HEAO A-2 and Einstein SSS spectra
completely consistent with the sample characteristics discussed
here.

3.0 BL LAC SPECTRA

The BL Lac sample at our disposal is less than half as large as
that for Sy I, and the initial results led to what probably represented a premature characterization of their average spectra.
Mushotzky et al. (1978) noted the hardness of the initial measurements of Mk421 and Mk501 (α ≈ 0.2), although there was some
evidence for steepening at low energies in both of these objects.
We later found (Mushotzky et al. 1979a) that Mk421 also exhibited
variability in the soft component on occasion, and that the hard
component could vary by at least a factor of 4 (see Figure 4).

     On the basis of several exposures to MK421, Mk501, PK0548-322,
2155-304 and 1219+305 from both HEAO A-2 and the Einstein SSS,
we now believe that all the BL Lac sources exhibit spectra which
can be synthesized from two power-law forms: a very hard component
(α ≈ 0.2) and a much softer (α ≈ 2) component. We have never
measured any evidence for absorption (N_H < 3 x 10^{21}), which may
or may not be related to the absence of optical emission lines,
and the hard component appears to be more highly variable than
does the soft component. In other words, an arbitrary BL Lac
spectral measurement is likely to yield evidence for a steep
spectral component more than half the time (and more often than
a hard component).

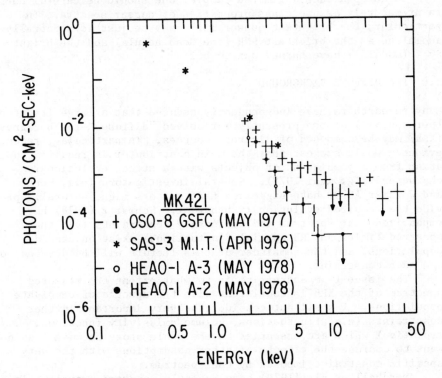

Fig. 4. Four separate measurements of Mk421.

4.0   QUASAR SPECTRA

The situation here is even more sparse than for BL Lac's. Several
exposures have been made to 3C273, but we have no correspondingly
deep exposures to any other quasars above a few keV except for
0241+62. The latter source is well-fit by a Crab-like spectrum
out to 50 keV (i.e. steeper than the Sy I average index), and
Fairall 9 is well-fit by a similar index below 4.5 keV with the
Einstein SSS. More SSS spectra of quasars will be available in
the near future, but only at energies commensurate with the 0.8 -
4.5 keV window of the SSS.

3C273 has been measured on several occasions by HEAO A-2.
It has exhibited 40% variations on timescales of 6 months, and
best-fit energy indices between 0.41 and 0.73 (the statistical
errors in the latter include the former, but not vice versa). The
deepest exposure measured the hardest overall spectrum out to $\sim$
60 keV (the steepest was the observation which was 40% lower in
intensity), but there was some marginal evidence for steepening
with lower energy (as well as evidence from another instrument on-
board HEAO-1 for steepening above 60 keV). A detailed description
of the HEAO A-2 3C273 observations is given in Worrall et al. (1979),
as well as appropriate references to the data at higher energies.

Based upon such a limited sample, one should be careful not
to prematurely assume class properties of quasar spectra. In
particular, the brightest quasar 3C273 may be just as spectrally
anomalous as the brightest SNR (the Crab nebula) and the brightest
Sy I (NGC4151) have turned out to be.

5.0   THE DIFFUSE BACKGROUND

Many researchers have independently deduced that a large fraction
(perhaps all) of the presently unresolved "diffuse X-ray background"
(XRB) may be composed of discrete sources. In particular, recent
evidence would suggest that the main contributor is required to
arise from a population of objects with a strong evolutionary
character (e.g. Avni 1978). Several investigators (e.g. Setti
and Woltjer 1979) have suggested that quasars might naturally provide the required evolutionary behavior. Schwartz (1978) has
emphasized that spectral and luminosity constraints (via the
observed limits on XRB spatial fluctuations) exist on candidate
populations, and the remainder of this lecture will be devoted to
one such constraint.

The issue I am addressing is to what extent the measured
spectrum of the XRB is consistent with the spectra of candidate
populations. I shall not consider the very important further
constraints (e.g. fluctuations, volume emissivity functions,
evolution) which are essential to the whole story because I do not
want to confuse the effects of other assumptions with the very
specific constraint imposed by the spectrum.

Marshall et al. (1979) have recently reported a systematic-

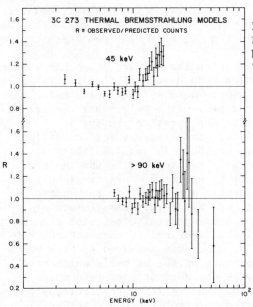

Fig. 5. Ratio of 3C273 spectrum measured from HEAO A-2 to expectation from bremsstrahlung continua.

free measurement of the XRB from HEAO A-2. Below ∿ 3 keV the galactic and extragalactic contributions cannot be unambiguously separated, and above ∿ 50 keV the experiment rapidly loses sensitivity. Between these two limits the spectrum is remarkably well-fit by thermal bremsstrahlung from a 40 ± 5 keV plasma. Field and Perrenod (1977) have discussed the possibility of a true inter-galactic plasma which might give rise to this spectrum, and approximately 1/3 the closure density would be required to match its intensity without clumping (a non-homogeneous plasma would require a smaller density). The quality of the fit does not necessarily demand a thermal bremsstrahlung origin for the XRB, or even a substantial fraction of it. It does, however, define the spectral conditions which the totality of the XRB contributors must meet.

3C273 is the best-measured of the quasars, and a comparison of its spectrum with the XRB is encouraging, as evidenced by Figure 5. A direct comparison fails, as shown by the top trace, but if we assume that quasars like 3C273 must be at Z > 1 we can achieve consistency at least over a limited dynamic range, as demonstrated in the lower trace. The fit is acceptable over a dynamic range of about 4 (it fails below 6 keV (not shown), but that may not be too serious because there are other XRB contributors and possible galactic complications there), as compared to a dynamic range of 15 in the HEAO A-2 XRB spectrum. It meets the necessary condition for being a consistent contributor to the XRB, but not the suffic-ient condition. Unfortunately, it cannot be tested in the crucial region near 40 keV where it should rapidly steepen. The positive detection of 3C273 by Cos B at γ-ray energies may indicate that

it cannot steepen rapidly enough at reasonable redshift, and Setti
and Woltjer (1979) have suggested, therefore, that 3C273-like quasars
would have to comprise no more than 5% of those from which the
XRB arises.

The other two quasars for which the Goddard group has spectra
are 0241+62 and Fairall 9, which both have exhibited power-law
spectra similar to that of the Crab nebula (the latter below 5 keV,
and the former between 2 and 50 keV). The only other spectra we
possess which may be relevant to the question are the Sy I and
BL Lac spectra discussed earlier, since the physical conditions
in their nuclei may be similar to those in quasars (there is no
evidence that their volume density evolves in the same manner,
but I prefer to restrict myself to spectral consistency only as
discussed above). The Sy I spectra are all steeper than the XRB
with the exception of NGC4151, and the γ-ray measurements of
NGC4151 constitute an even stronger spectral constraint than in
the case of 3C273 against objects identical to that one contributing substantially to the XRB. The only measured spectra which
are flatter are the hard components of the BL Lac's which are
occasionally measured along with the much steeper (and more
frequently observed) components. We are left, therefore, with
the difficulty that the measured spectra from those active
galactic nuclei which we expect to be most prototypical of those
which can make up the XRB are all steeper. See Figure 6, where
the ratio of the measured XRB spectrum to assumed power laws of
$\alpha = 0.4$ and $\alpha = 0.7$ are plotted, for an indication of how serious
this potential difficulty is even before steeper components arising
from other origins are subtracted. The same ratio of the measured
spectrum to thermal bremsstrahlung at 40 ± 5 keV is unity to
within a few percent over the whole range.

The new evidence which has seemingly given experimental
verification to the idea that the XRB is largely composed of the
contribution from evolved quasars is the detection of such objects
with the imaging instruments onboard the Einstein Observatory.
Extrapolations can be made which suggest that the XRB may be
thusly explained, but the spectral difficulty remains. Confining
my remarks to spectral considerations only, there are a few subtle
points worth considering. The first concerns the effective
energies at which these observations are made. Using the response
functions of the IPC and the HRI (Giacconi et al. 1979), the median
energy of detected photons is always less than 1 keV for incident
spectra steeper than an extrapolation to these energies of the
40 keV bremsstrahlung spectrum. The HRI median energy is always
about a factor of two lower than that of the IPC in its pre-launch
gain mode, so that it is always below $\frac{1}{2}$ keV, even for the flattest
candidate spectra. For the steep low energy components of Sy I
and BL Lacs, which may be representative of at least some of the
quasars, the median energy measured with the IPC is well below
$\frac{1}{2}$ keV. I shall assume, therefore, that the median energy which
characterizes the imaging results is no higher than $\frac{1}{2}$ keV. This

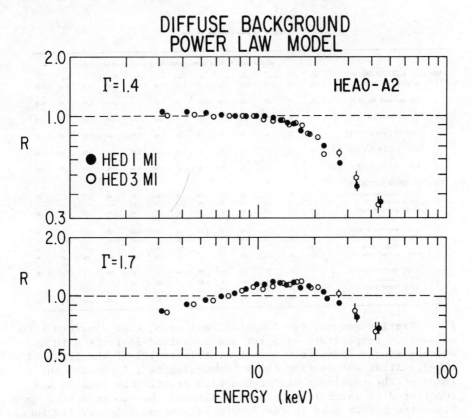

Fig. 6. Ratio of XRB spectrum measured from HEAO A-2 to expectation from power law continua.

means, therefore, that the imaging results are at an energy two orders of magnitude below the apparent thermal "peak" in the XRB, and two orders of magnitude above the peak in the optical cosmic spectrum which arises from starlight. We know that the starlight peak in the eV range has nothing to do with the X-ray background two orders of magnitude higher in energy, and the same orthogonality between the quasar images below 1 keV and the XRB at tens of keV may obtain if the quasar spectra are steep enough.

    A quantitative estimate is clearly in order at this point, and Table 1 summarizes such an attempt. The entries in the table are based on the following assumptions. I begin by assuming that a thermal bremsstrahlung spectrum characteristic of a 40 keV plasma is an adequate representation of the XRB everywhere (i.e. even below 3 keV and above 50 keV, where there is experimental evidence with which this assumption is inconsistent). I then subtract from this spectrum components which I am sure must contribute, although there is some uncertainty in the magnitude of this contribution;

TABLE 1

SYNTHESES OF THE X-RAY BACKGROUND SPECTRUM

		RATIO OF MODEL TO 40 keV THERMAL, NORMALIZED AT 5 keV, AT ENERGIES (keV):												
MODEL	DESCRIPTION (SEE TEXT)	.1	.2	.3	.5	1	2	3	5	10	20	30	50	100
1	40 keV thermal T(40)	1.00	1.00	1.00	1.00	1.00	1.00	1.00	1.00	1.00	1.00	1.00	1.00	1.00
2	0.4 index power law P(.4)	.88	.89	.89	.89	.90	.93	.95	1.00	1.13	1.46	1.87	3.08	10.75
3	0.7 index power law P(.7)	2.86	2.33	2.07	1.78	1.47	1.22	1.11	1.00	.92	.96	1.09	1.54	4.38
4	$T(40) - 0.1\{P(.6) + T(5)\}$	.84	.86	.88	.89	.92	.95	.97	1.00	1.03	1.04	1.03	.97	.61
5	$T(40) - 0.1\{P(.6) + T(5) + P(1)\}$	.41	.59	.68	.76	.85	.92	.96	1.00	1.05	1.06	1.05	.98	.55
6	$T(40) - 0.1\{P(.6) + T(5) + P(3)\}$	----	----	----	.21	.81	.93	.97	1.00	1.04	1.05	1.03	.97	.61
MODEL	FRACTION OF "TOTAL" FROM:													
4	P(.6): unevolved Seyferts	.14	.12	.11	.10	.09	.09	.08	.08	.08	.09	.11	.15	.35
	T(5): clusters	.10	.10	.10	.10	.09	.08	.07	.05	.02	----	----	----	----
	T(40): diffuse thermal	.76	.78	.79	.80	.81	.83	.85	.87	.90	.90	.89	.85	.65
5	P(.6): unevolved Seyferts	.10	.10	.10	.10	.09	.08	.08	.08	.08	.09	.11	.15	.34
	T(5): clusters	.08	.09	.09	.09	.09	.08	.07	.05	.02	----	----	----	----
	P(1): "flat" quasars	.26	.19	.16	.12	.09	.06	.05	.04	.03	.03	.03	.03	.05
	T(40): diffuse thermal	.56	.62	.65	.69	.73	.77	.80	.83	.86	.87	.86	.81	.60
6	P(.6): unevolved Seyferts	----	.02	.04	.07	.09	.08	.08	.08	.08	.09	.11	.15	.35
	T(5): clusters	----	.02	.03	.06	.09	.08	.07	.05	.02	----	----	----	----
	P(3): "steep" quasars	.97	.85	.67	.35	.09	.02	.01	----	----	----	----	----	----
	T(40): diffuse thermal	.02	.12	.26	.51	.74	.82	.84	.87	.90	.90	.89	.85	.65

I arbitrarily assume, for computational ease, that clusters with an average temperature of 5 keV and unevolved Seyferts with an average energy index of 0.6 each contribute 10% of the total at 1 keV. After subtracting these "contaminants", I compare the shape of the resultant difference with exactly the same 40 keV spectrum with which I began by normalizing the two at 5 keV. Since the fit is good in the "testable" region 3-50 keV (in fact, it is a better fit than is a power law of index 0.4 over the range 3-10 keV only), I reconstruct a synthesized XRB spectrum as the sum of a 40 keV thermal bremsstrahlung spectrum normalized to the same fraction of the original total 40 keV spectrum at 5 keV, plus the same cluster and Seyfert contributions, and compute the contributions of each component of the new total as a function of energy. No attempts are made at best temperature or fractional contribution fitting; I am interested here only in determining crude consistency or inconsistency.

I next go through exactly the same procedure, this time including a 10%-at-1 keV contribution from quasars with relatively flat Crab-like ($\alpha = 1.0$) or relatively steep ($\alpha = 3.0$) spectra in addition to the 20% already assigned to clusters and Seyferts. A compilation of the extent to which the unassigned fraction of the original 40 keV thermal bremsstrahlung spectrum is still consistent with a spectrum of the same shape is given in the top portion of Table 1 for each of the trials discussed. The bottom portion of the table indicates the fraction of the synthesized XRB spectrum which can then be ascribable to each presumed component. The important results are that the large fraction of the 3-50 keV XRB can still be consistently reconciled with a diffuse thermal

component for the trials I have chosen, even though the XRB below
1 keV may be dominated by the quasars if their spectra are steep
enough. It is important to note that the synthesized XRB spectra
all exceed the extrapolation of the 40 keV bremsstrahlung
spectrum below 3 keV (which may or may not be true), as well as
> 50 keV (which definitely is true; the unevolved Sy I contri-
bution I have assumed would dominate the measured XRB above
100 keV).

The arguments in this section have not been aimed at demon-
strating the necessity (or even consistency) of the XRB with a
substantial diffuse thermal component. In fact, I began by
attempting to satisfy the hypothesis that it could arise from
discrete evolved components. The apparent inconsistency between
the spectra of active galactic nuclei which have already been
measured and the XRB may be leading me astray if the evolution
extends to spectra as well as number density; if so, however, it
only makes the difficulty of inferring the > 10 keV contribution
from < 1 keV measurements even more acute. Such spectral evolution
(in particular, quasars at $Z > 1$ with spectra characteristic of
bremsstrahlung at $\gtrsim 100$ keV) can remove the requirement of a
substantial diffuse component, but there are no data of which I
am aware (or no measurements which can be made with Einstein that
I can imagine) which can unambiguously resolve the problem.

REFERENCES

Avni, Y., 1978, Astron. Ap. 63, L13.
Fabian, A.C., 1979, Proc. Royal Soc., in press.
Field, G. and Perrenod, S., 1977, Ap. J. 215, 717.
Giacconi, R. et al., 1979, Ap. J. 230, 540.
Ives, J.C., Sanford, P.W., and Penston, M.V., 1976, Ap. J. (Letters)
207, L159.
Marshall, F.E., Boldt, E.A., Holt, S.S., Miller, R., Mushotzky,
R.F., Rose, L.A., Rothschild, R.E. and Serlemitsos, P.J., 1979,
Ap. J. in press.
Mushotzky, R.F., Boldt, E.A., Holt, S.S., Pravdo, S.H., Serlemitsos,
P.J., Swank, J.H. and Rothschild, R.E., 1978, Ap. J. (Letters)
226, L65.
Mushotzky, R.F., Boldt, E.A., Holt, S.S. and Serlemitsos, P.J.,
1979a, Ap. J. (Letters), in press.
Mushotzky, R.F., Boldt, E.A., Holt, S.S., Marshall, F.E. and
Serlemitsos, P.J., 1979b, Ap. J., in press.
Schwartz, D.A., 1978, Proc. IAU/COSPAR Symp. X-ray Astron.
(Pergamon Press, Oxford).
Setti, G. and Woltjer, L., 1979, Astron. Ap. 76, L1.
Wilson, A.S., 1979, Proc. Royal Soc., in press.
Worrall, D.M., Mushotzky, R.F., Boldt, E.A., Holt, S.S., and
Serlemitsos, P.J., 1979, Ap. J. (Letters), in press.

X-RAY EMISSION FROM GALACTIC NUCLEI

M.J. Rees

Institute of Astronomy, Madingley Rd,
Cambridge, England

1. GENERAL REMARKS

The 2-10 keV X-ray luminosities of galactic nuclei range from
$L_x \leq 10^{36}$ erg s^{-1} for our own Galactic nucleus to $L_x \geq 10^{46}$ erg s^{-1}
for quasars. The X-ray spectra often appear hard and may be con-
sidered as quasi-power law continua, similar to the continuum
spectra observed in the radio and optical wavebands. The hard
X-ray and γ-ray luminosities may exceed the total emission from
the nucleus at other wavebands, as is the case with the nearby
galaxy Centaurus-A (NGC 5128) in which the spectrum has been confirmed above 1 MeV (Hall et al., 1976). X-ray variability has been
observed on timescales of days or less, as discussed by Lawrence, 1979
(see also Elvis et al., 1978). Such high luminosities, coupled
with rapid variations, strongly suggests that the X-rays originate
close to the power source in these galactic nuclei.
    It is important to emphasize, first of all, that there is no
reason to be surprised that such a large fraction of the energy
from active galaxies emerges in the X-ray band - indeed it is easy
to envisage a variety of possible mechanisms.
    Optical data on the broad emission lines imply that these arise
from clouds moving at speeds of several thousand km s^{-1}. Shocks
involving this kind of velocity automatically yield gas at temperatures of ≥ 10 keV: indeed, the clouds giving the optical emission
lines ($T_e \simeq 10^4$ °K) may be in pressure balance with a hot and more
rarified medium capable of emitting thermal X-rays via bremsstrahlung, or comptonisation of soft photons. We also observe directly
that quasars emit non-thermal continuum radiation, implying that
relativistic electrons are present. (These particles could have
been accelerated via shocks, or by some electromagnetic process
near a massive central object). These can give X-rays via synchrotron

339

R. Giacconi and G. Setti (eds.), X-Ray Astronomy, 339-354.
Copyright © 1980 by D. Reidel Publishing Company.

or Compton emission. The various possible processes can be
summarised in a flow diagram reproduced in Figure 1. This figure
is adapted from Fabian and Rees (1978), where some fuller quantitative discussion of the various possibilities may be found.

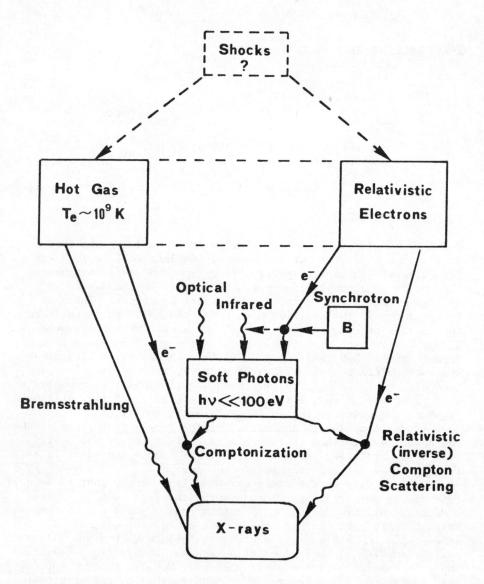

Fig. 1. Possible pathways for X-ray emission in galactic nuclei
(from Fabian and Rees (1978)).

The gas responsible for the broad emission lines - even for Lyman α - has densities $n_e \lesssim 10^{11}$ cm^{-3}; photoionization models then suggest that this gas lies $\gtrsim 10^{18}$ cm from the central continuum, and that it consists of enormous numbers of fast-moving clouds of individual dimensions $\lesssim 10^{15}$ cm (Blumenthal and Mathews, 1979; Carswell and Ferland, 1979). There is no physical reason why, still closer to the central non-thermal continuum, there should not be even denser ($n_e >> 10^{11}$ cm^{-3}) and more opaque clouds. Indeed, in a wide class of accretion models (Bergeron, these proceedings; Maraschi et al.,1979), infalling clouds could perhaps exist, with $(n_e T_e) \propto r^{-5/2}$, in pressure balance with a "hot-phase" medium at the virial temperature. At a distance r from the centre, the equivalent black body temperature is

$$T_{bb} \approx 3 \times 10^4 L_{46}^{1/4} \ (r/10^{15} \ cm)^{-\frac{1}{2}} \ K \tag{1}$$

(the non-thermal continuum luminosity, $10^{46} L_{46}$ erg s^{-1}, being assumed to come from a central region of size $r_{cont} < r$). In the high density limit, clouds located where $T_{bb} \gtrsim 10^4$ K would have $T_e \approx T_{bb}$ and would reprocess the incident continuum into approximately black body radiation. The form of the reprocessed continuum depends on the configuration and dynamics of the clouds - specifically, on how the covering factor depends on radius for $r_{cont} \lesssim r$ < r , where r is the radius such that $T_{bb}(r) \approx 10^4$ K. Recent IUE data on 3C 273 (Ulrich et al., 1979) reveal evidence for such a "thermal" continuum feature (see also de Bruyn, 1979). The dense clouds at $r << 10^{18}$ cm may be embedded in a hot (almost relativistic plasma) which radiates non-thermally or by comptonisation of the soft photons generated by the cool clouds (cf. Liang, 1979). The main uncertainties that one faces in attempting to quantify such models concern the influence of Compton heating and conductivity. The importance of the Compton effect depends on the X-ray spectrum; conductivity depends on the magnetic field geometry. All that can be stated with confidence is that a "snapshot" of the region within $\sim 10^{18}$ cm of the central source could reveal large numbers of streaky filaments (possibly aligned along the local magnetic field direction) embedded in a hotter medium; the survival or persistence of individual filaments is, however, more model-dependent.

To discriminate among the various mechanisms for the X-rays illustrated in Fig. 1, further extended spectral observations covering the hard X- and γ-ray bands are needed. These, combined with variability timescales, may eliminate some emission mechanisms. Highly variable sources seem most likely to emit via non-relativistic Compton scattering, the timescales for this process being shorter than for bremsstrahlung from the same electrons. The bulk of the material in the emitting region may be mildly relativistic ($T_e = 10^9 - 10^{10}$ K). Observations of X-rays from galactic nuclei should stimulate much needed study of processes such as: radiation processes in 'transrelativistic' plasmas where electron-electron bremsstrahlung and relativistic corrections are important;

production of e^+, e^- pairs and their effects on cooling and
opacity; effects of major differences between electron and ion
temperatures; comptonisation of spectra in inhomogeneous gas
clouds; and acceleration of ultrarelativistic particles by mildly
relativistic shocks.

   X-ray observations - along with studies of the non-thermal
optical continuum - offer the most direct evidence on physical
conditions close to the central "power house" (dimensions $\leq 10^{15}$ cm)
Relative to the continuum source, even the broad-line region is
diffuse "fuzz" out on the periphery, $10^2 - 10^3$ times further removed from the centre.

   The foregoing comments are, in effect, a summary of what can
be inferred from the X-ray data. In the remainder of this lecture,
I shall discuss in somewhat more detail the physics of the material
that emits the bulk of the continuum radiation, with the aim of
identifying the key parameters that determine its broad qualitative
features. I shall briefly conclude by mentioning the other con-
spicuous qualitative feature of nuclear activity in galaxies: the
production of directed outflow in collimated jets.

2. PHYSICAL CONDITIONS IN CENTRAL CONTINUUM SOURCE

The various components of the spectrum from active nuclei may be
emitted, via several different processes, from regions with a wide
range of length scales - indeed, the relevant dimensions stretch
all the way up to the sizes of extended radio sources. The primary
energy source may however have dimensions of only $r_{em} \leq 10^{15}$ cm;
in fact, the observed continuum variability in some BL Lac-type
objects sets a firm size limit of this order, unless bulk relativistic motions are involved. The high and variable optical polarization observed in some quasars and BL Lac objects (Angel et al.,
1978) then imposes - for these systems at least - severe constraints
on the physical conditions in the emitting region (cf. Blandford
and Rees, 1978). The occurrence of this polarization implies that
the emitting region is optically thin to electron scattering,
thereby setting an upper limit $\sim r_{em}^2/\sigma_T$ to the number of electrons
that can pervade the emission region - moreover, the amount of mass
associated with these electrons can then be compared with the mass-
equivalent of the radiation emitted in the light travel time across
the system, Lr_{em}/c^3. If the radiation mechanism is, e.g., synchrotro
or Compton emission, we can estimate the individual energies of the
relevant electrons; one then draws the general inference that the
radiation must come from relativistic particles that are reacceler-
ated at sites spread throughout the volume and re-used several times
during a timescale r_{em}/c. This conclusion is of course somewhat
sensitive to the geometry; to make it more realistic, we must allow
also for the likelihood that the material has an inhomogeneous
texture, e.g.,("multiphase" medium or a disc-plus-corona structure).

The synchrotron or Compton lifetime of any relativistic electron does indeed turn out to be, typically, much less than r_{em}/c; moreover the available electric and magnetic fields would be fully able to reaccelerate the particles in length-scales $<< r_{em}$ (cf. Blandford and Rees, 1978; Cavaliere and Morrison, 1979). The inferences outlined above do not entail any assumptions about the nature of the central power supply beyond an estimate of its characteristic size - it could, for the purposes of the argument, be either an accreting black hole or a rotationally-powered "spinar". My personal view is, however, that these considerations support the idea that at least the strong non-thermal active nuclei involve black holes. On this hypothesis, the luminosity comes predominantly from a region only a few times larger than the Schwarzschild radius r_s. Moreover, the material is there moving with speeds $\sim c$; there is no necessity, therefore, for the amount of material occupying this region at any time to be larger than is needed to supply the luminosity for a timescale $\sim r_s/c$. The ratio of the mass of luminous material to the central hole mass is enormously small - it is

$$\sim (r_s/c) \ \varepsilon^{-1} \ L/L_E \ t_s \tag{2}$$

where ε is the efficiency, $L_E = 4\pi GMm_p c/\sigma_T$ is the Eddington luminosity corresponding to the hole's mass, and t_s is the timescale (Salpeter 1964) over which an object radiating at the Eddington luminosity would convert its entire rest mass into radiation. If the central mass is M, this means that

$$\frac{\text{density of material in emitting region}}{(M/r_{em}^3)} \lesssim \frac{r_{em}/c}{t_s}$$

$$\simeq 3 \times 10^{-12} (r_{em}/10^{15} \text{cm}) \tag{3}$$

While this can naturally be fulfilled for accretion processes, where the bulk of the confining mass has already passed within the hole, it is a stringent constraint on spinar-type models, where one would expect, for $L \simeq L_E$, that radiation pressure would maintain an atmosphere, with a scale height $\sim r_{em}$, containing a significant fraction of the total mass.

(Note that this argument applies only to the most powerful non-thermal sources. Galactic nuclei which emit an apparently thermal continuum may involve a different process. On general astrophysical grounds, we would expect to observe precursors of massive black holes - e.g. spinars, dense star clusters, etc. - in some nuclei.)

The main thrust of the arguments outlined above, is that the plasma which emits polarized continuum radiation is characterised by an exceedingly low density relative to the amount of power which

it emits. We may elucidate this quantitatively by considering a
homogeneous slab of plasma; the results can then be straight-
forwardly extended to a quasar model with realistic parameters
and radial structure.

Consider a cube of dimension ct_{var}, filled with plasma of
mean electron density n_e. Suppose that the magnetic field B is
such that $B^2/8\pi = x_{mag} n_e m_p c^2$, and that the cube is emitting a
steady luminosity L. This luminosity can be parametrised in terms
of a dimensionless "luminosity per unit area", ℓ, such that

$$L = \ell \, \sigma_T^{-1} \, m_p c^2 \tau_{es} \, (ct_{var})^{-1} \tag{4}$$

where $\tau_{es} \simeq n_e ct_{var} \sigma_T$ $\tag{5}$

The significance of ℓ is that $\tau_{es}.\ell$ denotes the fraction
of the rest-mass energy which is radiated in the light travel
time across the emission region (assuming that the charge of the
electrons is balanced by ions). The thermal properties of the
plasma, if it is in a stationary state, can then be expressed in
terms of ℓ. In most ordinary thermal cosmic plasmas (or stellar
atmospheres), ℓ is much less than unity; in the quasar context
we may have $\ell \simeq 1$.

If $x_{mag} = 0$, and non-relativistic electron bremsstrahlung
is the only radiation process, then ℓ is restricted to low values:
for $T_e \lesssim 1$ Rydberg, $\ell \simeq (m_e/m_p)(e^2/\hbar c)^2$, even if reabsorption can
be neglected. The maximum value of ℓ before the electrons get
relativistic is $\ell \simeq (m_e/m_p)(e^2/\hbar c)$, for $\tau_{es} \lesssim 1$. If $\tau_{es} > 1$,
Compton scattering of bremsstrahlung photons raises ℓ only by a
logarithmic factor. Even when the electron energies become
relativistic $((\langle \gamma_e \rangle - 1) \gtrsim 1)$ ℓ only rises as $\langle \gamma_e \rangle$ for $\tau_{es} \lesssim 1$.

Consider now the effect of a magnetic field. If $kT \simeq m_e c^2$,
and reabsorption effects can be ignored, cyclotron cooling dominates bremsstrahlung by a factor $x_{mag}(e^2/\hbar c)^{-1}(m_p/m_e)$. Synchrotron
radiation is, of course, more efficient still by a further factor
γ_e^2. (The synchrotron timescale being only $\sim (m_e/m_p)\gamma_e^{-1} \ell \, t_{var}$).

These general arguments imply that the primary energy output
from sources with $\ell \gtrsim 1$ must come from relativistic electrons –
indeed, only a small fraction of the electrons need be relativistic
at any instant to carry away the energy.

Consider now a simple model for an accretion-powered quasar.
We introduce the parameter

$$\dot{m} = c^2 \dot{M}/L_E \tag{6}$$

The significance of m is that it calibrates the accretion rate in
terms of the value that would yield L_E if the mass-energy conversion were 100%. Another quantity of interest for accretion flows
is the inward drift velocity $v_{inflow}(r)$. For radial inflow this is

of order the free fall or Keplerian velocity $v_{Kep} \simeq c(r/r_s)^{-\frac{1}{2}}$; for disc-type accretion we have

$$v_{infall}/v_{Kep} \simeq \alpha(h/r)^2 \simeq \beta \tag{7}$$

where α is the now-conventional viscosity parameter introduced by Shakura and Sunyaev (1973) and h is the disc thickness.

If the accretion yields radiative efficiency ε, then the optical depth down to $r \simeq r_s$ is $\dot{m}\varepsilon^{-1}\beta^{-1}$ and the value of ℓ required near $r \simeq r_s$, where most of the energy release occurs, is

$$\ell = \dot{m}^{-1}\varepsilon\beta^2(h/r) \tag{8}$$

This relation (in conjunction with our earlier discussion) shows that a homogenous radial infall with $\beta = 1$ is likely (unless ε is very low) to require relativistic electrons and non-thermal emission mechanisms to radiate the energy released. The same is true for disc-type accretion if $\dot{m}$ is low: the disc will then go thermally unstable and heat up.

If there is a sufficiently wide range of scales between r_s and the (much larger) radius where the infalling mass is supplied, we may expect that some of the parameters will display a power-law dependence. For free fall accretion (or for a thick disc where α is independent of r) the density varies as $r^{-3/2}$, and the value of the parameter ℓ varies as r^{-1}.

A "snapshot" of the material in the emission region would reveal material in three distinct "phases".

I. Small dense clouds or filaments, at temperatures of min T_{bb} $\simeq 10^4 K$ (cf. equation 1). The individual clouds at a distance r from the centre would be very small - probably limited to $\sim (c_s/v_{free\ fall})r$, where $c_s \simeq 10^6(T/10^4 K)^{\frac{1}{2}}$ cm s^{-1} is the internal sound speed. Individual clouds need not persist or maintain their identity for even one dynamical timescale. This cool material will absorb, scatter and reprocess the radiation produced by the hotter phases.

II. Hot thermal plasma, in rough pressure balance with "phase I". The ion temperature will be of order $T_{virial} \simeq 10^{12}(r/r_s)^{-1} K$, but the electrons may be cooler. This idea of a two-temperature plasma was extensively discussed by Eardley et al. (1978) in the context of thick accretion discs. If $kT_e \gtrsim m_e c^2$, the effective cross-section for equalising the electron and ion temperatures is $\lesssim (m_e/m_p)^2\sigma_T$. There is therefore no reason to expect $T_i \simeq T_e$; in practice, Compton interactions would provide the dominant cooling for the electron component, constraining kT_e to remain $\lesssim m_e c^2$ even where the virial temperature and the ion temperature become higher than this.

III. Non-thermal relativistic electrons. Even though Coulomb interactions between electrons and ions may be ineffective for heating electrons to individual kinetic energies $\gtrsim m_e c^2$, this acceleration can of course happen via the collective processes associated with shock fronts, magnetic reconnection, etc. At any instant, there

may be many regions where relativistic electrons exist, having
been recently accelerated by some such process; the radiation
lifetime of these particles is so short (typically << r/c) that
they must be continually accelerated. A "snapshot" of the emission
region would show only a small fraction of the electrons in this
third phase; nevertheless, because relativistic electrons radiate
so efficiently the main primary radiation from active galactic
nuclei may come from "phase III".

The relative importance of these three phases depends on the
density, inflow timescale, etc; one may conjecture that the most
purely non-thermal phenomena will be found when the thermal cooling
is least effective, and the densities low. On the other hand, if
$\dot{m} \gtrsim 1$ and $\alpha << 1$, the cooling is efficient enough to guarantee that
most of the material is in phase I; the radiation then emerges
predominantly as a thermal ultraviolet continuum.

An obvious prerequisite for X-ray emission is that electrons
should be present with individual energies of at least several keV.
The foregoing general physical arguments confirm that this condition
should be fulfilled for most types of accretion flow, even though
the equivalent black body temperature T_{bb} (cf. equation (1)) would
correspond to less than $\sim$ 100 eV in the case of massive black holes.
The various pathways, summarised in Figure 1, whereby the different
gas phases can generate X-rays, may be classified under two headings:
(a) Relativistic (phase III) electrons can generate X-rays by
Compton scattering of soft photons; these may be either $\leq$ 100 eV
photons processed by phase I gas, or else synchrotron photons emitted by the relativistic electrons themselves. In the latter
case, the X-ray production is exceedingly sensitive to the field
strength and to the non-thermal electron spectrum. There is in
fact no reason why synchrotron radiation should not extend up to
the X-ray band: the only firm limit on the possible Lorentz factor
of non-thermal electrons is $\{(x_{mag} \cdot r_s)/(classical\ electron\ radius)\}^{1}$.
(b) Thermal bremsstrahlung, the other main alternative, requires,
for high efficiency, either $\tau >> 1$, or else $\alpha << .1$ so that the
available timescale greatly exceeds the free fall time. In the
former case comptonisation would modify the spectrum.

Only after detailed study of the spectral behaviour of variable
galactic nuclei will it be possible to decide among these options.

3.   THREE COMMENTS ON ACCRETION

3.1 Supercritical luminosities and accretion rates

The production of a steady luminosity exceeding L_E is obviously
peculiarly unlikely if accretion of gravitationally-attracted gas
provides the power supply - moreover some authors (e.g. Cowie et al.,
1978) have emphasised how "preheating" effects may inhibit the
accretion even at luminosities lower than L_E, particularly if the
emitted radiation is in the form of hard X-rays; also, radiation

pressure effects can be more important if the material is cool
enough for other opacity (e.g., photoionization) to dominate electron
scattering. (Other complications not yet adequately considered in
the literature involve the effects of pair production in enhancing
the number of electrons (and positrons) in compact sources of hard
radiation and thereby diminishing the effective value of L_E).

Even if the available mass supply, and the potential efficiency,
succeeds in providing a luminosity far exceeding L_E (and if the
mass can be supplied, for instance, by entire stars which are
essentially unaffected by radiation pressure until they are tidally
or collisionally disrupted), the main power output will be in kinetic
energy rather than in photons. If there is a wind with terminal
velocity v, we can define a radius r_{trap} (well within the photospheric radius) at which the bulk transport of radiation with the
flow is comparably important to the diffusive transport, with
speed c/τ. For $r < r_{trap}$, the matter and radiation can be treated
as a single fluid where radiation provides the main pressure.
Suppose that wind passes through a sonic point at a radius r_{sonic}
$< r_{trap}$. Then, between these two radii, the kinetic energy outflow
is almost constant, but the internal energy of trapped radiation
varies as $r^{-2/3}$. The ratio of the photon luminosity that escapes
to the kinetic energy outflow is then $(r_{trap}/r_{sonic})^{-2/3}$. When the
photon luminosity exceeds L_E, the parameters of the wind are likely
to be such that the kinetic energy flux is greater still.

An alternative possibility is that the power output remains
at a value $\sim L_E$ even though $\dot{m}$ becomes very large. This can happen
either through the flow adjusting so that ε becomes small, or by
more of the material being expelled before having got into a deep
potential well:

a) The flow pattern may adjust so that the high mass influx
can be accepted without the luminosity vastly exceeding L_E. This
can happen if a thick disc develops whose inner edge coincides
with a circular orbit of very small binding energy (i.e., $r \simeq 2r_s$ for
a Schwarzschild hole). The efficiency can then become arbitrarily
low, so that there is no limit to the value of $\dot{m}$ that can be swallowed
(Jaroszynski et al., 1979). This possibility requires that the
inward-spiralling material should be barytropic - in particular,
none acquires positive energy and escapes as a wind.

b) An alternative possibility, first outlined by Shakura and
Sunyaev (1973) and subsequently considered by others (Begelman and
Meier, (1980) and references cited therein), involves the production
of a radiation-driven wind which carries away the excess energy.
The efficiency per unit mass swallowed is comparable with the stan-
dard thin-disc value, and the luminosity is no more than $\sim L_E$.

Either of these last two possibilities (and many others as well)
is consistent only with a particular choice of the viscosity law;
until we know more about how the viscous dissipation depends on
local conditions we cannot decide between the various options. In
the case of a thick cool disc supported by radiation pressure,
specifying the isobars and flow pattern allows one to calculate the

net radiation flux out of each volume element and thus leaves no
freedom in the choice of the law governing viscous dissipation.

3.2 Disc-type accretion for very low $\dot{m}$

When $\dot{m} \ll 1$, spherically-symmetric inflow onto a black hole not
merely yields a low luminosity, but has low efficiency, because
the cooling timescales (except for electrons with very high γ)
will become much longer than the free fall time (cf. Shapiro, 1973).
One might have thought that, contrariwise, rotationally-inhibited
accretion onto black holes will always have high efficiency even
when $\dot{m}$ is very low, on the grounds that material has to work its
way in to the innermost stable orbit before being swallowed.
Interestingly, however, this may not be so.
      If $\dot{m}$ is low for a given α, cooling inefficiency at non-
relativistic electron temperatures will ensure that a disc thickens
up until $h \simeq r$. If the viscosity in a thick disc were primarily
due to magnetic effects, then, even though the appropriate value
of α is uncertain, there is no reason why this value should decrease
with $\dot{m}$. Therefore, the timescale for angular momentum redistribu-
tion (cf. equation 7) becomes independent of $\dot{m}$ even though the
cooling timescale varies as $\dot{m}^{-1}$. When $\dot{m}$ is sufficiently low,
angular momentum is redistributed, allowing some of the material
to approach the hole, in a time shorter than that over which
significant binding energy can be radiated. There is no inconsistency here: what must happen is that the inner part of the
disc resembles a "donut" in which pressure gradients in the radial
direction are important (cf. Abramovitz et al. 1978) extending in
almost to the location of the circular orbit of zero binding energy.
Material that loses sufficient angular momentum to get into such an
orbit can spill into the hole without having radiated significant
binding energy.
      This flow pattern may well be the norm in galactic nuclei
containing massive black holes ("dead quasars") but where the
fuelling rate has now fallen very low, corresponding to $\dot{m} \ll 1$.
The only significant radiation would be that due to a "non-thermal
tail" of relativistic electrons (phase III). The pressure distribution of the hot thermal plasma (phase II) in such a "donut" -
a cavity along the rotation axis, and $\rho \propto r^{-5/2}$ at large distances
from the hole - may be responsible for the collimation of jets and
beams from systems such as M 87. The high-entropy plasma escaping
in the beam could be produced by a more exotic process near the
hole, maybe involving electromagnetic effects (cf. Blandford and
Znajek, 1977).

3.3 Scaling laws

It is commonplace to draw analogies between stellar-mass accretion-
powered X-ray sources in our Galaxy (e.g., Cygnus X1) and the larger-
scale processes occuring in external galactic nuclei; but the

resemblance is indeed a remarkably close one. Suppose we consider
two black holes, of (respectively) $10M_\odot$ and $10^8M_\odot$. Provided that
the value of $\dot{M}$ scales with M, so that the parameter $\dot{m}$ (equation 6)
is the same, then the ratios of cooling and dynamical timescales
are identical. If the viscosity is primarily due to magnetic
reconnection, there is no reason why the (poorly known) value of
α should depend significantly on the mass of the hole. There is
thus no reason why the same kinds of flow pattern should not be
manifested on widely different scales (compare SS 433 and extragalactic double sources, for instance). The crucial parameter is
$\dot{m}$. The only features that do depend significantly on the mass
are the temperature T_{bb} of "cool phase" material ($\propto M^{-1/4}$ for a given
value of $\dot{m}$ and of (r/r_s)) and effects such as ionization and opacity
that depend on gas temperature.

4. BEAMS AND JETS

4.1 Observational evidence

There is now strong morphological evidence that active galactic
nuclei give rise to jet-like features, with lengths ranging from
a few parsecs to hundreds of kiloparsecs. Phenomena observed to
date include:
     a) Jets associated with double radio sources. These jets,
observed in NGC 6251, 3C 66B, 3C 219 and NGC 315, are linear
features with opening angles in the range of 3-15° and are often
one-sided.
     Of particular interest is the structure of the source associa-
ted with NGC 6251 (Waggett et al., 1977, Readhead et al., 1978).
A low-frequency radio survey recently revealed a very extended
double source, with very high energy content even though the
current radio power output is low. A five-gigahertz map made
with higher angular resolution revealed a straight jet, emerging
from the galaxy and pointing towards one of the components. VLBI
measurements show that, right in the nucleus itself, there is a
source only a few light years long pointing along the jet. The
nucleus of this galaxy thus contains a "cosmic blow-torch", generating a jet detectable out to a distance of half a million light
years, and presumably pumping energy continuously into the diffuse
extended structure. The fact that the jet is seen only on one
side of the galaxy might indicate that it is emerging relativistic-
ally; unless the axis is precisely in the plane of the sky, the
Doppler effect would strongly enhance the detectability of the jet
on the approaching side. So far, jets have only been reported
associated with weaker, comparatively close-by radio doubles.
While this is in part a selective effect, the failure to observe
jets in the most powerful doubles like Cygnus A is probably significant.

b) <u>Jets associated with compact radio sources</u> (e.g., 3C 273,
3C 345, 3C 147 and 3C 380 (see Readhead (1979) and references cited
therein).   Recent results from VLBI have shown that compact variable
radio sources are usually of the "core-jet" type in which an inhomogeneous jet emerges from an unresolved optically thick nucleus.
Apparent superluminal expansion is often observed between prominent
features in the jet and the core.   This is interpreted as evidence
for relativistic motion within the jets (Blandford, McKee and Rees,
1977).   Compact jets are usually roughly aligned with larger scale
radio features where these exist.   However it has recently been
discovered that the jets in the powerful sources are bent through
angles of up to $30°$ as they are observed with progressively larger
angular resolution (Readhead et al., 1979).   This bending is probably
exaggerated because the jet makes a small angle with the line of
sight, a condition that can also give rise to superluminal expansion (Scheuer and Readhead, 1979; Blandford and Konigl, 1979).
Extended double radio sources generally contain relatively faint
compact central components; the brightest of these (e.g., 3C 236 and
Cygnus A) have also been examined by VLBI and show linear structure.
These compact sources are much more closely aligned with the centre
radio components, suggesting the absence of a strong projection
effect.

c) <u>Optical jets</u>.   It is natural to include optical jets in
this list, most notably those associated with M87 and 3C 273 as they
too emanate from nuclei in which there is independent evidence of
activity.   Both of these examples are also radio jets.   In M87
there is a radio "counter jet" of power comparable to the radio
jet, but no optical "counter jet" (Turland, 1975).   Optical emission
has also been recently detected from 3C 66 and 3C 31 (van Breughel
et al., 1979).   A jet-like feature displaying emission lines blue-
shifted by 3000 km s^{-1} has been associated with DA 240, but this
bears no·relation to the radio structure (Burbidge, Smith and
Burbidge, 1975).

d) <u>X-ray jets</u>.   Centaurus A has recently been shown to contain
an X-ray jet located a few kpc from the nucleus and apparently
feeding the inner radio lobes (Schreier et al., 1979).   This may
be related to the optical jet reported earlier by van den Bergh.

e) <u>Galactic jets</u>.   The moving emission lines from the galactic
object SS 433 have been modelled kinematically as a pair of anti-
parallel precessing jets (Abell and Margon, 1979) with a velocity
of $\sim 0.27c$.   There is also evidence for radio jets from SS 433 which
may suggest an analogy with Sco X-1.

4.2   The Jet Velocity

Only in the cases of SS 433 have convincing direct measurements
of jet speeds been reported.   In most cases the jet speed V (not
to be confused with the expansion speed of the double radio source)
can range from the escape velocity of an elliptical galaxy (~ 300
km s^{-1}) to the speed of light.   The evidence is conflicting.   The

existence of superluminal expansion and the preponderance of one-sided jets (easily explained in terms of "Doppler favouritism" (e.g., Shklovsky, 1977; Rees, 1978a) suggests that $V \simeq c$, as might be expected if the jets originate at the bottom of a relativistic potential well. However, there is the problem of "waste energy" first emphasized by Longair et al.(1973). In many sources a lower bound on the momentum flowing through the jet can be obtained by multiplying the minimum pressure in the hot spots by their apparent area. Typical values are in the range $10^{34} - 10^{35}$ dyne. Further multiplications by $V/2$ gives the jet power L_o. A large value of V often leads to powers much greater than appear to be dissipated as heat in the radio components. Furthermore, dynamical and radiative arguments applied to individual jets in less powerful sources such as 3C 31 and NGC 1265 (Begelman et al., 1979) and 3C 449 (Perley et al., 1979) yield speeds ranging from $\sim$ 500 km s^{-1} to $\sim$ 10,000 km s^{-1}. If a jet emerges from a collimation region with a speed $\sim$ c and is subsequently decelerated to a speed $V \ll c$ by entraining surrounding gas, then conservation of linear momentum (approximately valid for a highly supersonic jet) implies that the power in the collimation region exceed the final power by $\gtrsim (c/V)$. The "waste energy" that is difficult to dissipate invisibly in the radio components has then got to be similarly processed in the nucleus. For a strong source like Cygnus A, a relativistic jet power is 10^{46} erg s^{-1}, much greater than the known power in the nucleus.

There is the opposite problem with the mass flux. If the jet is decelerated by entrainment of surrounding material, then the mass loss must <u>increase</u> by $\sim (c/V)$. Again, using Cygnus A as an example, a jet speed of only 1000 km s^{-1} requires 100 $M_\odot$yr^{-1} and, over the lifetime of the radio source, at least $10^{10}M_\odot$ of gas has to be injected into the radio components - much more than is believed to be present within an elliptical galaxy. (Most of this mass would in any case have to be hidden in cool, dense filaments in order not to lead to excessive Faraday depolarization). So, high jet speeds seem to require too much energy and low jet speeds too much mass for the galaxy to provide.

We can make a further independent estimate of the momentum flowing in the jet if we believe that the jet is "free", i.e. unconfined by external pressure. This requires that the jet Mach number M exceed θ^{-1} where θ is the jet opening angle. For an observed jet like NGC 6251, where a minimum equipartition pressure p_j at a distance r along the jet can be estimated, this leads to a momentum estimate $p_j r^2$. Such estimates are consistent with the values obtained from the radio components, and Mach numbers as large as 30 are suggested for the most collimated sources.

4.3 Collimation

Collimation of the relativistic plasma must involve some asymmetry in the galactic nucleus, most likely due to rotation. For instance, if the central power-source is embedded in a rotationally-flattened cloud, the plasma will tunnel its way out by excavating nozzles along the rotation axis. This "twin exhaust" configuration (which can create directed beams even if the plasma is generated with an isotropic pressure) could be established on any scale: in particular, there is no reason why the nozzles should not be at ≤ 100 r_s if the outflowing plasma originates very near the hole (Blandford and Rees, 1979; Rees, 1978b).

The nozzle walls could be confined by gas supported by pressure of radiation, ions or magnetic fields. Hot inward-spiralling material orbiting the hole would form a thick disc or "donut", centrifugally inhibited from draining into the hole. The natural "walls" of such a channel are set by the (roughly paraboloidal) surface within which a particle would have so little angular momentum that the hole could swallow it.

If the black hole is spinning, the Lense-Thirring "dragging of inertial frames" influences material sufficiently close to it. An orbit of radius r and period t_{orb} that does not lie in the hole's equatorial plane precesses on a timescale $t_{prec} \simeq t_{Kep}$ x $(J/J_{max})^{-1}(r/r_s)^{3/2}$, J being the hole's angular momentum and J_{max} the maximum angular momentum of a Kerr hole. If the material drifts inward only slowly (i.e., $t_{inflow}(r) \gg t_{Kep}(r)$) then the precession builds up over many orbits. As Bardeen and Petterson (1974) pointed out, there is then a critical radius r_{BP} at which $t_{prec} \simeq t_{inflow} \simeq \beta^{-1}t_{Kep}$. For r ≤ r_{BP}, the gas flow is axisymmetric with respect to the hole, <u>irrespective of</u> its original angular momentum. Thus, if the beams are collimated at a radius ≤ r_{BP}, their orientation is, in effect, stabilised by the hole; they are then impervious to jitter resulting from short-term fluctuations in the flow pattern of the surrounding gas.

The value of r_{BP} is ≫ r_s only if the viscosity is low enough to ensure that the inflowing material makes many orbits, thus allowing the Lense-Thirring precession to have a cumulative effect. For instance, if the collimation occurs in a thick "donut", then we require α ≤ 0.01 if black hole's alignment is to affect the flow out to ∿ 50 r_s.

4.4 Some further questions regarding the physics of jets

The obvious key question concerns the flow pattern in the region where the jets are formed, and the nature of the collimation mechanism. This may occur in the relativistic domain near a black hole, but it is unclear whether it is basically a fluid-dynamical process or whether a crucial role may be played by electromagnetic effects, anisotropic radiation pressure, etc. The jet could conceivably comprise electron-positron pairs rather than ordinary matter; it may contain a magnetic field, or even carry a current (Lovelace, 1976; Benford, 1978). The radio emission - even the mill arc-second structures seen by VLBI techniques - involve scales vastly larger than the primary energy source. If some of the extreme

cases of optical variability involve directed relativistic outflow
(Blandford and Rees, 1978; Scheuer and Readhead, 1979) they may
provide clues to jet behaviour on rather smaller scales.

Even though the jets may be moving at a high (or even relativistic) speed the material in them could cool down to a low temperature, yielding a high Mach number and the possibility of unconfined
"free jets". Adaptation of the arguments given in section 2 show
that a jet emanating from near $r \simeq r_s$ with an energy flux of order
L_E could certainly cool to a non-relativistic value of T_e. But
such a jet would only be directly observable if:

(i)   internal magnetic field energy were dissipated and
partially converted into relativistic electrons;

(ii)   friction and entrainment at the jet boundaries (or interaction with "obstacles" in the jet's path) dissipated some of its
bulk kinetic energy; or

(iii)   the ejection velocity were variable, and thus led to the
development of internal shocks, which accelerated relativistic
electrons. Even the thermal component of the beam material may
have a large enough emission measure to be detected if instabilities
cause it to form into clouds with a small filling factor.

The radio observations of curvature in jets indicate that the
physics can be further complicated by such effects as transverse
motion of an external medium, buoyancy, misalignment of the galaxy
with the spin axis of its central black hole, and possibly precession of the hole itself.

REFERENCES

Abell, G.O. and Margon, B., 1979, Nature 279, 701.
Abramowitz, M., Jaroszynski, M. and Sikora, M., 1978, Astron. &
Astrophys. 63, 221.
Angel, R. et al. 1978, in Proc. Pittsburg Conference on BL Lac
Objects, ed. A.M. Wolfe.
Bardeen, J.M. and Petterson, J.A., 1974, Astrophys.J.(Lett), 195,
L.65.
Begelman, M.C. and Meier, D., 1980 in preparation.
Begelman, M.C., Rees, M.J. and Blandford, R.D., 1979, Nature, 279, 770.
Benford, G., 1978, Mon.Not.R.astr. Soc. 183, 29.
Blandford, R.D. and Konigl, A., 1979, Astrophys.J., 232, 34.
Blandford, R.D., McKee, C.F. and Rees, M.J., 1977, Nature, 267, 211.
Blandford, R.D. and Rees, M.J., 1974, Mon.Not.R.astr.Soc., 169, 395.
Blandford, R.D. and Rees, M.J., 1978 in Proc.Pittsburg Conference
on BL Lac Objects, ed. A.M. Wolfe.
Blandford, R.D. and Znajek, R.L., 1977, Mon.Not.R.astr.Soc., 179, 433.
Blumenthal, G. and Mathews, W.G., 1979, Astrophys.J., 233, 479.
Burbidge, E.M., Smith, H.E. and Burbidge, G.R., 1975, Astrophys.J.
(Lett), 196, L137.
Carswell, R.F. and Ferland, G., 1979, Mon.Not.R.astr.Soc. (in press)
Cavaliere, A. and Morrison, P., 1979, Astrophys.J. (in press)

Cowie, L., Ostriker, J.P. and Stark, A.A., 1978, Astrophys.J.,226, 1041.

De Bruyn, G., 1979, in "Highlights of Astronomy",Vol. 5 (D. Reidel Publishing Co., Dordrecht), p. 631.

Eardley, D. et al. 1978, Astrophys.J.,224, 53.

Elvis, M. et al. 1978, Mon.Not.R.astr.Soc. 183, 129.

Fabian, A.C. and Rees, M.J., 1978 in Proc. IAU/Cospar Symposium on X-ray Astronomy,ed. W.A. Baity and L.E. Peterson (Pergamon), p.381.

Hall, R.D. et al., 1976, Astrophys.J.,210,631.

Jaroszynski, M., Abramowitz, M. and Prezynski, B., 1979, Acta Astron. (in press)

Lawrence,A.,seminar presented at the School, Erice.

Liang, E., 1979, Astrophys.J.(Lett.), 231, L111.

Longair, M.S., Ryle, M. and Scheuer, P.A.G., 1973, Mon.Not.R.astr. Soc., 164, 243.

Lovelace, R.V.E., 1976, Nature,262, 649.

Maraschi, L. et al. 1979, Astron. & Astrophys. (in press)

Perley, R.A., Willis, A.G. and Scott, J.S., 1979, Nature,281, 437.

Readhead, A.C.S., 1979, Proc. IAU Symposium 92, ed. G.O. Abell and P.J.E. Peebles, p. 165.

Readhead, A.C.S., Cohen, M.H. and Blandford, R.D. 1978, Nature, 272, 131.

Readhead, A.C.S. et al.,1979, Astrophys.J., 231, 299.

Rees, M.J., 1978a, Mon.Not.R.astr.Soc.,184, 61P.

Rees, M.J., 1978b, Nature,275, 516.

Salpeter, E.E., 1964, Astrophys.J., 140, 796.

Scheuer, P.A.G. and Readhead, A.C.S., 1979, Nature,277, 182.

Schreier, E. et al. 1979, Astrophys.J.(Lett.), 234, L39.

Shakura, N.I. and Sunyaev, R.A., 1973, Astron. & Astrophys., 24, 337.

Shapiro, S.L., 1973, Astrophys.J.,180, 531.

Shklovski, I.S., 1977, Soviet Astron.A.J., 21, 401.

Turland, B.D., 1975, Mon.Not.R.astr.Soc.,170, 281.

Ulrich, M.H. et al., 1979, Mon.Not.R.astr.Soc.,(in press)

van Breughel, W., Miley, G. and Butcher, H., 1979, Astrophys.J., (in press)

Waggett, P.C., Warner, P.J. and Baldwin, J.E., 1977, Mon.Not.R. astr.Soc.,181, 405.

GAS CLOSE TO THE RADIATIVE CONTINUUM SOURCE(S) IN ACTIVE NUCLEI

J. Bergeron

European Southern Observatory, Geneva, Switzerland

I. BROAD LINE REGION: BLR

The BLR is identified through its broad optical and UV lines (velocity of order 10^4 km s^{-1}) and has a temperature of order 10^4K.

a. Mass of the BLR

The equivalent width of Hβ is roughly constant (dispersed by a factor of 5 around 70 Å) for a large spread in absolute luminosities (factor 10^3). This is expected for optically thick nebulae covering a large fraction of the sky around the central continuum source (Searle and Sargent 1968). We will see below, however, that this argument is not confirmed by the UV observations of nearby active nuclei. Furthermore the inferred variation of EW (Hβ) with the steepness of the spectrum is not present (Bergeron 1979). This may be explained either by the difficulty to separate the stellar component from the non-thermal one, or by a lack of continuity between the spectral dependence in the optical and UV.

From the line intensity, L_ℓ, and emissivity one deduces the product of the gas density n and its mass M. Recombination lines have an emissivity only slowly varying with the gas temperature T. However for high optical depth in the Lyman lines and high density (see below), the emissivities of the Lyman and Balmer lines may differ from the pure radiative case by up to an order of magnitude (Krolik and McKee 1978). This is suggested by the low ratio Lyα/Hβ. Therefore the value of nM as inferred from Hβ is uncertain by a factor 3 to 10.

Using EW(Hβ) $\propto L_\ell \propto L_{opt}$, and assuming a proportionality between the optical and UV luminosities, one gets

R. Giacconi and G. Setti (eds.), X-Ray Astronomy, 355-362.

$$nM \propto L_\ell \propto L_{UV} \tag{1}$$

One finds for NGC 4151 a mass of $3\,n_9^{-1}$ $M_\odot$, where n_9 is the gas density in units of 10^9 cm^{-3}, and for 3C 273 a value 10^3 times larger.

b. Emission line spectrum

The emission line spectrum is very similar for most active nuclei (nearby galaxies and quasars).

Radiative heating of the gas is dominant over collisional heating for most of the BLR. This is suggested by the spread in degree of ionization and by the larger energy density for the photons than for the matter.

Let us define first the ionization parameter

$$I = \frac{L_{UV}}{nr^2} \tag{2}$$

where r is the overall size of the BLR and L_{UV} is the product of the total flux at the Lyman limit by the Lyman frequency.

The range of values for I, in erg cm s^{-1}, are

$$10^{-3} \lesssim I \lesssim 10^{-1}.$$

The corresponding range of ionization, for the optically thin case, is

$$5 \times 10^{-2} \gtrsim H^o/H^+ \gtrsim 2.5 \times 10^{-4}.$$

The lines of lower excitation (SiII, MgII, FeII) could then be formed only in regions optically thick to the Lyman continuum. This requires column densities larger than 10^{21} cm^{-2}.

The ratio of energy densities in radiation and matter may be written

$$\frac{\xi\nu}{\xi m} \left(\propto \frac{I}{fT} \right) \simeq 10^{-2} f^{-1} T_4^{-1} > 1, \tag{3}$$

where T is the gas temperature given in units of $10^4 K$ and f the gas filling factor. The latter is usually smaller than 10^{-3} (see next section).

The possible density domain is determined by the presence or absence of forbidden lines. The critical density n_c for a forbidden line is defined by the equality between the spontaneous transition rate and the collisional deexcitation rate. The line emissivity becomes independent of the electronic density n_e for $n_e \gg n_c$. It is then reduced compared to recombination line emissivity by n_c/n_e.

The absence of broad $[OIII]$ and $[FeII]$ lines implies $n_e > 10^8$ cm^{-3} { $n_c([FeII]) = 10^7$cm^{-3}}. The presence of CIII] is compatible with $n_e \simeq 10^9 - 10^{11}$ cm^{-3} (Nussbaumer and Schild 1979, with a correction of their results at low temperature). The upper value is much higher than the critical density for this line { n_e (CIII]) $= 3 \times 10^9$ cm^{-3}}. It reflects the higher contribution to the absolute intensity of a line for regions with $n_e \gtrsim n_c$.

c. Filling factor and spatial coverage.

The filling factor can be deduced from eqs (1) and (2)

$$f \propto (\frac{L_\ell}{L_{UV}})\ L_{UV}^{-\frac{1}{2}}\ I^{\frac{3}{2}}\ n^{-\frac{1}{2}} . \tag{4}$$

For NGC 4151 $f = 10^{-3}\ n_9^{-\frac{1}{2}}$ and is of order $10^{-3}\ n_9^{-\frac{1}{2}}$ for high redshift quasars. The average column density of matter on the line of sight, $\bar{n}r$, would be similar for all objects (direct consequence of the small spread of values for EW(Hβ) and I) only if the spatial coverage was constant (independent of L_{UV}, i.e. of f).

Spherical blobs of matter with radius x, spread uniformly throughout the sphere of radius r, cover a solid angle roughly given by

$$\Omega \simeq \pi\ f\ \frac{r}{x}$$

Thus for a central continuum source, $\Omega \simeq 4\pi$ requires $x \lesssim rf/4$. This critical size is proportional to $n^{-1}I$, and is independent of absolute luminosities. It is equal to $2 \times 10^{13} n_9^{-1}$ for NGC 4151.

Ω is probably much smaller than 4π as suggested by (i) the rareness of a low X-ray energy turn-over in the spectrum of Seyfert galaxies and nearby quasars (may be with the exception of low luminosity active nuclei, see lecture by S. Holt), (ii) the absence of a Lyman discontinuity at the emission redshift for large z quasars (Smith 1978).

If the emissive matter is in the form of a disk, f is a measure of its thickness h. The disk must be thin, especially for bright quasars for which h/r $\simeq 10^{-3}$. In this alternative the column density on the line of sight must be very high when looking through the disk. A few active nuclei must then absorb low energy X-ray up to 10-30 keV, if the overall size of the BLR is much larger than the X-ray continuum source (as suggested by the faster variation time-scales for the X-ray continuum than for the optical emission lines, by more than an order of magnitude).

If the matter is clumpy Ω will be a decreasing function of L_{UV} for a clump size increasing with the overall dimension of the BLR.

Such a case could arise, e.g. for a clumpy thick atmosphere of a
thin disk in which the blob size would be related to the disk radius.

A decrease of $\Omega/4\pi$ with increasing luminosity would provide
an explanation for the decrease of EW(CIV) with increasing luminosity found by Baldwin (1977) for high redshift radio-quasars. At
the low luminosity end this effect should disappear when Ω approaches 4π. Even when $\Omega/4\pi > 0.1$, it could already be masked by the
spread of values in n and I. Thus the different behaviours of the
EW(CIV) for high redshift quasars and of the EW(Hβ) for nearby
active nuclei, with absolute luminosities, are not inconsistent.
Indeed not only are the nearby active nuclei of much lower luminosity than the Baldwin sample of quasars, but also a non-purely-
radiative recombination origin of the Balmer lines could introduce
a large scatter in the relationship. The correlation between EW(Hβ)
and L_{opt} thus becomes less meaningful.

Furthermore the conclusion Ω close to 4π found by Searle and
Sargent (1968) for Seyfert galaxies must be relaxed. It was deduced
assuming $I(Ly\alpha)/I(H\beta) \simeq 23-40$, which overestimated the gas radiative
losses by up to an order of magnitude. The same conclusion is also
derived by Philipps (1978) from a study of the FeII lines, but he
invokes a specific origin for these lines (pure fluorescence
mechanism) which is in conflict with the lack of a low energy X-ray
turn-over for 11 nearby active nuclei out of 13.

The minimum column density for individual blobs is of order
10^{22}cm^{-3}. The optical depth in the Lyman continuum for one such
clump of matter is large, but it can be penetrated by X-ray photons
of energy greater than 2 keV. They heat and partly ionize the gas,
$H^+/H^O \gtrsim 0.1$, for $I \gtrsim 10^{-2}$ and an average spectral index between the
optical and X-ray energies of order unity.

If the FeII lines are emitted by the same blobs as the Balmer
lines, they can arise only from regions shielded from the 2000 Å
continuum. Indeed, the UV continuum can excite the Fe^+ ground level,
populate the excited levels and ionize substantially Fe^+ via these
excited levels. Column densities of $10^{22}- 10^{24}$cm^{-2}are necessary
to retain a substantial fraction of Fe^+ (Philipps 1978). For higher
column densities of $10^{24}- 10^{25}$cm^{-2}, the FeII lines could be formed
under collisional excitation for temperatures in the range 7000 -
1200 K (Collin-Souffrin et al. 1979). The latter regions are optically
thick at photon energies smaller than 7 - 15 keV.

The EW(FeII) increases with the optical luminosity for Seyfert
galaxies, with however a large scatter towards small EW (Bergeron
1979, Kunth and Sargent 1979). Possible explanations are either
an increase of energy input with increasing luminosity or an increase of the FeII emissive volume relative to the volume responsible for the Balmer lines. These effects must overbalance that
due to the decrease of $\Omega/4\pi$ with luminosity.

Clumpy spherical models with a clump size increasing with the
overall dimension of the BLR may explain the different behaviours

of the EW(CIV) and the EW(FeII) with absolute luminosity. The CIV emission comes from regions of moderate optical depth in the Lyman continuum (column densities at most of 10^{21} cm^{-3}, leading to dimensions always smaller than x . In this picture the corresponding emitting volume for a given clump would be proportional to x^2. The FeII emitting volume would be proportional to x^3 until the clump column density exceeds 10^{25}- 10^{26} cm^{-2}. For larger column densities it would become, as for CIV, proportional to x^2.

II. MASS FLUX

If the BLR is in an inflow regime towards a compact object, one could estimate whether its accretion rate is large enough to account for the observed radiative output.

In the case of spherical accretion at a velocity v(r), the mass flow rate may be written

$$\dot{M} = 4\pi\ r^2\ n\ m_H\ f\ v , \tag{5}$$

which, using eqs (1), (2), and (4), becomes

$$L_t \propto \dot{M} \propto (\ \frac{L_\ell}{L_{UV}}\ )^{\frac{1}{2}}\ L_\ell^{\frac{1}{2}}\ I^{\frac{1}{2}}\ n^{-\frac{1}{2}}\ v . \tag{6}$$

An upper limit to the accretion velocity is given by the full width of the broad lines, $v \lesssim$ FWZI/2. For Seyfert galaxies this latter varies at most by a factor of 3 for a range of 10^3 in absolute luminosities.

The virial velocity v_V is proportional to the mass of the central compact object. Thus if the flow velocity is proportional to v_V one gets

$$v \propto M_c/r \propto I.$$

This implies that v is independent of L_{UV}, for similar duration of the active phase, in agreement with the observations.

The mass flow rate necessary to achieve a luminosity L, given in units of 10^{45} erg s^{-1}, with a conversion efficiency factor ε, is given by

$$\dot{M} = 1.7\ x\ 10^{-2}\ \varepsilon^{-1}\ L_{45}\ M_\odot\ yr^{-1} .$$

The mass flux deduced from the observations for the Seyfert galaxy NGC 4151 (L_t = 1 x 10^{45}erg s^{-1} including the soft γ-ray emission) and the quasar 3C 273 (L_t = 5 x 10^{46}erg s^{-1}) are equal to

$$\dot{M} \text{ (NGC 4151)} = 0.3 \, n_9^{-\frac{1}{2}} \, v_3 \, M_\odot \, yr^{-1},$$

$$\dot{M} \text{ (3C 273)} = 10 \, n_9^{-\frac{1}{2}} \, v_3 \, M_\odot \, yr^{-1},$$

where v is given in units of 10^3 km s^{-1}. If a substantial fraction of the observed line width is due to a radial inflow, there is enough matter in the BLR to power the observed radiative energy output.

For active nuclei with similar L_ℓ/L_{UV} and I a consequence of eq. (6) is the decrease of L_t/L_{UV} for increasing L_{UV}. This is not confirmed by X-ray observations in the keV range. Both nearby active nuclei and quasars have roughly equal luminosities in the optical (or UV) and keV ranges. Yet this may be in agreement with the flattening of the hard X-ray spectrum up to MeV energies for NGC 4151 ($L_t/L_{UV} \simeq 10^2$, with a power mainly emitted in the soft γ-ray range) and no such effect, e.g., for 3C 273 ($L_t/L_{UV} \simeq 3$). More observations are needed beyond 50 keV to confirm this possible effect.

III. INTERCLOUD HOT PHASE

The 7 keV Fe line is marginally detected in NGC 4151. If this is a recombination line and not a K line, one obtains, for a temperature of 3×10^7 K,

$$n \, M \text{ (IC)} \simeq 0.1 \, \left(\frac{Fe \, / \, H}{4 \times 10^{-5}}\right) n \, M \text{ (BLR)}.$$

In the assumption of an homogeneous medium, the density of the hot phase is given by

$$n \text{ (IC)} = 10^7 \, r_{17}^{-3/2} \, cm^{-3},$$

and its mass is then large

$$M \text{ (IC)} = 30 \, r_{17}^{3/2} \, M_\odot \,,$$

$$= 10 \, n_9^{-1} \, M(BLR) \,.$$

The thermal velocity of the gas is 800 km s^{-1}, smaller by nearly an order of magnitude than the FWZI/2 of the BLR.

If the ionization equilibrium constraint, $Fe^{25+}/Fe^{26+} \gtrsim 1$, is used to estimate the hot gas density, one finds, for photoionization by the observed X-ray flux,

$$n_7 \; (IC) \; r_{17}{}^2 \gtrsim 1 \; .$$

This is in agreement with the density obtained for the homogeneous case.

The X-ray continuum emission of the hot gas is equal to

$$L_{ff} = 4 \times 10^{-23} \; n^2 \; V \; T_8{}^{\frac{1}{2}} \; erg \; s^{-1} \; , \hspace{1cm} (7)$$

where V is the volume occupied by the hot gas. For NGC 4151 L_{ff} = 1 x 10^{42} erg s^{-1} for T_8 = 0.3. The observed luminosity is L(2–10 keV) = 6 x 10^{42} erg s^{-1}.

Pressure equilibrium between the two phases requires n(BLR)$\simeq$ (1–3) x 10^{10} cm^{-3}. For lower density of the BLR, pressure equilibrium would still be possible if the Fe recombination line intensity was at least 1% of the observed intensity. In this case the observed line would mainly be a K line arising in a colder gas, possibly by BLR itself.

For a hot phase with a temperature higher than 10^8 K, Fe would be entirely ionized, and the signature of the IC would be only its thermal X-ray emission.

In the case of pressure equilibrium, there is a strong conduction between the hot gas and the clumps of the BLR. The mass flow is estimated following Zel'dowich and Pikel'ner (1969) and Bergeron and Gunn (1977). For NGC 4151, the evaporation rate is equal to m $\simeq$ 10^{13}– 10^{14} cm^{-2} s^{-1} for a hot phase temperature of 10^8 K, a radius for the BLR of 10^{17} cm and Compton heating, by the hard X-ray photons, of the hot gas and ionized transition layer between the two phases.

The minimum mass which will evaporate is estimated by considering the free fall time as characteristic time-scale. The mass of the central object in NGC 4151 is equal to 2 x 10^7 M$_\odot$ for a life time of the active phase of 10^8 yr, an efficiency factor of 10% and a constant power output during activity of 10^{45} erg s^{-1}. The free fall time in then equal to 1.5 x 10^8 $r_{17}{}^{3/2}$ s. Column densities of 10^{21} – 10^{22} cm^{-2} can evaporate in such a time. The evaporation rate will be reduced if a magnetic field is present, with the lines of force tangled, e.g., by turbulent motions. Transverse to a magnetic field $B_\perp$ the coefficient of thermal conductivity is reduced by about the factor $(r_c/\ell)^2$, where r_c is the radius of gyration and ℓ the mean free path of the hot electrons. This reduction factor is equal to

$$(r_c/\ell)^2 \simeq 10^{-15} n_{10}(BLR) T_8^{3/2}(IC) B_\perp^{-1}. \hspace{1cm} (8)$$

A field of a strength similar to galactic interstellar fields
(a few μ gauss) is enough to prevent heat conduction transverse
to the lines of forces.

If the hot gas thermal pressure is much smaller than the
thermal pressure in the clumps, their confinement could be
achieved by magnetic pressure if the field strength is larger than
1 gauss.

REFERENCES

Baldwin, J., 1977, Astrophys.J., 214,679.
Bergeron, J., 1979, Stars and Stars systems, ed. B.E. Westerlund,
     p. 67.
Bergeron, J., and Gunn, J.E., 1977, Astrophys. J., 217,892.
Collin-Souffrin, S., Dumont, S., Heidmann, N., and Joly, M., 1979,
     Astron.Astrophys., in press.
Krolik, J.H., and McKee, C.F., 1978, Astrophys.J. Suppl., 37,459.
Kunth, D., and Sargent, W.L.W., 1979, Astron.Astrophys., 76,50.
Nussbaumer, H., and Schild, H., 1979, Astron.Astrophys., 75,L17.
Philipps, M.M., 1978, Astrophys.J., 226,736.
Searle, L., and Sargent, W.L.W., 1968, Astrophys.J., 153,1003.
Smith, M.G., 1978, Vistas in Astron., 22,321.
Zel'dovich, Ya.B., and Pikel'ner, S.B., 1969, Soviet Phys. JETP,
     29,170.

CATEGORIES OF EXTRAGALACTIC X-RAY SOURCES

Martin J. Rees

Institute of Astronomy,
Madingley Road,
Cambridge, England.

This lecture is concerned with the various classes of extragalactic sources that may appear in X-ray surveys, and the implications for the X-ray background.

1. QUASARS AND THE BACKGROUND

I have summarized in a separate contribution some astrophysical considerations relating to quasars and active galactic nuclei. The following remarks deal solely with the collective importance of quasars, etc. (whatever astrophysical mechanisms may be involved) to the X-ray background.

Preliminary data from the first deep surveys carried out with the Einstein Observatory, covering only about one square degree, and reported by Giacconi elsewhere in these proceedings, show that most quasars down to $\sim$19th magnitude are detectable as X-ray sources. The limiting sensitivity is $\sim 10^{-14}$ erg cm^{-2} s^{-1} in the 1 - 3 keV band, $\sim 10^3$ lower than earlier surveys. This is not a surprising result – it was predictable from the scanty previous data if the ratio of X-ray and optical luminosities for the typical quasars was similar to that of 3C273 and Seyfert nuclei. Setti and Woltjer (1973, 1979) and others have pointed out that only a modest extrapolation from the Einstein Observatory limiting sensitivity would account for all the X-ray background (at $\lesssim$ 5 keV): indeed, the argument can now be inverted, and the X-ray background used to set constraints on how far, and how steeply, the quasar counts can be extrapolated to fainter magnitudes (Fabian and Rees, 1978; Setti and Woltjer, 1979). See Figure 1 and caption for further explanation. This sets contraints on the evolution with z of the quasar population, and on extrapolations of the luminosity function towards fainter objects (Seyfert galaxies, etc.).

R. Giacconi and G. Setti (eds.), X-Ray Astronomy, 363-375.
Copyright © 1980 by D. Reidel Publishing Company.

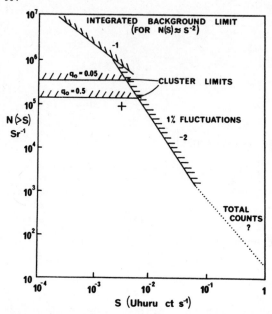

Fig. 1. Integrated background and fluctuation constraints plotted
in Uhuru flux units ($\approx 1.1 \times 10^{-11}$ erg cm^{-2} s^{-1} at 1 - 3 keV). These
constraints depend in detail on the actual source counts, and thus
are only approximate; the lines drawn (logarithmic slopes -1 and
-2 respectively) do however indicate the limits set by the integrated background (assuming $N(>S) \propto S^{-2}$), and the absence of detected fluctuations in the background, respectively (Fabian and
Rees, 1978). The total surface density for rich clusters out to
$z \simeq 3$, assuming a number density 4×10^5 (c/H$_o$)$^{-3}$, is indicated
for $q_o = 0.5$ and $q_o = 0.05$: it is clear that rich clusters, however
they may evolve, cannot reach the integrated background line without contradicting the fluctuation constraints. The cross denotes
a point on the source counts estimated by Giacconi et al.(1979)
from the Einstein Observatory's first deep surveys. Most of these
sources may be quasars. This point lies on an extrapolation, with
logarithmic slope -1.5, of the counts at high flux densities. This
is, however, fortuitous: most of the Uhuru-level extragalactic
sources are clusters, whose counts probably have a slope flatter
than -1.5; the quasar counts may be as steep as -2, in which case
only a modest further extrapolation to fainter fluxes suffices to
account for the entire X-ray background.

    The parameter α_{ox} - denoting the spectral index that is obtained
by interpolating a power-law spectrum between the optical (2500 Å)
and X-ray (2 keV) flux densities - shows no obvious correlation
with luminosity or redshift (Tananbaum et al.,1979; Avni et al.,
1979). This gives us some confidence in assuming that the X-ray

counts do indeed have the same slope as the optical counts. However it will be important, both for cosmological applications and for our physical understanding of quasars and their evolution, to test this by studying the counts and redshift distribution for an *X-ray selected* sample. (There should soon be a large enough sample to permit an analysis similar to Schmidt's (1970) well-known comparison of an optically-selected and a radio-selected sample).

The Einstein Observatory data refers to relatively low energy X-rays well below $\sim$ 4 keV. Indeed, for a power-law quasar spectrum, most of the photons counted are below 1 keV unless there is a low energy cut-off due to absorption. One needs some spectral information before one can estimate how much quasars contribute to the $\gtrsim$ 10 keV X-ray background.

At the time of writing, the best evidence on the X-ray spectra of active galactic nuclei comes from the work of Mushotsky et al. (1979) using the HEAO-A2 experiment, which covers the energy range 2 - 50 keV. Evidence on the hard X-ray spectra of bona fide quasars is even more sparse, being restricted essentially to 3C 273 (Worrall et al.,1979). This object shows $\sim$ 40% variations on timescales of 6 months, with α in the range 0.4 - 0.7 between 5 and 50 keV. The spectrum seems steeper both at lower and at higher energies (there is a Cos B measurement of $\sim$ 100 Mev γ-rays from 3C 273 (Swanenburg et al.,1978)).

These data, and their relevance to the X-ray background at higher energies, are reviewed by Holt elsewhere in this volume.

According to Tananbaum et al.(1979) α_{ox} varies between 0.94 and 1.86. The mean is 1.27 (though this refers to a sample biased in favour of radio-selected quasars). The fact that the X-ray spectral index is flatter than α_{ox} has the astrophysically-interesting consequence that the luminosity of quasars is concentrated in the optical/ UV band or in the hard X-ray band. As far as the X-ray background is concerned, the X-ray spectral index of quasars seems flat enough to ensure that - given that they are a major contributor at $\lesssim$ 5 keV - they may also contribute most of the background at higher energies: indeed the fact that the γ-ray background is not stronger allows us already to exclude the possibility that most active nuclei have spectra like NGC 4151.

2. CLUSTERS OF GALAXIES

2.1 The Present State of Cluster Sources

In clusters such as Coma, there now seems little doubt that most of the X-rays below $\sim$ 10 keV result from thermal bremsstrahlung. The X-ray emission comes predominantly from the core of the cluster, since the emissivity per unit volume is proportional to n_e^2. The following three inferences about conditions in the core seem fairly uncontroversial:

(i)   The electron temperature T_e is close to the virial temperature T_v, so that (sound speed) $\simeq$ (virial velocity).

(ii) The gas mass in the core amounts to only $\sim$ 2 per cent of the total virial mass, which is partly contributed by the galaxies, but is largely in the form of non-gaseous "dark matter" (cf. White and Rees, 1978).

(iii) The data are consistent with an iron abundance in the hot gas which is $\sim \frac{1}{2}$ the solar value (Culhane,1979).

Statements (i) and (ii) imply a cooling time such that $t_{cool} \simeq t_{Hubble} \gg t_{dynamical}$. As a first approximation, the present state of the intracluster gas can thus be fitted by a quasi-static model, evolving slowly due to infall, cooling, conduction, mass loss (and energy input) from galaxies, etc.

Note that - in order that they should indeed be "uncontroversial" - the above statements (i) - (iii) are worded in a cautious and limited way: they refer only to the cores of rich clusters at the present epoch. We know much less about the much larger amount of gas residing in the outer parts of clusters, because its emissivity is low. This gas may not be in equilibrium (it could be flowing either in or out), and its temperature and composition are very uncertain. The same statements do not generally apply to small groups (where T_v is lower), nor to rich clusters at earlier times. The X-ray luminosity of rich clusters seems to be positively correlated with virial velocity, cluster richness and central galaxy density, and negatively correlated with the fraction of spirals in the cluster. The latter is in accord with ideas that the gas is stripped from galaxies in "dynamically old" clusters; in unrelaxed clusters such as Virgo, X-ray emission with a softer spectrum is associated with some individual galaxies (Forman et al.,1979).

Some relevant questions can be summarised under the following headings:

a) Is there evidence for inflow into central massive galaxies? Giant CD galaxies might have gained mass by infall of cooling material which then condenses into stars (Silk,1976); Fabian and Nulsen (1977) have discussed the gas dynamics of this inflow, and considered its possible relevance to (e.g.) the X-ray emission and optical filaments of NGC 1275.

b) Is T_e (or the core radius for the gas distribution) significantly different from that implied by the distribution of galaxies? This is not of course a straightforward question if the cluster is non-spherical (Binney and Strimpel,1978); nor is it obvious that the galaxies should be accurate tracers for the disstribution of the missing mass (Cavaliere and Fusco-Femiano,1976). This question is however relevant to the balance between cooling and heating for the core gas.

c) Are the X-ray properties of clusters correlated with their radio properties? If so, this might indicate a heat input or dynamical effect from relativistic particles or, conversely, the lifetime or luminosity of diffuse non-thermal radio sources might depend on the external gas pressure, which is in turn correlated with the bremsstrahlung X-ray emissivity (Lea and Holman,1978).

Note that inverse Compton scattering of the microwave background
(by relativistic electrons associated with current or past radio
outbursts) may be a significant contribution to the X-ray flux
from clusters above 10 keV even if it is swamped by bremsstrahlung
at lower photon energies.

   d) What are the temperatures, densities and gas flows in the
outer parts of clusters? This question will probably be answered
when HEAO-2 provides X-ray maps with adequate angular resolution
in several energy bands. It will be particularly interesting to
compare such maps with the new radio maps now coming from the
Very Large Array - the diffuse structure of radio sources is of
course another tracer of the gas pressure and flow pattern. Relevant but indirect information may come from searches for the
Sunyaev-Zel'dovich (1972) "microwave dip", or from the orientation
of radio trail sources on the edges of clusters (cf. Valentijn,
1979).

   e) Another basic question concerns the Fe emission from
clusters. At the moment, it is unclear whether the inferred high
abundance of iron extends beyond the core of the cluster (though
the fact that the emission line has similar equivalent widths in
several clusters precludes the likelihood that the iron is restricted to individual active galaxies). The required iron abundance
in cluster cores could have been plausibly supplied by "sweeping"
of processed material lost from stars in the individual galaxies
(Takahara and Ikeuchi,1977). Binney in his contribution to this
volume, offers the interesting suggestion that the presence of
heavy elements in clusters is a natural corollary of the absence
of G-dwarfs with below $\sim$ 1/3 of solar abundance in the disc of
our Galaxy. If the iron were to pervade the outlying parts of
clusters, this would perhaps favour more radical ideas involving
a pregalactic burst of "Population III" nucleosynthesis. (One
would then, however, need to assume that population II objects -
globular clusters, etc. - are coeval with population III and themselves date from any era prior to the formation of galaxies as
single units).

   f) Is there a high energy tail to the X-ray spectrum from
clusters? Apart from the possibility of inverse Compton X-rays,
some gas may be heated to temperatures $> T_v$ by explosive outbursts
from galaxies related to radio sources (Lea and Holman,1978).
Clusters may conceivably contain a population of unresolved point
X-ray sources.

2.2 Evolution of Gas Content of Clusters

   Some consideration of how clusters evolve is a prerequisite
not only for understanding their present properties, but also for
interpreting the data on large-redshift clusters that HEAO-2 is
now providing.

   Two contrasting hypotheses have been explored for galaxy and
cluster evolution:

A. Galaxies (and dark matter) may have condensed from primordial
gas before clusters assemble. Galaxies then gradually aggregate
into clusters in a manner that can be simulated by gravitional
N-body computations. The present gas in clusters then results from
either: infall of the small fraction of material ($\sim 10\%$) that escaped incorporation into the first generation of gravitationally-
bound systems; or ejection from galaxies, via stellar winds and
planetaries, supernovae, "sweeping" of interstellar matter by ram
pressure of pre-existing cluster gas, etc. (Binney, these proceedings).

B. Protoclusters (i.e. gas clouds of $\sim 10^{14} M_\odot$) may have been the
first objects to have condensed out of the expanding universe;
they then cool and fragment into galaxies. In this scheme, advo-
cated particularly by the Moscow group (Doroshkevich et al.,1974;
Zel'dovich,1977), the present intracluster gas represents material
that has not (yet) cooled and condensed.

Type-A hypotheses yield clusters whose X-ray luminosities
increase with time; on the other hand, hypothesis B would predict
that clusters were brightest when they had just virialised but
were still predominantly gaseous, but that the X-ray luminosity
would thereafter decline as the initial gas gradually cooled and
condensed into galaxies.

Observations of distant clusters with the Einstein Observatory
offer our best hope of discriminating between these two basic
schemes. In practice, a range of schemes between the extremes of
A and B seems possible within the limits of current observation.
Preliminary data on a small sample of clusters with $z \gtrsim 0.5$ can
merely rule out extreme X-ray evolution of either sign (Henry et
al.,1979).

In a detailed study of hypothesis A, Perrenod (1978) has considered the X-ray emission from the gas that accumulates in a re-
alistically evolving cluster potential. He considers two alternative origins for the gas (t_{cool} is in general $\gtrsim H_\rho^{-1}$): either the
gas was distributed like the galaxies at the epoch of cluster
'turnaround', and falls in (being shock-heated) as the cluster
virialises; or the gas results from mass-loss from galaxies. In
this case, stellar-evolutionary considerations suggest that the
loss rate would have been higher in the past, going roughly as t^{-1}.

Perrenod presents numerous alternative models in these two
categories. The general findings are that L_x may increase by a
factor up to ~ 10 between the epoch corresponding to $z \simeq 1$ and the
present. This is a combined consequence of the cumulative build-up
of gas, and the deepening of the cluster potential well as virial-
isation proceeds.

Even though the evolutionary effects are large, Perrenod
points out that there may be grounds for optimism about the prospects of using X-ray clusters for "geometrical" cosmology. This
is because the X-ray evolution involves physical processes that
seem more straightforward and tractable than those bedevilling

optical and radio tests. (Perrenod claims that once a cluster has
turned around, its evolution depends solely on its dynamical age,
which is related to z in a manner that depends on q_o. Uncertainties
on the heat and mass input from galaxies, and their dependence on
time, would complicate the picture, so that prospects of clear
cosmological tests seem remote).

The X-ray emission from clusters may turn on more suddenly
than in Perrenod's schemes: for instance, gas could perhaps be
retained in galactic halos (at temperatures $kT \lesssim 1$ keV) until the
galaxies fall together into a cluster and the halo material is
stripped away and shock-heated to the cluster virial temperature
(Norman and Silk,1979). One difficulty stems from the fact that
the hypothesised halo gas has a short cooling time; on the other
hand, the Butcher-Oemler (1978) evidence for a sharp evolutionary
change in galaxy colours (indicating a sudden quenching of star
formation) supports this general idea, at least qualitatively.

The models so far investigated do, however, cover only a
small fraction of the tenable options. For instance, any of the
following three schemes would lead to higher values of L_x in the
past.

(i)   According to hypothesis B, a newly virialised gaseous
protocluster could have $t_{cool} < t_{Hubble}$, and could then radiate
much of its binding energy on a timescale $\sim t_{cool}$. (The actual
amount of energy radiated would depend on the fraction of gas
which was shock-heated to X-ray temperatures rather than being
squeezed into dense cool "pancakes").

(ii)   Even in hypothesis A, infalling gas might overshoot
and reach $T_e \simeq T_{escape} \simeq 10 \, T_{virial}$ before bouncing and establishing
a quasi-static distribution.

(iii) If cluster gas is ejected from galaxies via supernova-
driven winds, the velocities and temperatures may be so high that
only a small fraction of the gas remains gravitationally confined
even within a cluster; the bulk may escape and constitute an
ultra-hot inter-cluster medium (cf. section 3.1).

2.3 The Sunyaev-Zel'dovich (1972) "micro-wave dip"

Another probe for the gas distribution in clusters of galaxies
is the Sunyaev-Zel'dovich effect: Compton scattering of microwave
background photons passing through the cluster causes a reduction

$$\frac{\Delta T}{T} = -2\tau_{es}\left(\frac{kT_e}{m_e c^2} \right) \qquad (1)$$

in the temperature on the Rayleigh-Jeans part of the curve, and
an enhancement on the exponential tail. Detections of this effect
are as yet only marginal - the latest data were reported by
Birkenshaw and by Lake at IAU Symposium 92. In principle, however,
combination of such data with X-ray maps of clusters of galaxies

will provide evidence on the gas distribution and on the shape of
the potential well.

Two other implications of this effect deserve mention.

(i) Several authors (e.g. Gunn,1978; Cavaliere et al.,1978;
Silk and White,1978) have pointed out that, in principle, combined
microwave and X-ray measurements of gas in a cluster of galaxies
of known redshift can yield a direct measurement of the Hubble
constant H_o, bypassing all intermediate steps normally involved
in building up a cosmic distance scale. The principle can be understood if we consider a homogeneous cloud of angular size θ, temperature T_e and density n_e. The X-ray data yield values of T_e and
of $n_e^2(cz/H_o)\theta^3$; they therefore yield a value of n_e that depends
on $H_o^{\frac{1}{2}}$. The microwave dip amplitude yields $n_e\theta(cz/H_o)$; and thus a
value of n_e depending on H_o. Consistency between these results
therefore – at least in principle – pins down H_o. Extension of
the same method to large redshifts can also in principle determine
q_o in a straightforward "geometric" way, without evolutionary uncertainties entering in. It may be a long time before this method
can be applied in practice; it requires good microwave and X-ray
data with high angular resolution; proper allowance must be made
for the density and temperature profile, and also for inhomogeneity, non-sphericity, etc. (Other "direct" methods of measuring
H_o, incidentally, might involve supernovae; or variable compact
radio sources observed with VLBI techniques, if the kinematics
ever become properly understood and prove sufficiently standardised).

(ii) Sunyaev and Zel'dovich (1979) have recently pointed out
some "higher order" microwave effects whereby the peculiar velocity
of clusters can be determined: a radial velocity v_{pec} relative to
the mean Hubble flow at the cluster's location yields an effect
of order $(v_{pec}/c)\tau_{es}$ – smaller by a factor $(m_e/m_p)^{1/2}(v_{pec}/v_{virial})$
than the effect (1), if the gas is at the virial temperature;
transverse peculiar velocities show up as a polarization effect,
with the electric vector perpendicular to the velocity, of magnitude $(v_{pec}/c)^2\tau_{es}$.

Rich clusters probably contribute 5 – 10% of the X-ray background. One can rule out the possibility that they contribute all
the background, in whatever fashion they may evolve, because the
small angular scale fluctuation limits imply that the total number
of contributing sources must exceed the number of rich clusters
out to $z \simeq 3$ (Fabian and Rees,1978).

3. OTHER POSSIBLE SOURCES OF EXTRAGALACTIC X-RAY EMISSION

3.1 Ultra-hot Inter-cluster Gas?

The gas temperatures in clusters are ∿5 keV, implying an exponential fall-off in their contribution to the background at higher
energies. One possible contributor to a genuinely diffuse hard
X-ray background might be thermal bremsstrahlung from a very hot

gas between the clusters. As has recently been emphasised by the
Goddard group (Marshall et al.,1979) the background spectrum is,
empirically, very closely fitted by a 35-40 keV bremsstrahlung
spectrum. The excellence of this fit may well be fortuitous: indeed the fit could be destroyed by merely substracting off the
contributions from other source of background (e.g. quasars and
clusters), which are known to be substantial below 10 keV and do
not have this spectrum.

Field and Perrenod (1977) discussed the possibility that
intergalactic gas is indeed responsible for the background. If
such gas is reheated at redshift z_{heat} to a temperature such that
$kT = 35(1+z_{heat})$keV, then it can emit the entire X-ray background
if its density corresponds to

$$\Omega_{gas} \simeq \frac{1}{3} \{(1+z_{heat})^{1.6}-1\}^{\frac{1}{2}} h_{100}^{-3/2} \qquad (2)$$

Of course, less gas would be needed if it were clumped; but at
such high temperatures it would be uninfluenced by the gravitational field fluctuations arising from clusters of galaxies.

The main difficulty with this hypothesis, as Field and Perrenod
realised, is the energy requirements for heating such a large mass
of gas. Subsidiary problems relate to the long electron-ion coupling time and the Compton distortions of the microwave background
spectrum that this ultra-hot gas would cause (Wright,1979); also,
conductivity and pressure balance considerations might prejudice
the existence and survival of low-density HI clouds around galaxies
and in the Local Group.

3.2 Compton Scattering in Remote Radio Sources

Inverse Compton scattering of microwave background photons
by relativistic electrons was long ago suggested as a contributor
to the X-ray background (Felten and Morrison,1966). This process
would tend to be more efficient at large redshifts, for two reasons:
the energy density of the microwave background varies as $(1 + z)^4$;
and the many suppliers of relativistic electrons - radio source,
quasars, etc. - were more prolific at early epochs. As a corollary
of this effect, Compton scattering may "snuff out" extended sources
at large redshifts (Rees and Setti,1968).

X-rays observed in the range 1 - 10 keV are generated predominantly by electrons with Lorentz factors in the range $(1 - 3) \times 10^3$.
The radio sources with the highest inverse Compton X-ray luminosities are therefore those with the largest energy content in
the form of such electrons. These sources will not necessarily be
those with the highest radio luminosities: the maximum stored
energy is inferred to exist in very extended "giant" sources and
in the low surface-brightness "bridges" joining the components of
some strong double sources. On the assumption of equipartition,
the energy stored in $\gamma \simeq 10^3$ relativistic electrons in NGC 6251

and 3C 236 is $\geq 10^{60}$ erg (Waggett et al.,1977; Willis et al., 1974). If the magnetic field is weaker than its equipartition strength, the inferred energy stored in relativistic electrons is even larger. Some of the X-ray seen from Centaurus A are probably inverse Compton emission from the inner radio lobes (Schreier et al.,1979).

A radio source with a typical spectra index ~ 0.7 containing $10^{60}\varepsilon_{60}$ erg of relativistic electrons with $\gamma \geq 10^3$, and at redshift z, will emit an inverse Compton X-ray power of $\sim 3 \times 10^{43} (1+z)^4$ ε_{60} erg/s. We do not know the appropriate values of ε_{60}, but it would seem quite probable that there may be diffuse objects at $z \simeq 2$, a few hundred kpc in extent, emitting $\sim 10^{46}$ erg of 1 - 10 keV X-rays. These sources would generally not be particularly powerful radio sources (objects such as Cygnus A or 3C 9 are powerful radio sources because they have a strong magnetic field, rather than because they have an exceptional energy content), but they may feature in deep radio surveys. Several Westerbork radio sources have already been identified in Eistein Observatory deep surveys, though it is not yet clear whether the X-rays come from the radio lobes or from the active nucleus (Giacconi et al.,1979).

3.3 Young Galaxies

According to some theories, galaxies pass through a bright early phase when the rate of star formation, supernova out-burst, etc. is ~ 100 times higher than in a present-day galaxy (e.g. Meier,1976; Ostriker and Thuan,1975 and Binney (these proceedings)). The properties of such systems, and their potential detectability, depend on many uncertainties; but we can readily see that a young galaxy where supernova outbursts were frequent could give rise to thermal or non-thermal X-rays with a much higher power than a present-day galaxy.

Winds from young galaxies. A possible mechanism that could generate a 40 keV bremsstrahlung-type spectrum without demanding such a high mass of gas might be supernova-driven winds in young galaxies. Mathews and Baker(1971), Mathews and Bregman(1978) and others have discussed galactic winds. The mass is supplied by supernova ejecta, with characteristic velocities of $\sim 10^4$ km/sec (~ 100 keV per particle), and by more gentle processes such as stellar winds and planetary nebulae. Supernovae give the main energy input, though not necessarily the main mass supply. Two simple requirements for a wind are

$$M_{SN}/(M_{SN} + M_{other}) \geq \frac{\text{(escape energy, per proton, from galaxy)}}{100 \text{ keV}} \quad (3)$$

and

$$t_{cool} \geq t_{outflow} \quad (4)$$

(Note that these constraints could perhaps be eased in a more complex, though may be more realistic, case when the gas has a multiphase structure).

For a given value of the ratio (3), the inequality (4) is more easily satisfied now than in the past. However, galaxies might have experienced an initial bright phase when supernovae provided the main mass loss. There would then have been a very fast ($\sim 10^4$ km/sec) hot (~ 100 keV) wind emanating from young galaxies. For instance the supernova rate within the core region of a young galaxy may be high enough to sustain, for $\sim 10^7$ yrs, a wind kinetic energy output $\sim 3 \times 10^{46}$ erg/sec. Even though (3) is fulfilled by an unduly wide margin, such a wind would yield $\sim 5 \times 10^{44}$ erg/sec, per galaxy, in hard X-rays (Bookbinder et al., 1979). If their upper-main-sequence stars have a Population I composition, young galaxies may produce stronger integrated X-ray emission than present-day galaxies because metal-poor stars spend longer in the part of the H-R diagram that permits strong radiation-driven winds, so the lifetime of a typical X-ray binary is correspondingly prolonged.

3.4 The Background Spectrum

The well-known problem of accounting for any sharp break or feature in the background spectrum has been reassessed by De Zotti et al. (1979). If various categories of source contributing to the background have different spectra, there is of course no reason why the degree of isotropy should be similar at different X-ray energies (cf. Rees,1973): the small scale fluctuations due to individual discrete sources will depend on photon energy, and the larger scale anisotropies may be energy-dependent too if different source populations evolve differently with redshift.

REFERENCES

Avni, Y., et al., 1979, Astrophys.J. (Letters), in press.
Binney, J. and Strimpel, O., 1978, Mon.Not.Roy.Astr.Soc., 185, 473.
Bookbinder, J., Cowie, L.L., Krolik, J., Ostriker, J.P., and Rees, M.J., 1979, Astrophys. J., in press.
Butcher, H., and Oemler, A., 1978, Astrophys.J., 219, 18.
Cavaliere, A., Danese, L., and De Zotti, G., 1978, X-ray Astronomy, W.A. Baity and L.E. Peterson (eds.), Pergamon Press.
Cavaliere, A., and Fusco-Femiano, R., 1976, Astron.Astrophys., 49, 137.
Culhane, L., 1979, Phil.Trans.R.Soc., in press, and references cited therein.
De Zotti, G. et al., 1979, Objects of High Redshift, Proc.IAU Symp. No. 92, in press.
Doroshkevich, A.G., Sunyaev, R.A., and Zel'dovich, Y.B., 1974, Confrontation of Cosmological Theories and Observational Data, M.S. Longair (ed.), D. Reidel Pub. Co., Dordrecht-Holland.
Fabian, A.C., and Nulsen, P.E.J., 1977, Mon.Not.Roy.Astr.Soc., 180, 479.

Fabian, A.C., and Rees, M.J., 1978, Mon.Not.Roy.Astr.Soc., 185, 109.
Felten, J.E., and Morrison, P., 1966, Astrophys.J., 146, 686.
Field, G.B., and Perrenod, S.C., 1977, Astrophys.J., 215, 717.
Forman, W., Schwarz, J., Jones, C., Liller, W., and Fabian, A.C.,
1979, Astrophys.J.(Letters), 234, L27.
Giacconi, R., Bechtold, J., Branduardi, G., Forman, W., Henry, J.P.,
Jones, C., Kellog, E., van der Laan, H., Liller, W., Marshall, H.,
Murray, S.S., Pye, J., Schreier, E., Sargent, W.L.W., Seward, F.,
and Tananbaum, H., 1979, Astrophys.J.(Letters), 234, L1.
Gunn, J.E., 1978, Observational Cosmology, Proc. Saas Fee School,
Geneva Observatory Publications.
Henry, J.P.,Branduardi, G., Briel, U., Fabricant, D., Feigelson, E.,
Murray, S., Soltan, A., and Tananbaum, H., 1979, Astrophys.J.(Let-
ters), 234, L15.
Lea, S.M., and Holman, G.D., 1978, Astrophys.J., 222, 29.
Marshall, F.E., Boldt, E.A., Holt, S.S., Miller, R., Mushotsky, R.F.,
Rose, L.A., Rothschild, R.E., and Serlemitsos, P.J., 1979,
Astrophys.J., in press.
Mathews, W.G., and Baker, J.C., 1971, Astrophys.J., 170, 241.
Mathews, W.G., and Bregman, J.N., 1978, Astrophys.J., 244, 308.
Meier, D., 1976, Astrophys.J., 207, 343.
Mushotsky, R.F., Boldt, E.A., Holt, S.S., Marshall, F.E., Pravdo,
S.H., Serlemitsos, P.J., and Swank, J.H., 1979, Astrophys.J., in
press.
Norman, C., and Silk, J.I., 1979, Astrophys.J.(Letters), 233, L1.
Ostriker, J.P., and Thuan, T.X., 1975, Astrophys.J., 202, 353.
Perrenod, S.C., 1978, Astrophys.J., 226, 566.
Rees, M.J., 1973, X- and Gamma-Ray Astronomy, H. Bradt and
R. Giacconi (eds.), D. Reidel Publ. Co., Dordrecht-Holland, p. 250.
Rees, M.J., and Setti, G., 1968, Nature, 217, 326.
Schmidt, M., 1970, Astrophys.J., 162, 371.
Schreier, E.J., Feigelson, E., Delvaille, J., Giacconi, R., Grindlay,
J., Schwartz, D.A., and Fabian, A.C., 1979, Astrophys.J.(Letters),
234, L39.
Setti, G., and Woltjer, L., 1973, X- and Gamma-Ray Astronomy,
H. Bradt and R. Giacconi (eds.), D. Reidel Publ. Co., Dordrecht-
Holland, p. 208.
Setti, G., and Woltjer, L., 1979, Astron.Astrophys., 76, L1.
Silk, J., 1976, Astrophys.J., 208, 646.
Silk, J., and White, S.D.M., 1978, Astrophys.J.(Letters), 226, L103.
Sunyaev, R.A., and Zel'dovich, Y.B., 1972, Comm.Astrophys.Sp.
Phys., 4, 173.
Sunyaev, R.A., and Zel'dovich, Y.B., 1979, Mon.Not.Roy.Astr.Soc.,
in press.
Swanenburg, B.N., Bennet, K., Bignami, G.F., Caraveo, P., Hermsen,
W., Kanbach, G., Masnou, J.L., Mayer-Hasselwander, H.A., Paul, J.A.,
Sacco, B., Scarsi, L., and Wills, R.D., 1978, Nature, 275, 298.
Takahara, F., and Ikeuchi, S., 1977, Prog.Theor.Phys., 58, 1728.
Tananbaum, H., Avni, Y., Branduardi, G., Elvis, M., Fabbiano, G.,
Feigelson, E., Giacconi, R., Henry, J.P., Pye, J.P., Soltan, A.,

and Zamorani, G., 1979, Astrophys.J.(Letters), 234, L9.
Valentijn, E.A., 1979, Astron.Astrophys., 78, 367.
Waggett, P.C., Warner, P.J., and Baldwin, J.E., 1977, Mon.Not.Roy.
Astr.Soc., 81, 465.
White, S.D.M., and Rees, M.J., 1978, Mon.Not.Roy.Astr.Soc., 183,
341.
Willis, A.G., Strom, R.G., and Wilson, A.S., 1974, Nature, 250,
625.
Worrall, D.M., Mushotsky, R.F., Boldt, E.A., Holt, S.S., and
Serlemitsos, P.J., 1979, Astrophys.J., 232, 683.
Wright, E.L., 1979, Astrophys.J., 232, 348.
Zel'dovich, Y.B., 1977, The Large-Scale Structure of the Universe,
M.S. Longair and J. Einasto (eds.), D. Reidel Publ. Co., Dordrecht-
Holland, p. 409, and references cited therein.

THE X-RAY BACKGROUND AS A PROBE OF THE MATTER DISTRIBUTION ON VERY LARGE SCALES

Martin J. Rees

Institute of Astronomy,
Madingley Road,
Cambridge, England

The X-ray background obviously holds important clues to the evolutionary history and spatial distribution of its sources, be they quasars, hot gas, or some as yet unknown population. This lecture deals with just one aspect of this subject: the sensitivity of the X-ray background as a probe for density irregularities on scales larger than superclusters. Whatever the precise origin of the background may be, it is plausible to suppose that any large scale inhomogeneities in the overall distribution of gravitating matter will induce associated inhomogeneities in the spatial density of X-ray sources. (This relation may not be a strict proportionality – the X-ray source density or emissivity may depend on a higher power of the overall density – but this would do no more than change a numerical coefficient of order unity in the following expressions, where we consider only small fractional perturbations.)

The covariance function data on the distribution of galaxies become imprecise on scales exceeding 20 Mpc, but the universe definitely appears more homogeneous on scales $\gtrsim$ 100 Mpc than on any scales $\lesssim$ 10 Mpc. There is good radio and optical evidence that the distribution on the sky of all kinds of luminous objects becomes smoother as we look deeper – we are not in a Charlier-style hierarchical universe where the density contrasts are the same on every scale.

Fluctuations on larger scales ($\lambda \gtrsim$ 100 Mpc, say) are thus only of small amplitude (certainly $\delta\rho/\rho \lesssim 1$). Their influence may nonetheless be significant: the velocity perturbations that they induce are of order

$$V_{pec} \simeq c\Omega(\delta\rho/\rho)(\lambda/\lambda_H) , \tag{1}$$

and the gravitational field perturbations are

377

R. Giacconi and G. Setti (eds.), X-Ray Astronomy, 377-384.
Copyright © 1980 by D. Reidel Publishing Company.

$$\Delta\phi \simeq c^2\Omega(\delta\rho/\rho)(\lambda/\lambda_H)^2; \tag{2}$$

this means that peculiar velocities are dominated by the largest scales unless $\delta\rho/\rho$ falls off more steeply than λ^{-1} ($\propto M^{-1/3}$) and that the largest scales cause the dominant perturbations in the gravitational field unless $\delta\rho/\rho$ falls off steeper than λ^{-2}($\propto M^{-2/3}$). Anisotropies in the microwave background on small angular scales are due to gravitational and Doppler effects on the 'cosmic photosphere' (the gravitational effect dominating for scales exceeding the horizon size at the epoch of last scattering (Sunyaev and Zel'dovich, 1970).

Constraints are placed on large-scale inhomogeneities by the isotropy of any class of extragalactic discrete sources; but these are of limited value. Inferences drawn from optical counts of galaxies are bedevilled by the possible effects of patchy Galactic obscuration; for radio sources, absorption is negligible but the problem here is the broad luminosity distribution (such that intrinsically faint nearby sources and powerful remote ones appear in surveys in comparable numbers at the same flux density). The best limits amount to $\lesssim 5\%$ on scales of $\sim 1/3$ the Hubble radius (Webster, 1977; Fanti, 1979).

Given that the X-ray background is predominantly from cosmological distances ($z \gtrsim 1$), and that the Galactic contribution, away from the plane, is only a small contamination, X-ray data can provide useful evidence on the distribution of matter, if we assume that the X-ray emission per unit volume (at a given epoch) scales with the overall matter density. The present isotropy limits have a precision of about 1 percent on large angular scales. Warwick et al.(1979), in an analysis of Ariel V data (2 - 18 keV), obtained an upper limit of better than 1 percent to any 24h effect, after subtracting off a component depending on galactic latitude. The limits on angular scales of $\lesssim 20°$ are no better than a few percent, but there are prospects of significant improvements from larger-area detectors, and from measurements at higher energies where the Galactic disc contributes relatively less.

Suppose that the input into the X-ray background at a redshift z amounts to a power per unit comoving volume of $\varepsilon(z)$. The background intensity then involves an integral of the form

$$\mathcal{J} = \frac{c}{4\pi} \int \varepsilon(z) (1 + z)^{-\alpha} \frac{dt}{dz} dz , \tag{3}$$

where α is the effective spectral index, and dt/dz depends on the cosmological model. It is convenient to define a number

$$x = \mathcal{J}/(\frac{c}{4\pi} \varepsilon(0)H_o^{-1}) \tag{4}$$

The significance of this number is that a static non-evolving Euclidean universe of radius cH_o^{-1} would emit x times the actual X-ray background. The value of x is of order unity: the $(1+z)^{-\alpha}$ term tends to reduce it below unity; on the other hand, any evolutionary

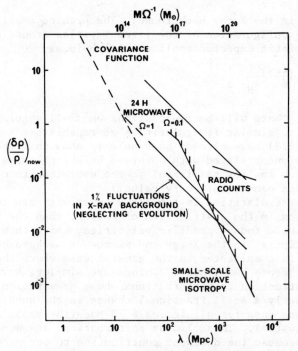

Figure 1. Constraints on density perturbations ($\delta\rho/\rho$) on various length scales λ exceeding $\sim$ 10 Mpc (adapted from Fabian et al., 1979). The lines labelled 'covariance function' and 'radio counts' assume that the galaxy and radio source distributions mimic the underlying mass distribution. The 'microwave velocity' lines give δv_G = 600 km s^{-1} and assume that we lie at the edge of such a perturbation. The radio count limits can perhaps be improved to 3 percent on scales of 1 Gpc, but depend upon the radio luminosity function, etc. The microwave background observations imply upper limits to the Doppler and gravitational perturbations at the epoch of last scattering, and these yield approximately the limits indicated. Limits of $\sim$ 1% on the isotropy of the X-ray background would yield, apart from an evolutionary correction (cf. equation 5), the line shown: note that this is potentially the most sensitive constraint on scales 100 - 1000 Mpc; on larger scales the microwave limits are likely to remain the best.

effect which enhanced the contribution from large z would tend to raise it.

Clumping of the sources, on any scale, would cause a corresponding anisotropy or 'graininess' in the observed X-ray background. Suppose, for illustration, that the clumping has a characteristic (comoving) scale λ (<< the present Hubble radius) and that the amplitude of the variations is $\delta\rho/\rho$. In general ($\delta\rho/\rho$) will be a function of z. The inhomogeneities nearest to us will cause

anisotropies in the X-ray background. The precise amplitude depends on the configuration of the irregularities around us, but
the characteristic expected amplitude is obviously

$$\left(\frac{\Delta I}{I}\right)_x \simeq \left(\frac{\delta\rho}{\rho}\right)\left(\frac{\lambda}{\lambda_H}\right)\frac{1}{x} \tag{5}$$

Additionally, there will be fluctuations on small angular scales
$\sim (\lambda/\lambda_H)$ due to similar irregularities at redshifts $z \gtrsim 1$. (Turner
and Geller (1979) have already been able to show that the X-ray
background is uncorrelated with observed bright galaxies, to a
precision which implies sufficient source evolution to make x twice
as large as its non-evolutionary value.)

If the irregularities in the X-ray emissivity are related to
inhomogeneities in the distribution of matter, then the gravitational effects should induce peculiar velocities, which themselves show
up as '24h effects' in the X-ray and microwave background. This
relationship is complicated in the general case where the perturbations have become non-linear. Things are simpler, however, if
we restrict attention to perturbations whose gravitational influence makes merely a small fractional change in the Hubble flow, as
seems to be the case for all scales $\gtrsim$ 20 Mpc (the scale of the
Local Supercluster). These linear perturbations are nevertheless
interesting because the dominant contribution to our peculiar mo-
tion could arise from perturbations with $\lambda \gg$ 20 Mpc and $(\delta\rho/\rho) \ll 1$
(cf. (1)).

The peculiar velocity induced at the boundary of a lump of
scale λ and amplitude $\delta\rho/\rho$ actually depends on the value of Ω.
The problem is discussed by Silk (1974), who presents graphs for
V_{pec} more accurate than the crude relation (1), as a function of
$\delta\rho/\rho$ for different choices of Ω. When $\lambda \ll \lambda_H$, our peculiar velocity will give a 24 hour X-ray anisotropy of amplitude

$$\left(\frac{\Delta I}{I}\right)_x = (3 + \alpha)\frac{V_{pec}}{c} \tag{6}$$

(when $\lambda \simeq \lambda_H$ the situation is more complicated because the sources
of a substantial fraction of the X-ray background are themselves
given a peculiar velocity: see below). The microwave background,
if observations are made on the Rayleigh-Jeans portion of the
spectrum ($\alpha = -2$), would yield

$$\left(\frac{\Delta I}{I}\right)_{mic} = \frac{V_{pec}}{c}. \tag{7}$$

Comparison of (5) – (7) shows that these measurements yield
an estimate of Ω: if the microwave background anisotropy is induced by inhomogeneities on scale λ ($\ll \lambda_H$), then the X-rays should
show an effect $(3 + \alpha)$ times larger due to our motion (equation 6),
and an effect due to the clumping of sources themselves which is

larger by a factor $\sim x\Omega^{-1}$. In principle, therefore, comparison of
X-ray and microwave data can yield constraints on Ω (Fabian and
Warwick, 1979); if Ω is very small the Galactic peculiar velocity
of ~ 600 km s^{-1}, reported by Smoot et al. (1977) and Corey and
Wilkinson (1976), could not be induced by a large-scale inhomogeneity on any scale between ~ 100 Mpc and the Hubble distance,
without the corresponding anisotropy in the X-ray source distribution exceeding what is observed.

Figure 1 shows the relative sensitivity of the microwave,
X-ray and other limits in restricting the inhomogeneity of the
universe on large scales. The amplitude of the inhomogeneities
on a scale λ is expressed in terms of the present value of $(\delta\rho/\rho)$.
The large scales ($\lambda > 100$ Mpc) would have grown since entering the
horizon unimpeded by radiation pressure effects, so for them $(\delta\rho/\rho)$
is related to the curvature fluctuation by

$$\approx \left(\frac{\delta\rho}{\rho}\right)\left(\frac{\lambda}{\lambda_H}\right)^2 \tag{8}$$

(Note that Ω does not enter into this expression.) The upper limit
on any 24h component in the X-rays sets a limit shown as a line in
Fig. 1. The peculiar velocity induced by the corresponding inhomogeneity shows up as an extra contribution depending on Ω.

We see from this diagram that X-ray isotropy observations with
~ 1 percent precision provide our best limits on the spectrum of
inhomogeneities on scales between ~ 100 Mpc and ~ 1000 Mpc; on
scales exceeding 1000 Mpc the most sensitive limits come from the
microwave background − more specifically, from limits to the gravitational potential fluctuations around the surface of last scattering. On scales $\gtrsim \lambda_H$ other gravitational effects must be allowed
for (cf. Rees and Sciama, 1968; Dyer, 1976). The interpretation
of the experimental limits to $\Delta T/T$ on smaller angular scales depends on the sharpness of the last scattering surface of 'cosmic
photosphere', the influence of early reheating, etc. However,
these uncertainties do not enter on large scales, and on the
smaller scales the X-ray limits may be better in any case.

In compiling Figure 1, inhomogeneities of characteristic am-
plitude $(\delta\rho/\rho)$ were assumed to pervade the whole universe − $(\delta\rho/\rho)$
measures the amplitude of the Fourier components over a particular
range of length scales. If, on the other hand, the universe contains isolated 'lumps' on large scales, embedded in a much smoother
general background, then, as Fabian and Warwick (1979) have pointed
out, there may be detectable consequences for the X-ray background,
for observations with presently attainable sensitivity, even
though the influence of the 'lump' may be undetectable as far as
the microwave background is concerned. Of particular interest is
the possibility that there may be isolated inhomogeneities on
scales fully comparable with the observed part of the universe.

The naive discussion leading to (5) and (6) needs some modification for inhomogeneities with scales comparable with the

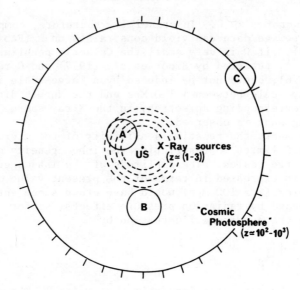

Figure 2. This diagram is intended to illustrate schematically the
effect of a single isolated 'lump' of dimensions comparable with
the Hubble radius, placed in various positions around us. The
radial coordinate is the 'comoving r' of the Robertson-Walker metric; the 'cosmic photosphere' is assumed to be at z ≃ 1000, and
the sources of the X-ray background at z ≤ 3. (Note that, in terms
of the coordinate r, z ≃ 3 is about half-way to z ≃ 1000 if Ω = 1.
If Ω < 1 it is less than half way; moreover, volumes then increase
faster than ∝ r³.) The effects on the X-ray and microwave background are as follows (cf. Fabian and Warwick, 1979):
A. (i) 24h effects in microwave and X-ray backgrounds due to our
peculiar motion; (ii) 12h effect of similar magnitude in X-ray
background owing to shear induced by lump; (iii) 24h effect, $Ω^{-1}$
times larger than others, due to excess sources in region of lump.
B. As compared with case A, effect (ii) is more important relative
to (i); but (iii) is not absent because the lump lies beyond the
X-ray sources.
C. There will in this case be a small 12h effect in the X-rays;
our peculiar velocity shows up in the microwave background but not
significantly in X-rays, since all the 'sources' are falling at
almost the same rate as us. The dominant effect, however, would
be in the microwave background, due to the gravitational perturbation on the cosmic photosphere.

Hubble radius λ_H ($\lambda \gtrsim 1000$ Mpc, say). A general treatment of cosmological models containing inhomogeneities on scales $\gtrsim \lambda_H$ would be prohibitively complicated. However, drastic simplifications ensue if we restrict attention to very small amplitudes - this is a justifiable restriction for observational cosmology because we know, from the microwave background, that the Universe is not far from being accurately "Robertson-Walker".

Provided that the curvature fluctuations are indeed small, we can consistently take a Friedmann model, with a certain overall curvature (or, equivalently, a well-defined mean density parameter Ω_0 at the present epoch t_0). The nature of the fluctuations can then be specified by defining the density at each point (comoving coordinate $\underline{r}$) as $\rho(1 + \delta(\underline{r}, t))$; at each location $\underline{r}$, δ increases with t according to the standard expression for the growth of linear density perturbations in a Friedmann model of density parameter Ω_0. Under the restriction that $\delta \ll 1$, one can extend relation (1) for the peculiar velocity to the case $\lambda \simeq \lambda_H$. For an observer at the edge of a "Swiss cheese" perturbation one obtains

$$V_{pec} = \frac{\delta}{2} c\Omega\left(\frac{\lambda}{\lambda_H}\right) \left(\frac{2(1-\Omega)^{\frac{1}{2}} - 1}{2(1-\Omega)^{\frac{1}{2}} - \Omega}\right) \tag{9}$$

The effect on the X-ray background can be analyzed into three components
  (i) A 24h effect due to our motion towards the 'lump'. (Reduced below (6) because the sources are themselves 'falling' as well.)
  (ii) A 12h quadrupole effect, with a minimum along the axis of symmetry towards the 'lump', due to the shear induced within the volume whence the bulk of the X-ray background originates.
  (iii) A 24h effect due to the enhanced density of sources in the 'lump'. (This effect would not exist if the lump lay beyond the redshift at which the X-ray background originates.)
These effects are illustrated in Figure 2.
  Finally, we note that the X-ray background can set constraints on cosmological anisotropies on scales $\gg \lambda_H$ (Wolfe, 1970). An ever-expanding Friedmann universe ($\Omega \lesssim 1$) could contain small amplitude "ripples" in the curvature of the hypersurfaces of homogeneity, on scales exceeding the present horizon. These would show up as 12h and 24h anisotropies in both the microwave and the X-ray background. In the limit where $\lambda \gg \lambda_H$, growing modes would yield comparable amplitudes in X-ray and microwave backgrounds; on the other hand, decaying modes of anisotropy would show up much less strongly in the X-ray background, since this comes from smaller redshifts than the microwaves.

Acknowledgements: I am particularly grateful to A.C. Fabian for helpful and relevant discussion.

REFERENCES

Corey, B.E., and Wilkinson, D.T., 1976, Bull.Am.Astr.Soc., 8, 351.
Dyer, C.C., 1976, Mon.Not.Roy.Astr.Soc., 175, 429.
Fabian, A.C., and Warwick, R.S., 1979, Nature, 280, 39.
Fabian, A.C., Warwick, R.S., and Pye, J.P., 1979, Physica Scripta, in press.
Fanti, C., 1979, Objects of high redshift, Proc. IAU Symp. No. 92, to be published.
Rees, M.J., and Sciama, D.W., 1968, Nature, 217, 511.
Silk, J.I., 1974, Astrophys.J., 193, 525.
Smoot, G.F., Gorenstein, M.V., and Muller, R.A., 1977, Phys.Rev. Letters, 39, 898.
Sunyaev, R.A., Zel'dovich, Y.B., 1970, Astrophys.Sp.Sci., 7, 3.
Turner, E., and Geller, M., 1979, Astrophys.J., in press.
Warwick, R.S., Pye, J.P., and Fabian, A.C., 1979, Mon.Not.Roy.Astr. Soc., in press.
Webster, A.S., 1977, Radio Astronomy and Cosmology, D.L. Jauncey (ed.), D. Reidel Publ. Co., Dordrecht-Holland, p. 269.
Wolfe, A.M., 1970, Astrophys.J., 159, 61.

DEEP SURVEYS WITH THE EINSTEIN OBSERVATORY

Riccardo Giacconi

Harvard/Smithsonian Center for Astrophysics
Cambridge, Massachusetts USA 02138

The very first rocket experiment which discovered X-ray radiation
from extrasolar sources in 1962 also discovered the existence of
an apparently isotropic background of radiation in the 2-8 keV range of
unknown origin.

       Subsequent rocket flights devoted to the study of this radia-
tion discovered that the night sky was aglow with X-rays of energies
as low as 0.25 keV. The soft X-ray background (< 1 keV) was found
to be anisotropic, strongest at the galactic poles than in the galactic
plane. This initially suggested an extragalactic origin of the radiation.
However, it soon became apparent that a substantial fraction of this
radiation had to originate in the galaxy. The detailed comparison between observed galactic latitude dependence and expected galactic
absorption did not support an extragalactic origin (Bowyer et al, 1968).A
substantial flux remained for instance even in the galactic plane where
an extragalactic flux would be completely absorbed. The brilliant
experiment by Kraushaar and his collaborators to observe a shadowing
of the radiation by the Small Magellanic Cloud produced a null result
(McCammon et al, 1971). Finally, several regions of enhanced emission
and large angular extent (several degrees) appeared to be connected with
old supernova remnants and other galactic features. Thus a large fraction of the soft X-ray background is certainly galactic. As to its origin,
there appears to be both a component due to low luminosity unresolved
stellar sources and a truly diffused component which, if thermal, is
characterized by a temperature of order of 10^6K. The detailed study of
X-ray luminosities for stars spanning the entire range of stellar spectral
types which is being carried out by Einstein, combined with the known

R. Giacconi and G. Setti (eds.), X-Ray Astronomy, 385-398.

distribution of stars in the galaxy will ultimately provide a great deal
of information on the discrete source contributions to the soft background. Study of diffused and extended emission features with Einstein
imaging detectors will also provide an observational basis for the study
of the diffused component.

Most interesting, however, from the point of view of cosmological studies is the X-ray background at higher energies (> 2 keV).
The first extensive studies carried out from UHURU revealed a perva-
sive, intense background of radiation which shows no evidence of large
scale anisotropy. Figure 1 shows the appearance of the background as
seen by the UHURU detectors. Clearly except for individual strong
sources, the flux of this background radiation is the dominant feature
of the X-ray sky.

The absence of correlation with galactic features and the
high degree of isotropy can only be reconciled with an extragalactic
origin of the radiation. Once this is established, it can easily be
shown that a large fraction of the radiation must originate at redshifts
greater than 1 (Schwartz and Gursky, 1974). Even in the extreme assumption of a uniform distribution of sources in a Euclidean universe,
this fraction is greater than 20%. Therefore, the study of the cosmic
X-ray background can give us information on the early epochs of formation of the universe and particularly in the range of redshifts intermediate between those accessible to optical observations ($Z \lesssim 3$) and the
very large redshifts (~ 1000) from which the observed microwave
background radiation appears to originate.

Prior to using the X-ray background as a tool in cosmological research, we must however understand its origin. UHURU and
OSO-8 data on the fluctuations and spectrum of the background showed
the following: (a) the observed granularity of the background (~ 3%)
could be reconciled with the superposition of a large number of discrete
sources, (b) the spectrum of the background could be described as a
combination of power law spectra, or as a thermal bremsstrahlung
spectrum of temperature ~40 keV, and (c) no known class of extragalactic X-ray emitters could explain the totality of the background without
taking into consideration evolutionary effects; however, once evolutionary effects were taken into account, several classes of objects could
yield a large fraction - if not all - of the background (for a review
and bibliography, see Schwartz, 1979). The data thus did not provide
sufficient constraints to decide on whether the background was due to
discrete sources or to truly diffuse processes and a number of equally
plausible theories on its nature were advanced.

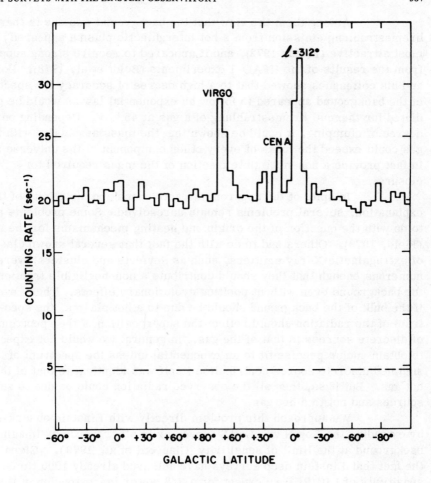

Figure 1

Among them the view that the background was due to thermal bremsstrahlung emission from a hot intergalactic plasma appeared most attractive (Field, 1972), and it appeared to receive strong support from the results of the HEAO-1 experiments (Boldt et al, 1978). Boldt and his colleagues showed that to a high degree of accuracy the spectrum of the background appeared to follow an exponential law as would be predicted for thermal bremsstrahlung of a gas at 45 keV. Depending on the degree of clumping, it could be shown that the mass associated with the gas could exceed the mass of every other component of the universe and in fact provide a non-negligible fraction of the mass required for closure.

In spite of the apparent attractiveness and simplicity of this explanation, several problems remain unresolved. Some problems had to do with the question of the origin and heating mechanisms for the gas (Field, 1979). Others had to do with the fact that several known classes of extragalactic X-ray emitters, such as Seyferts and clusters, were numerous enough that they should contribute a non-negligible fraction of the background even without positing evolutionary effects. Thus, even if the bulk of the background should be due to a hot plasma, the spectrum of the radiation should reflect the superposition of the spectrum of discrete sources to that of the gas. In general we would not expect to obtain such a precise fit to an exponential unless the spectrum of the summed discrete source contribution itself was identical to that of the hot gas. But if so, then all the observed radiation could be due to such sources and not to a hot gas.

We approach this problem directly with Einstein observations by extending the Log N vs. Log S curve, or in other words by imaging the background at the limit of sensitivity (Giacconi et al, 1979). Given the fact that Einstein deep surveys have achieved already 1000 times the sensitivity of UHURU, we expect for a 3/2 power law extension of the number intensity relation found with UHURU to be able to observe several million sources in the sky or several tens of sources per square degree. At this source density imaging is essential to avoid source confusion. We point at selected regions of the sky where we have optical and radio coverage and we cover the region with a mosaic of exposures as shown in Figures 2 and 3. Given the fact that we are navigating in unknown seas, we observe the fields at least twice with the IPC and use cross-correlations to reassure ourselves about the validity of our automated detection techniques.

The preliminary results of the surveys are summarized in Tables 1 and 2 and in Figures 4, 5, 6 and 7. In these figures the circle of the X-ray source positional uncertainty is shown superimposed on deep plates of the fields obtained by W. Sargent at Kitt Peak and by W. Liller at CTIO.

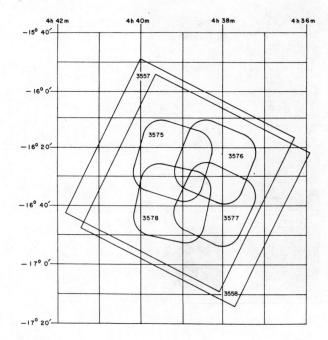

Figure 2

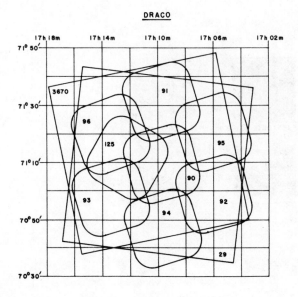

Figure 3

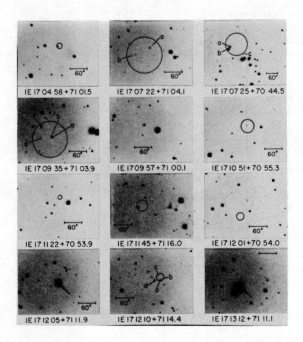

Figure 4

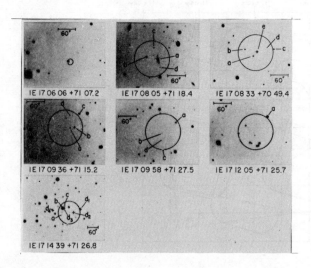

Figure 5

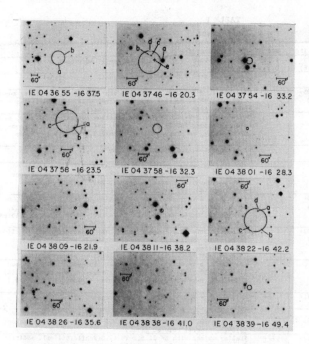

Figure 6

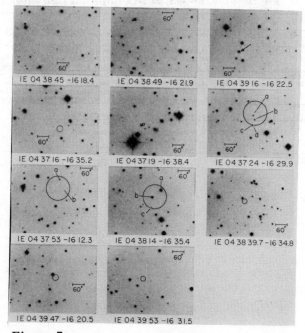

Figure 7

TABLE 1

DRACO SURVEY

Name	IPC 29	IPC 3670	Av. IPC	HRI	Optical Candidate
1E 17 04 58 / 71 01.5			(.53+ .16)	1.2+ 0.3	B=17.5, B-V=+0.3, QSO Z=2.0 (Sargent), no proper motion.
1E 17 06 06 / 71 07.2				1.11+ .30	B=16.0, B-V=+0.7,G-type star (Sargent), UX-Ari-type(?)
1E 17 07 22 / 71 04.1*	.98+ .26	1.16+ .29	1.05+ .20		(a)B=18.2, B-V= +0.9. Late type star (Sargent), no proper motion; (b) B=19, neutral colored galaxy with reddish stellar nucleus. Also, 4-6 faint stellar objects, B from 21-22.
1E 17 07 25 / 70 44.5	1.73+ .39				(a) B=16.0, B-V= +1.0, stellar spectrum approx. KO(Liller), no proper motion; (b) B=14.9, B-V=1.0, V=7000km s^{-1}(Liller),compact galaxy, no proper motion;(c)B=18.2,B-V= +0.6,no proper motion.
1E 17 07 57 / 70 54.9	1.79+ .40				Poor position, positional error >60".
1E 17 08 05 / 71 18.4	.85+ .26				(a) Blue Sb galaxy, B=17.5,Z=0.0039,λ3727[O II]in emission (Sargent), no proper motion;(b)B=18.4, B-V= +0.6, metal poor star(Sargent), no proper motion;(c) B=19.8, B-V= +0.6, no proper motion; (d) B≈21, B-V= +1.0.
1E 17 08 33 / 70 49.4			.59+ .16		(a)B=17.5, B-V= +1.0, no proper motion;(b)B=17.7,B-V= +0.4, no proper motion;(c)B=17.3,B-V=+0.9, no proper motion;(d)B=19.5, B-V=+0.6;(e)Radio source 2.8+ 0.4 mJ @21 cm;Two neutral colored galaxies; B=20. One stellar object, B=20.5.B-V= +0.3.
1E 17 09 35 / 71 03.9*	.94+ .25	1.21+ .29	1.03+ .19		(a)B=18.0,B-V= +0.4 compact galaxy,small redshift λ3727[O II] in emission (Sargent), no proper motion (b)B=17.6,B-V=0.5,no, proper motion (c)B=18.7,B-V= +0.5, no proper motion; (d)B=19.8,B-V= +0.8,small diffuse object, no proper motion.
1E 17 09 36 / 71 15.2	.87+ .24				(a)B=19.1, B-V= +0.6, late type star (Sargent), no proper motion;(b)B=18.1, B-V= -0.1, A type (Sargent), no proper motion;(c)B=16.7, B-V= +0.7, no proper motion; (d)B=20.5, B-V = +0.7 object.
1E 17 09 57 / 71 00.1	1.90+ .30	1.51+ .31	1.71+ .21	.46+ .12	B=12.0,B-V= + 0.5, G-type star (Sargent)
1E 17 09 58 / 71 27.5		1.46+ .30			(a)B=18.1,B-V= +0.5, G-type subdwarf (Sargent), no proper motion;(b)B=20.8,B-V= +0.8;(c)B=18.9,B-V= +0.8,no proper motion; (d)B=21, neutral colored galaxy.
1E 17 10 51 / 70 55.3*	1.17+ .27	1.10+ .28	1.13+ .20	.77+ .29	Stellar object with B= 21.5,B-V= +1.0(Schild,Liller). Additional faint galaxies (~ 22-23m) (Kristian-Young-Westphal).
1E 17 11 22 / 70 53.9*		1.84+ .33		.41+ .14	B=19.1, B-V= 0.7, featureless spectrum (Sargent). No proper motion. BL Lac (?).
1E 17 11 45 / 71 16.0				.78+.23 .70+.24	B=17.5, B-V= +0.2, QSO Z=1.6 (Sargent), no proper motion.
1E 17 12 01 / 70 54.01		.99+ .30		.73+ .21	B=20.8, B-V= +0.7
1E 17 12 05 / 71 11.9	2.84+ .33	2.50+ .33	2.67+ .23	.99+ .23	B=10.6, B-V= +0.4, stellar spectrum;~ F2(Liller), no proper motion.
1E 17 12 5 / 71 25.7		.86+ .28			(a)B=15.6,B-V= +0.8, no proper motion. Six other objects (brighter than 17.1, +0.8> B-V >+0.4, no proper motion). Two objects, B=17.9 and B=20.1,both with B-V= +0.2,no proper motion.
1E 17 12 10 / 71 14.4*	2.63+ .33	2.82+ .35	2.69+ .24		(a)B=18.0,B-V= +0.6,G-type star(Sargent),no proper motion; (b)B=19.6,B-V= +0.9,M-type star(Sargent),no proper motion; (c)B=20.5, B-V= +0.1 compact galaxy(Sargent),no proper motion; (d) radio source 62.0± 0.9mJ @ 21 cm, 99± 4mJ @49cm,extended
1E 17 13 12 / 71 11.1	3.81+ .39	3.75+ .39	3.77+ .28	2.93+.35 2.66+.58	SAO 8737. B=9.7,B-V= +0.8, FO; proper motion 0."05/year
1E 17 14 39 / 71 26.8	1.28+ .37				(a)B=20.1,B-V= -0.3;(b)B=13.0,B-V= +0.3,no proper motion; (c)B=19.0,B-V= +0.3,no proper motion;(d) 4 radio sources: d1 3.4+ .4mJ @ 21cm; d2 1.8+ .4mJ @ 21cm; d3 3.4 ± .4mJ @ 21cm; d4 20.3±1.4mJ @ 21 cm,39± 2 mJ @ 49 cm.

TABLE 2

Eridanus Survey

NAME	IPC 3556	IPC3557	Av.IPC	HRI	Optical Candidate
1E 04 36 55 -16 37.5	1.66± .44	2.58± .63	1.96± .36		(a)B=19.2,B-V= +0.1,blue featureless spectrum,extension to UV (WD,QSO?);(b) B=19.8,B-V= +0.8,plus 6 to 8 faint stellar ob - jects at B from 21 to 22.
1E 04 37 16 -16 35.2				.87± .25	B=21.5, B-V< +0.6, UV bright stellar object.
1E 04 37 19 -16 38.4				.41± .11	B=17.3, B-V= +0.4 featureless spectrum,no proper motion
1E 04 37 24 -16 29.9	1.36± .37				(a)B=19.6, B-V= +0.1,(b)B=19.4,B-V= +0.7,no proper motion; (c) B=19.1,B-V= +1.3,no proper motion; (d)B=21.2,B=V= +0.5 diffuse object, UV bright.
1E 04 37 46 -16 20. 3°		2.63± .51			(a)B=13.8,B-V= +0.5,Late F star(Liller),no proper motion; (b) B≥ 22,(c)B≈ 21 neutral color galaxy;(d)B=20.6,B-V= +0.4; (e) radio source ⩬3 mJ @6 cm
1E 04 37 53 -16 12.3	1.60± .43				(a)B≈19.5,B-V≈0.5, diffuse object; (b) B=21.8,B-V= +0.8,stel- lar object,featureless spectrum,UV bright.Two additional faint objects, B≈22).
1E 04 37 54 -16 33.2			(.84± .22)	.93± .29	B=13.3,B-V= +0.8, G5 star, no proper motion
1E 04 37 58 -16 23.5°	2.10± .40	1.73± .42	1.92± .29		(a)B=18.4,B-V= +0.9, no proper motion; (b) B=20.4,B-V= +0.8, (c)B≈22, B-V< +0.5, UV bright.
1E 04 38 00 -16 32.3	1.11± .34			.78± .30	Empty field
1E 04 38 01 -16 28.3		1.67± .41		.55± .15	Empty field
1E 04 38 09 -16 21.9°	2.29± .40	.93± .36	1.54± .27	.82± .17	B≈21.5, B-V< +0.6, blue featureless spectrum
1E 04 38 11 -16 38.2		(2.13± .58)		.32± .10	B=17.8,B-V=-0.2, QSO Z=1.96 (Liller, Smith),UV bright; no proper motion
1E 04 38 14 -16 35.4		1.56± .40			(a)B=14.4,B-V= -0.4,no proper motion;featureless spectrum. (b)B=16.9,B-V= +0.4; (c) B=20.3,B-V= +1.4. Also 4-6 stellar objects, B from 21 to 22.
1E 04 38 22 -16 42.2°	1.94± .38	1.25± .39	1.60± .27		(a)B=19.2,B-V= +0.8,stellar,no proper motion; (b) B=19.8, B-V= +0.9 stellar,no proper motion; (c)B≈19.5,diffuse blue object; (d)B=19.9,B-V= +1.0.
1E 04 38 26 -16 35.6°	4.07± .47	4.42± .57	4.21± .36	1.16±.15 .83±.14 .50±.15	B=19.8,B-V= +0.2, QSO Z=0.5 (Smith), UV bright Radio source 10± 2mJ @ 6 cm
1E 04 38 38 -16 41.0	1.83± .37	2.99± .51	2.23± .30	1.05±.15 .90±.17	B=13.1, B-V = +0.7, G0 Star, small proper motion
1E 04 38 39 -16 34.8				1.03± .25	B=17.1,B-V= +0.5,stellar object, no proper motion. Featureless spectrum, jet extending to NE seen on both blue and red PS plates; not on UV 4-m plate
1E 04 38 39 -16 49.4		2.07± .46		.58± .21	B=21.1,B-V= +0.3, stellar object with fainter, possibly diffuse companion
1E 04 38 45 -16 18.4		.89± .36		.80±.21 .64±.20	B=19.1,B-V= +0.3,featureless spectrum; no proper motion, UV bright
1E 04 38 49 -16 21.9		1.28± .39		.50±.15 .26±.10	B=21.0, B-V= +0.4, possibly diffuse object; radio source 27±2 mJ @ 6 cm
1E 04 39 16 -16 22.5			(.72± .22)	.67± .16	B=15.8,B-V= +1.2 K8-K9 star; no proper motion
1E 04 39 47 -16 20.5				.91± .26	B=13.7,B-V= +0.7, Late F(Liller), no proper motion
1E 04 39 53 -16 31.5				.65± .21	Empty field

The flux limit of the survey is 1.3×10^{-14} erg cm^{-2} s^{-1} or about 10^{-7}
Sco X-1. In the two surveys thus far completed, we detect 43 X-ray
sources in about 1/2 square degree of sky. Our positional accuracy
in locating sources is dependent on the detector used and the source
intensity. In the case of the IPC we have adopted a positional uncertainty
of 60 arc sec radius mainly due to systematic uncertainties in converting
detector coordinates to sky positions. For the HRI sources the location
accuracy varies from about 5 to about 20 arc sec radius, which depends
mainly on the statistical uncertainty due to the small number of source
counts.

Since we are concerned with the contribution of discrete
sources to the extragalactic X-ray background, we first limit our
band to the 1 to 3 keV range (for conversions of count rate to flux we
assume that the sources have the spectrum of the background) and then
we separate the contribution of stars from the objects detected. This is
done by direct measurement of the redshifts of candidate objects where
possible, or by the optical and radio morphology of potential counterparts. To accomplish this we have been assisted by many optical and
radio astronomers whose contributions to our study are greatly appreciated. We have also measured the B magnitude and B-V colors for optical candidates as well as their proper motions over a 25-year baseline.
In making these observations we have identified new quasars, compact
galaxies, and possible BL Lac objects. In some instances we have found
X-ray error boxes which appear completely empty at the plate limit of
available optical material. In one such case, additional observations using
CCD cameras at Mt. Hopkins (Gursky and Schild, 1979) and Mt. Palomar
(Kristian, Westphal and Young, 1979) have revealed the presence of objects fainter than $M_R \sim 21.5$. We conclude that about 1/3 of the X-ray
sources are associated with stars and we eliminate these from further
consideration.

Of the remaining source we select, for the purposes of estimating the background contribution, only those within the central 32 x
32 arc min region of the survey field and which are observed at the
5 sigma level of significance or higher. These constraints reduce possible errors due to statistical uncertainty in source intensity and geo -
metrical corrections to the flux. The results are shown in Figure 8
which is a plot of the number-flux observations for X-ray sources. Included in this figure are the previous distributions from the UHURU
survey (Murray, 1977), and the HEAO-1 flux limit, our new result and
the Einstein flux limit for an exposure of 5×10^5 seconds. We also indicate the background limit for a number-flux relation which has the form
$N (> S) S^{-1.5}$.

The Einstein deep survey point is at a flux of 2.6×10^{-11} erg
cm^{-2} s^{-1} and corresponds to 6.3×10^4 sources ster^{-1}. This lies just

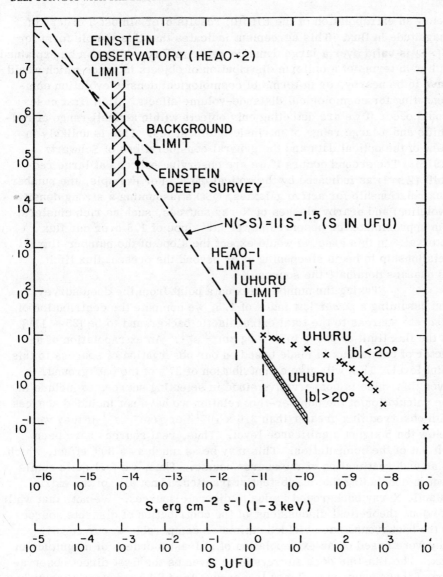

Figure 8

below an extrapolation of the UHURU results over almost 3 orders of magnitude in flux. This agreement indicates that this simple form for N(> S) is valid over a large dynamic range in flux. This can be explained either in terms of a uniform distribution of objects in space which would have to be nearby, or in terms of cosmological density evolution compensating for cosmological distance-volume effects. This first case would occur if we are detecting only objects within a small range of redshifts and a large range of intrinsic luminosities. This is unlikely in view of the optical data and the general considerations of Schwartz (1978). The second occurs if we are observing sources at large redshift ($Z \gtrsim 1$) as indicated by the optical data. For example, the number-flux relationship for active galaxies, QSO's (assuming a strong density evolution) and nearby classes of X-ray sources, such as rich clusters, can approximate the observed power law slope of 1.5 over our flux interval. In this case we would expect the slope of the number-flux relationship to begin steepening at or beyond the present flux limit as quasars dominate the sources.

Taking the number-flux data point from the deep survey, and assuming a power law index of 1.5, we compute the contribution of discrete sources to the total extragalactic background to be $(25 \pm 11)\%$ at the flux limit of 2.6×10^{-14} erg cm^{-2} s^{-1}. An extrapolation of a factor of two in flux is made based on our observation of sources to this faint level. This will give a contribution of 37% of the background. Even this value is conservative since in selecting sources to include in the calculation of the number-flux relation we have not included sources with observed flux greater than 2.6×10^{-14} erg cm^{-2} s^{-1} if they were below the 5 sigma significance level. Thus, real sources have been left out of the computation. This may be as much as a 30% effect, which is an underestimation of the source counts. Thus, we feel quite confident in concluding that a substantial fraction, if not all, of the extragalactic X-ray background is due to discrete sources. We note that while previous theoretical discussions of the contribution of discrete sources to the background have yielded estimates comparable to our result, they were based on an extrapolation of at least 3 orders of magnitude in flux. The Einstein deep surveys have given us the first direct observation of these sources at fluxes low enough to yield a substantial contribution.

One might be concerned by the fact that this contribution by discrete sources has been established only at the relatively low energies (1 - 3 keV) to which Einstein is sensitive and that the situation may be quite different at higher energies. It should be noted, however, that although this may well be the case, the thermal bremsstrahlung explanation cannot be selectively used to interpret data within a restricted range of energies. The finding that at low energy much of the background is due to discrete sources severely constrains the fraction of the background that can be due to thermal emission from a hot plasma at all energies.

The final result is that the Einstein deep survey and quasar investigations (described by Tananbaum elsewhere in this book) converge to the conclusion that the X-ray background is composed principally of unresolved sources and that they are probably QSO's. Sources of the type seen in the Deep Survey themselves account for about 37% of the background. Assuming QSO evolution models, as suggested by optical observations, the remainder of the background could easily be accounted for. This apparently leaves little room to observe a possible residual intergalactic source of X-rays that is truly diffuse. However, uncertainties in the X-ray flux and spectrum of QSO's and their evolution will probably always make it impossible to exclude a contribution of up to about 10% from an intergalactic plasma, and an even larger contribution cannot be rigorously excluded at the present time. Yet the existence of such a plasma at a level that produces only a small fraction of the X-ray background would still represent more mass than has been found previously in the universe. How could one detect such a plasma if it is not possible to observe it directly in fact of the overwhelming contribution by the unresolved QSO's? The answer may lie, as pointed out by Gorenstein, in the study of the X-ray structure of clusters of galaxies. The spatial distribution of hot gas near the outer boundaries of a cluster of galaxies may be affected by the pressure exerted by a gas in the intergalactic medium that may be even hotter. External pressure will be manifested in subtle, but possibly detectable effects upon the shape of the X-ray surface brightness distribution.

Future study of the sources of the background by more advanced X-ray observatories with increased sensitivity, positional accuracy and spectral coverage will provide definite conclusions to the question of their origin. It will also provide information on the evolution of its sources at a very early epoch which will be of fundamental importance to cosmology.

REFERENCES

Boldt, E. A., 1978 NASA Technical Memo 78106 (March)
Bowyer, C. S., Field, B. G. and Mack, J. E., 1968 NATURE 217, 32
Field, G., 1962 Ann. Rev. Astron. Astrophys. 10, pp. 227-260.
Field, G., 1979, submitted to Astrophys. J. for publication
Giacconi, R., 1979 Astrophys. J. 230, 540.
Gursky, H. and Schild, R., 1979 private communication
Kristian, J., Westphal, J. and Young, P., 1979 private communication

McCammon, D., Bunner, A. N., Coleman, P.L. and Kraushaar, W.L.,
        1971 Astrophys. J. (Letters) 178, L33
Murray, S. S., 1977 CFA preprint #680
Schwartz, D. and Gursky, H., from "X-ray Astronomy", 1974, pub-
        lished by D. Reidel Publishing Co., Dordrecht, Holland
        (eds. R. Giacconi and H. Gursky), pp. 359-388.
Schwartz, D., 1979 (COSPAR) X-ray Astronomy (eds. W. A. Baity
        and L. E. Peterson), Pergamon Press Oxford and New York

SUBJECT INDEX